12 Springer Series in Solid-State Sciences

Edited by Peter Fulde

Springer Series in Solid-State Sciences

Editors: M. Cardona P. Fulde H.-J. Queisser

S. Nakajima Y. Toyozawa R. Abe

The Physics
of Elementary
Excitations

With 114 Figures

Springer-Verlag Berlin Heidelberg New York 1980

Professor Dr. *Sadao Nakajima*
Professor Dr. *Yutaka Toyozawa*

The Institute for Solid State Physics, The University of Tokyo,
Roppongi, Minato-ku,
Tokyo 106, Japan

Professor Dr. *Ryuzo Abe*

College of General Education, The University of Tokyo, Komaba, Meguro-ku,
Tokyo 153, Japan

Series Editors:

Professor Dr. Manuel Cardona
Professor Dr. Peter Fulde
Professor Dr. Hans-Joachim Queisser

Max-Planck-Institut für Festkörperforschung, Heisenbergstrasse 1
D-7000 Stuttgart 80, Fed. Rep. of Germany

Revised translation of the Japanese edition:
Soreiki No Butsuri by Sadao Nakajima, Yutaka Toyozawa, and Ryuzo Abe.
© by Sadao Nakajima
Originally published by Iwanami Shoten, Publishers, Tokyo (1978).
English translation by Sadao Nakajima, Yutaka Toyozawa, and Ryuzo Abe.

ISBN-13: 978-3-642-81442-6   e-ISBN-13: 978-3-642-81440-2
DOI: 10.1007/978-3-642-81440-2

Library of Congress Cataloging in Publication Data. Nakajima, Sadao, 1923-. The physics of elementary excitation. (Springer series in solid-state sciences ; 12) Translation of Soreiki no butsuri which was published as vol. 2 of Bussei. Includes bibliographies. 1. Exciton theory. 2. Solid state physics. 3. Liquids. I. Toyozawa, Yutaka, 1926- joint author. II. Abe, Ryuzo, 1930- joint author. III. Title. IV. Series. QC176.8.E9N3413   530.4'1   80-11692

Preface

This book is an introduction to the physics of elementary excitations in condensed matter with emphasis on basic concepts and their mathematical representations.

The nature of the book is mainly determined by the fact that it was originally written, in Japanese, as one volume of *Iwanami Series of Fundamental Physics* supervised by Professor H. Yukawa. Our task was to portray the theory of condensed matter from a unified point of view for the student looking for his own research field and also for more senior readers interested in fundamentals of contemporary physics. As our point of view, we chose the concept of elementary excitation, which we believe to be one of the most fruitful concepts discovered by the quantum theory of matter.

The present English edition has been translated by the authors themselves from the second, revised Japanese edition published in 1978, six years after publication of the first edition. In translating, we have introduced no major modifications; only the list of references has been made more suitable to overseas readers.

In the English as well as in the Japanese editions, Chaps. 1, 4, and part of 6 were written by Nakajima, Chaps. 2, 5, and 7 by Toyozawa, and Chaps. 3 and part of 6 by Abe.

Finally we should like to thank Professor P. Fulde for kind help and Dr. H. Lotsch, Springer-Verlag, for patient cooperation in making this English edition a reality.

Tokyo, July 1980
S. Nakajima
Y. Toyozawa · R. Abe

Contents

Part II   Interaction Between Elementary Excitations

Part I

Families of
Elementary Excitations

1. Crystals and Phonons

The purpose of this introductory chapter is to describe the basic ideas of elementary excitations and their mathematical representations by taking the phonon—the well-known excitation in solids—as a typical example. All concepts related to the phonon are fairly simple, so that one can without much difficulty understand them both intuitively and mathematically. Historically, too, the phonon was the first concept of elementary excitation introduced in the description of excited states of condensed matter.

1.1 Order and Elementary Excitations

Before going into the details of phonons, let us briefly think of the reason why it is often possible to describe excited states of a macroscopic system in terms of elementary excitations.

As statistical mechanics tells us, the motion of microscopic particles in a macroscopic system is more or less random and its degree of randomness is measured by the entropy. As the temperature approaches absolute zero (with other macrovariables such as volume being kept constant), however, the entropy also vanishes (the third law of thermodynamics). Thus the macroscopic system at absolute zero is not only in the state of least energy, but also in the state of *perfect order*. The type of order which manifests itself in each particular system depends on charge, mass, and spin of constituent particles as well as on the particle density and the nature of the interaction between particles.

From the microscopic point of view, what we usually observe as a macroscopic property is nothing else but the physical response which the macroscopic system shows against a certain external perturbation we apply to the system. For example, we observe the thermal energy absorbed by a solid when it is brought in contact with an appropriate heat reservoir (specific heat), or we observe the thermal energy flowing through the solid (thermal conductivity). We may also place our system in a macroscopic field of external force such as an electric field produced by a condenser or a magnetic field produced by a solenoid, and observe the electric or magnetic moments produced in the system. When the external field is weak, the response will be proportional to the field strength, and the proportionality constant (a tensor in general) defines the corresponding material constant such as dielectric constant, magnetic susceptibility, etc. The external field may be oscillating in space and time, in which case the material constant may

depend on wavelength and frequency. To emphasize this dependence, we sometimes use the notion of response functions; for instance, the dielectric function
instead of the dielectric constant.

We can get a variety of further information by observing the effect which our
system exerts upon an external beam of particles. We thus observe scattering
and/or absorption of laser, x-ray, electron, and neutron beams by a macroscopic
system.

Now, when the external perturbation is applied, our system will in general be
brought into excited states, in which the order is no longer perfect. If the imperfection thus produced is "weak", however, we would expect an approximate
superposition principle to hold in the sense that a linear combination of imperfections with various wavelengths should also represent a possible imperfection.
In other words, the weak disorder may look like a wave. In view of the wave–
particle duality in quantum theory, the macroscopic system in low excited states
lying near the ground state may thus be regarded as a collection of "particles".
These particles are what we call *elementary excitations* (or *quasiparticles*).

A simple example is the oscillation of atoms in a solid crystal. For simplicity
we consider the system of inert atoms such as Ar. Let us first apply classical
mechanics to their translational motion. The ground state is then such that all
atoms are at rest in the configuration of a minimum energy of the interatomic
interaction. But the energy of the interaction in a macroscopic system of similar
particles is a minimum when particles sit at regular lattice sites. Hence the order
in this case means a crystal, and the imperfection which weakly destroys this
order is the oscillation of atoms around their equilibrium sites. When the amplitude of the oscillation is small, we may expand the interatomic potential into
powers of atomic displacements from lattice sites and ignore terms of third and
higher order (*anharmonic* terms). In this *harmonic approximation*, the force acting
on each atom is linear in displacements and the motion is a superposition of
harmonic oscillations called normal modes. The nomal mode is a collective
motion, in which all atoms oscillate with a single frequency and with definite
relative phases. In the long-wavelength limit, it reduces to the ordinary elastic
wave propagating through the solid.

In actuality the motions of atoms obey quantum mechanics so that oscillations should be "quantized". By applying quantum mechanics to the atomic
oscillations in a crystal, we obtain *phonons* as sound wave quanta. The situation
is similar to the case of photons which we obtain by quantization of electromagnetic oscillations in a cavity. The fundamental difference is, however, that
we cannot take the phonon out of the solid.

Another example is the *magnon* in a ferromagnetic crystal. The crystal contains magnetic atoms, whose magnetic moments are, in the ground state, aligned
along a certain direction by the quantum-mechanical exchange interaction between magnetic moments themselves, without the aid of the external magnetic
field. The weak imperfection of this orientation order propagates through the
crystal as the *spin wave*, whose quanta are called magnons (Sect. 4.3).

We thus expect that there should exist a variety of elementary excitations corresponding to various types of ordering in macroscopic systems. We further notice that the superposition principle of imperfections is only approximate, so that there should always exist a certain interaction between elementary excitations. For example, anharmonic terms ignored by the harmonic approximation of the lattice vibration give rise to the interaction between normal modes (mode–mode coupling), which means the interaction between phonons in quantum theory. In the case of the ferromagnet, the exchange interaction between magnetic moments depends on the interatomic distance and therefore is modulated by the lattice vibration. This leads to the interaction between phonons and magnons.

In short, the physics of elementary excitations is the investigation of families of elementary excitations and their interactions.

1.2 One-Dimensional Model

For simplicity let us consider the atomic oscillation in a simple one-dimensional model. As we will see in Chap. 4, one- and two-dimensional crystals can not actually be stable in the *thermodynamic limit*, where the size of the crystal, and therefore the number of atoms, becomes infinite with the density of atoms being kept constant. For the present we ignore such a sophisticated problem.

1.2.1 One-Dimensional Lattice

Suppose that we have N similar atoms on the x axis and let their coordinates be $x_1, x_2, \ldots, x_N$ from left to right, where N may be as big as 10^8. Assume that the force acting between two atoms is attractive for larger distances, but becomes strongly repulsive at a short distance of a few Å. We denote the interaction potential as a function of the interatomic distance r by $u(r)$. Figure 1.1 shows $u(r)$ for two argon atoms as an example.

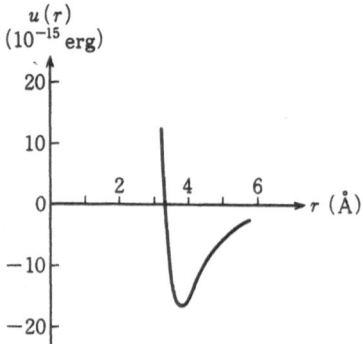

Fig. 1.1. Interaction potential between two Ar atoms

For simplicity we take into account only the interaction between nearest neighbors and write the total interaction potential as

$$U = u(x_2 - x_1) + u(x_3 - x_2) + \cdots + u(x_N - x_{N-1}) \,. \tag{1.2.1}$$

Taking the derivative, we obtain the force acting on the nth atom, except for two end atoms with $n = 1$ and N, as $u'(x_{n+1} - x_n) - u'(x_n - x_{n-1})$, which vanishes when atoms form a one-dimensional lattice defined by $x_2 - x_1 = x_3 - x_2 = \cdots = x_N - x_{N-1}$ $(\equiv a)$. The *lattice constant a* is determined by the boundary condition $u'(a) = 0$, which means that no external force acts on two end atoms.

The periodic structure may be disordered by displacing atoms slightly from their equilibrium sites. We denote the displacement of the nth atom by ξ_n and assume that the relative displacement $\xi_n - \xi_{n-1}$ of neighboring atoms is small compared with the lattice constant. In the harmonic approximation, the increase of the potential (1.2.1) from the minimum is given by

$$\Delta U = \frac{1}{2} f[(\xi_2 - \xi_1)^2 + (\xi_3 - \xi_2)^2 + \cdots + (\xi_N - \xi_{N-1})^2] \tag{1.2.2}$$

where $f = u''(a)$ is the spring constant.

When $\xi_n - \xi_{n-1} = a\varepsilon$, where the constant ε satisfies $|\varepsilon| \ll 1$, we have the homogeneous dilatation (or contraction depending on the sign of ε) of the lattice constant from a to $a(1 + \varepsilon)$ and ε is called the strain in the theory of elasticity. The force acting on the atom inside the crystal vanishes also in this case, but we should apply external forces $\pm fa\varepsilon$ to the two end atoms to keep them in equilibrium. The ratio of this external force to the strain is usually called Young's modulus.

1.2.2 Lattice Vibration

From (1.2.2), we obtain the equations of motion

$$M \frac{d^2 \xi_n}{dt^2} = -f(2\xi_n - \xi_{n+1} - \xi_{n-1}) \tag{1.2.3}$$

where M is the atomic mass. The form of equations for two end atoms is slightly different. To keep the same form for all atoms, we apply the *periodic boundary condition*. We thus assume (1.2.3) for all integers n and impose the condition that exactly the same motion shall repeat itself whenever n increases by N:

$$\xi_{N+n} = \xi_n \tag{1.2.4}$$

We will frequently make use of this type of boundary condition when we deal with bulk properties of a macroscopic system, which do not depend on the choice of boundary condition in the thermodynamic limit.

Fig. 1.2. A model of harmonic oscillations in the one-dimensional lattice

The normal mode is defined as a solution of (1.2.3), in which all atoms oscillate with one and the same frequency. Since (1.2.3) are linear differential equations with real coefficients, the real (or imaginary) part of a complex solution is also a solution. We write the normal mode as $\xi_n = \mathrm{Re}\{A_n e^{-i\omega t}\}$, where $\omega/2\pi$ is the frequency and $i = \sqrt{-1}$. Substituting in (1.2.3), we obtain equations to determine complex constants A_n:

$$M\omega^2 A_n = f(2A_n - A_{n+1} - A_{n-1}) .\tag{1.2.5}$$

The solution

$$A_n = A \exp(ikna)\tag{1.2.6}$$

satisfies the periodic boundary condition (1.2.4) when the wave numbers are given by

$$k = \frac{2\pi l}{Na} \quad (l = 0, \pm 1, \pm 2, \ldots) .\tag{1.2.7}$$

Neighboring atoms have the phase difference ka, so that $2\pi/|k|$ is the wavelength. Substituting (1.2.6) in (1.2.5), we obtain the frequency as

$$\omega(k) = 2\left(\frac{f}{M}\right)^{1/2}\left|\sin\frac{1}{2}ka\right| .\tag{1.2.8}$$

As is shown in Fig. 1.3, (1.2.8) and therefore $\exp[i(kna - \omega t)]$ are invariant when k is replaced by $k + K$, where $K = 0, \pm(2\pi/a), \pm(4\pi/a), \ldots$ are called (one-dimensional) *reciprocal lattice vectors*. In other words, k and $k + K$ define one and the same normal mode. To obtain *independent* normal modes, we may restrict the wave vector k within the *first Brillouin zone* defined by

$$-\frac{\pi}{a} < k \leqq \frac{\pi}{a} .\tag{1.2.9}$$

Then k is called the *reduced* wave vector. Since l in (1.2.7) runs from $-N/2$ to $N/2$, the number of reduced wave vectors is just equal to N, as it should be.

When $|ka| \ll 1$, the frequency given by (1.2.8) is inversely proportional to the wavelength and the proportionality constant defines the velocity of sound propagating along our chain of atoms,

$$\omega(k) = c_s |k|, \quad c_s = \left(\frac{fa^2}{M}\right)^{1/2}.$$ (1.2.10)

In accordance with the theory of elasticity, the sound velocity c_s is equal to the square root of the elastic constant fa divided by the mass per unit length M/a. Thus our chain of atoms behaves as if it were a continuous elastic medium in the long-wavelength limit.

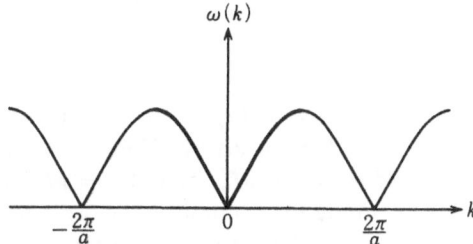

Fig. 1.3. The frequency spectrum of the one-dimensional solid

1.2.3 Diatomic crystals

Let us now turn to the diatomic system, in which two types of atoms, A and B, are alternately arranged along the x axis. Again we will take into account only the interaction between neighboring A and B atoms. The total energy of interaction is then a minimum when atoms form a one-dimensional crystal, whose lattice constant a is now twice as long as the nearest neighbor distance. The unit cell of our crystal is the interval of the length a and contains two atoms A and B. The whole crystal is a repetition of the unit cells and may also be regarded as being composed of two sublattices, one formed by A atoms and the other formed by B atoms. In this subsection, we denote the total number of atoms by $2N$, so that N is the total number of unit cells.

We now write displacements of A atom and B atom in the nth unit cell respectively as $A \exp[i(kna - \omega t)]$ and $B \exp[i(kna - \omega t)]$. Amplitudes A and B satisfy

$$M_a \omega^2 A = 2fA - f[1 + \exp(-ika)] B$$
$$M_b \omega^2 B = 2fB - f[1 + \exp(ika)] A$$ (1.2.11)

where M_a, M_b are atomic masses and f is the spring constant between neighboring atoms. Nontrivial solutions of (1.2.11) exist when the determinant formed by the coefficients vanishes. This gives a quadratic equation of ω^2 and therefore two solutions $\omega_{\pm}^2 (k)$ for each value of k (Fig.1.4).

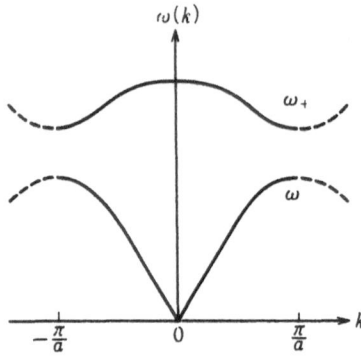

Fig. 1.4. The frequency spectrum of a diatomic crystal with acoustic branch ω_- and optic branch ω_+

In the long-wavelength limit $|ka| \ll 1$, we have $\omega_-(k) \simeq c_s|k|$, $\omega_+(k) \cong (2f/M_r)^{1/2}$, for which $A \cong B$ and $M_a A \cong -M_b B$, respectively. Here c_s is the velocity of sound given by (1.2.10) with $M = M_a + M_b$, and $M_r = M_a M_b \times (M_a + M_b)^{-1}$ is the reduced mass. The ω_- mode in this limit is the ordinary elastic wave, in which two atoms in the unit cell oscillate in phase. This mode is therefore called the *acoustic branch*. When the ω_+ mode is excited, on the other hand, two atoms in the unit cell oscillate in antiphase in the long-wavelength limit. In the case of an ionic crystal such as NaCl, the unit cell contains two ions possessing electric charges of equal magnitude and opposite sign and therefore acquires the electric dipole moment oscillating with the same frequency as the ω_+ mode. Thus the ω_+ mode is strongly coupled with the light of the same frequency and called the *optic branch* (see Chap. 2).

Since $\omega_- = (2f/2M_a)^{1/2}$, $\omega_+ = (2f/M_b)^{1/2}$ at $|ka| = \pi$ and for $M_a > M_b$, we see that $\omega_+(k)$ is almost independent of k in the limit $M_a \gg M_b$. This is exaggerated in Fig. 1.6, in which the k dependence of the optic branch is completely neglected (the *Einstein model*). The dispersion of the acoustic branch, on the other hand, is approximated by the linear relation $\omega_-(k) = c_s|k|$ (the *Debye model*).

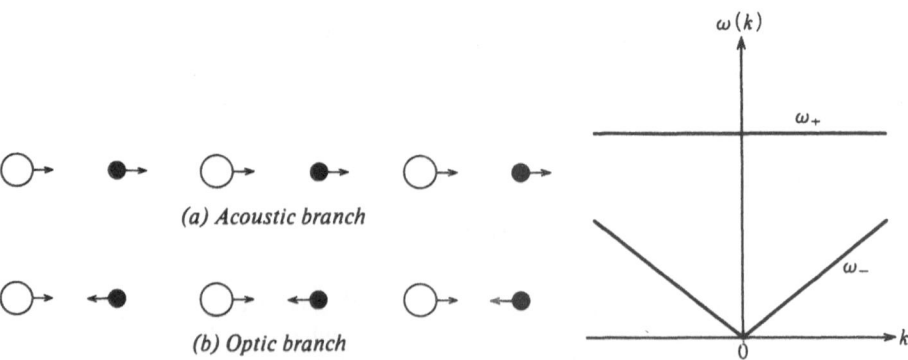

(a) Acoustic branch

(b) Optic branch

Fig. 1.5. Phases of oscillations in a diatomic crystal

Fig. 1.6.
Debye (ω_-) and Einstein (ω_+) models

1.3 Three-Dimensional Crystals

1.3.1 Lattice and Reciprocal Lattice

In a three-dimensional system of atoms, the energy of interaction is also a minimum when atoms form a crystal. In order to specify its fundamental periods, however, we now need three vectors a_1, a_2, a_3 which are not on a plane. With use of a set of three integers n_1, n_2, n_3, we may write the crystal period in general as

$$n = n_1 a_1 + n_2 a_2 + n_3 a_3 \tag{1.3.1}$$

which we call a *lattice vector*. The set of end points of all lattice vectors drawn from a common origin defines a *space lattice* (Fig. 1.7). The parallelepiped spanned by a_1, a_2, a_3 may be taken as its unit cell.

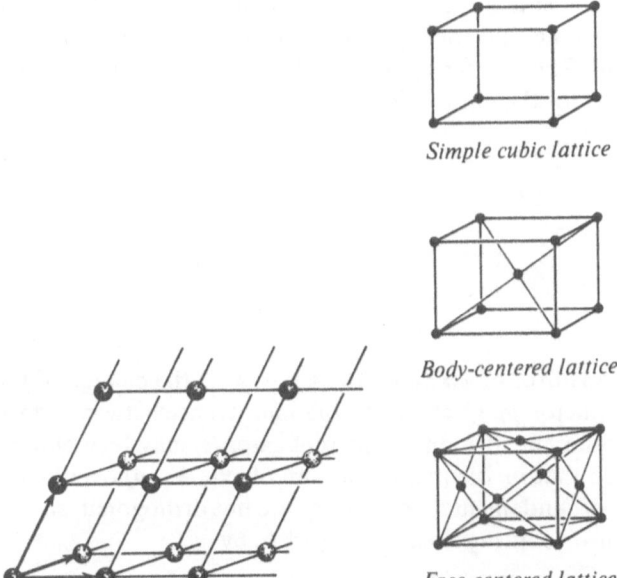

Simple cubic lattice

Body-centered lattice

Face-centered lattice   **Fig. 1.7.** Space lattices

Take one of the unit cells and suppose that it contains s atoms at r_1, r_2, ..., r_s when their oscillations are ignored. Then we find at $r_\sigma + n$ the atom which is identical with the one at r_σ, where n is any lattice vector and $\sigma = 1, 2, ..., s$. The equilibrium position of each atom may therefore be written as $R(n,\sigma) = r_\sigma + n$.

Precise information on the crystal structure is obtained by the analysis of the diffraction pattern when a wave of the wavelength comparable with the atomic spacing, i.e., of the order of 1 Å, is scattered by the crystal. The wavelength of 1 Å means the x-ray of 10^4 eV in energy, the electron of 10^2 eV, the neutron of 10^{-1} eV, etc. Let us take the case of the neutron as an example. The energy of the neutron is of the same order of magnitude as the kinetic energy of the crystal atom, so that one can easily observe the inelastic scattering by which the neutron changes its kinetic energy. We can thus obtain information about the dynamical behavior of atoms. For the present, however, we restrict ourselves to the elastic scattering and regard atoms in the crystal as fixed scattering centers against the incident neutron.

For simplicity we assume that each unit cell of the crystal contains one and the same atom, without even isotope impurities. Let R_j be the position of the jth atom and r be the position of the neutron. We write the interaction potential between the neutron and the atoms as

$$\mathscr{H}' = \sum_{j=1}^{N} v(r - R_j) . \tag{1.3.2}$$

Suppose that the neutron enters the crystal with momentum p and leaves it with momentum p'. We assume that the scattering potential is weak and apply the Born approximation. The scattering probability is proportional to the square of the absolute value of the matrix element

$$\langle p' | \mathscr{H}' | p \rangle = \int dr \, \mathscr{H}' \exp\left[\frac{i}{\hbar}(p - p') \cdot r\right]$$

$$\equiv v_k \rho_k$$

$$\rho_k = \sum_{j=1}^{N} \exp(ik \cdot R_j) . \tag{1.3.3}$$

Here v_k is the Fourier transform of $v(r)$ and $\hbar k = p - p'$ is the change of the neutron momentum. The factor ρ_k gives rise to the interference between waves scattered from atoms at different sites. In the present case, R_j may be identified with the lattice vector n, so that ρ_k is equal to the sum of $\exp(ik \cdot n)$ over all n.

In general the vectors of fundamental periods, a_j, are not orthogonal, so that it is convenient to introduce "contragradient" vectors b_j by

$$a_j \cdot b_l = 2\pi\delta_{jl} = \begin{cases} 2\pi & (j = l) \\ 0 & (j \neq l) . \end{cases} \tag{1.3.4}$$

Explicitly $b_1 = (2\pi/V_0)a_2 \times a_3$, etc., where $V_0 = a_1 \cdot (a_2 \times a_3)$ is the volume of the unit cell of the original lattice. The lattice whose fundamental periods are defined as b_1, b_2, b_3 is called the reciprocal lattice. The reciprocal lattice vectors are defined by

$$K = K_1 b_1 + K_2 b_2 + K_3 b_3 \tag{1.3.5}$$

where K_1, K_2, K_3 are integers. In accordance with this, we write the wave vector k in (1.3.3) also as

$$k = k_1 b_1 + k_2 b_2 + k_3 b_3 \, . \tag{1.3.6}$$

Then ρ_k takes the form $L(k_1)L(k_2)L(k_3)$, where $L(k_j)$ is the sum of $\exp(2\pi i k_j n_j)$ over n_j. Suppose that our crystal is a parallelepiped spanned by Ga_1, Ga_2, Ga_3, where G is an integer of the order of 10^8. Thus the integer n_j runs from 0 to G. We can easily see that $L(k_j)$, being of the order of unity in general, takes the exceptionally big value G when k_j is an integer. If we ignore G^{-1} against unity, therefore, we have

$$N^{-1} \sum_n \exp(i k \cdot n) = \begin{cases} 1 & (k = K) \\ 0 & (k \neq K) \end{cases} \tag{1.3.7}$$

where K is given by (1.3.5). Thus waves scattered by each atom interfere constructively with one another when the *Bragg condition*

$$p = p' + \hbar K \tag{1.3.8}$$

is satisfied. We have $p = p'$ in the case of the elastic scattering, so that the neutron is reflected like light by the plane bisecting the reciprocal lattice vector K (Fig.1.8).

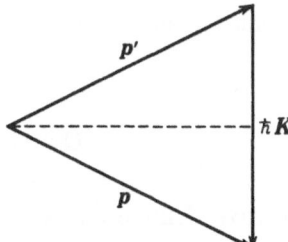

Fig. 1.8. The Bragg condition

1.3.2 The Hamiltonian of the Harmonic Approximation

We may deal with the lattice vibration in the three-dimensional crystal in the same way as in the one-dimensional case. Let $\xi(n,\sigma)$ be the displacement of the atom whose equilibrium position is $R(n,\sigma)$. The number of possible lattice vectors n is equal to the number of unit cells in the crystal, and σ runs from 1 to s, where s is the number of atoms in the unit cell. The total number of Cartesian components $\xi_j(n,\sigma)$, $j = x,y,z$ is equal to $3N$ as it should be, where N is now the number of atoms. We denote these components by $q_1, q_2, \ldots, q_{3N}$ and canonically conjugate momenta by p_1, p_2, ..., p_{3N}. In the harmonic approximation, the interaction potential measured from its minimum is a homogeneous quadratic function of q_n, so that the Hamiltonian takes the form

$$\mathscr{H} = \sum_n \frac{1}{2M_n} p_n^2 + \frac{1}{2} \sum_{nn'} f_{nn'} q_n q_{n'} \tag{1.3.9}$$

where M_n are atomic masses and $f_{n'n} = f_{nn'}$ are spring constants.

Note that the matrix formed by spring constants $f_{nn'}$ is symmetric and positive definite. The symmetry is due to the fact that sping constants are equal to second order partial derivatives of the interaction potential at the equilibrium configuration of atoms. The positive definiteness means that this configuration is a stable one corresponding to a minimum of the potential. This is not necessarily the absolute minimum, but may be one of the relative minima. The interaction potential can have a number of minima corresponding to various crystal forms and separated from one another by maxima. This is why many crystals undergo the structural transition from one crystal phase to another. It is even possible that a relative minimum corresponds to an amorphous solid, in which no long range periodicity can be found in the atomic arrangement.

Canonical equations of motion are

$$\frac{dq_n}{dt} = \frac{\partial \mathscr{H}}{\partial p_n}, \quad \frac{dp_n}{dt} = -\frac{\partial \mathscr{H}}{\partial q_n}. \tag{1.3.10}$$

Substituting (1.3.9), we obtain $p_n = M\dot{q}_n$ from the first equation and therefore $M_n \ddot{q}_n = -\sum_{n'} f_{nn'} q_{n'}$ from the second equation. The normal mode is defined as $q_n = (V_n/M_n^{1/2})\exp(-i\omega t)$, where

$$\omega^2 V_n = \sum_{n'} F_{nn'} V_{n'}, \quad F_{nn'} = \frac{f_{nn'}}{(M_n M_{n'})^{1/2}}. \tag{1.3.11}$$

Thus ω^2 is equal to one of $3N$ eigenvalues of the positive definite and symmetric matrix defined by $F_{nn'}$. We denote these eigenvalues by ω_λ^2 and components of the corresponding $3N$-dimensional eigenvectors by $V_{n\lambda}$ which may be complex. These eigenvectors form a complete orthonormalized system:

$$\sum_\lambda V_{n\lambda} V_{n'\lambda}^* = \delta_{nn'}, \quad \sum_n V_{n\lambda}^* V_{n\lambda'} = \delta_{\lambda\lambda'} \tag{1.3.12}$$

where * means the complex conjugate. Since $F_{nn'}$ are real, $V_{n\lambda}^*$ also form an eigenvector which belongs to the same eigenvalue as $V_{n\lambda}$. Assuming these two eigenvectors to be independent of each other, we define the parameter $-\lambda$ by $V_{n-\lambda} = V_{n\lambda}^*$.

Since our equations of motion are linear, the general solution is the superposition of normal modes:

$$q_n = \sum_\lambda M_n^{-1/2} V_{n\lambda} Q_\lambda, \quad p_n = \sum_\lambda M_n^{1/2} V_{n\lambda} P_\lambda . \tag{1.3.13}$$

With use of (1.3.12) we may write the inverse transformation as

$$Q_\lambda = \sum_n V_{n\lambda}^* M_n^{1/2} q_n, \quad P_\lambda = \sum_n V_{n\lambda}^* M_n^{-1/2} p_n \tag{1.3.14}$$

so that

$$Q_\lambda^* = Q_{-\lambda}, \quad P_\lambda^* = P_{-\lambda} . \tag{1.3.15}$$

We can easily write equations of motion for Q, P. It is more convenient to introduce

$$B_\lambda = \frac{1}{2^{1/2}} \left(Q_\lambda + \frac{i}{\omega_\lambda} P_\lambda \right) \tag{1.3.16}$$

which satisfies

$$\frac{dB_\lambda}{dt} = - i\omega_\lambda B_\lambda . \tag{1.3.17}$$

Hence B_λ is proportional to $\exp(-i\omega_\lambda t)$. We may also introduce real variables

$$q_\lambda = \frac{1}{2^{1/2}} (B_\lambda + B_\lambda^*), \qquad p_\lambda = \frac{i}{2^{1/2}} \omega_\lambda (B_\lambda^* - B_\lambda) \tag{1.3.18}$$

which satisfy

$$\frac{dq_\lambda}{dt} = p_\lambda, \qquad \frac{dp_\lambda}{dt} = - \omega_\lambda^2 q_\lambda . \tag{1.3.19}$$

They are canonical equations of motion which we obtain from the Hamiltonian

$$\mathcal{H} = \sum_\lambda \left(\frac{1}{2} p_\lambda^2 + \frac{1}{2} \omega_\lambda^2 q_\lambda^2 \right) . \tag{1.3.20}$$

This can be obtained also from the Hamiltonian (1.3.9) by successive transformations (1.3.13,16,18). In our harmonic Hamiltonian (1.3.20), we have no cross terms which cause energy transfer from one mode to another (the mode–mode coupling): each mode is a simple harmonic oscillation independent of others.

1.3.3 Periodic Crystals

The formalism given in the preceding paragraph may be applied to a crystal disordered by impurities or even to an amorphous solid, since we have nowhere made use of the periodicity of the crystal. We now apply it to a perfect crystal in which spring constants depend only on relative lattice vectors $n - n'$, and atomic masses only on the parameter σ. Hence $F_{nn'} = F_{\alpha\alpha'} (n - n')$, where α stands for the set of the index σ to specify atoms in the unit cell and the index j to specify Cartesian components x, y, z. We may thus write the solution of (1.3.11) as

$$V_n = v_\alpha \exp(i\boldsymbol{k} \cdot \boldsymbol{n}) \tag{1.3.21}$$

where v_α is a solution of the $3s$-dimensional eigenvalue equation

$$\left.\begin{aligned} \omega^2 v_\alpha &= \sum_{\alpha'} G_{\alpha\alpha'}(\boldsymbol{k})v_{\alpha'} \\ G_{\alpha\alpha'}(\boldsymbol{k}) &= \sum_{n} F_{\alpha\alpha'}(\boldsymbol{n})\exp(-i\boldsymbol{k} \cdot \boldsymbol{n}) \end{aligned}\right\} \tag{1.3.22}$$

We again impose the periodic boundary condition that (1.3.21) shall be invariant against $n \to n + Ga_j$. Hence $\boldsymbol{k}$ is given by (1.3.6), where

$$k_j = \frac{2\pi}{G} l_j, \qquad l_j = 0, \ \pm 1, \ \pm 2, \ldots . \tag{1.3.23}$$

Since (1.3.21,22) are invariant when $\boldsymbol{k}$ is displaced by a reciprocal lattice vector, we may restrict $\boldsymbol{k}$ to a unit cell of the reciprocal lattice to obtain independent normal modes. For instance $-G/2 < l_j < G/2$ in (1.3.23). The number of wave vectors thus restricted is equal to G^3, the number of unit cells in the crystal. For each one of them we have $3s$ eigenvalues of (1.3.22). We distinguish them by $r = 1, 2, \ldots, 3s$. The set of $\boldsymbol{k}$ and r correspond to λ and $-\boldsymbol{k}$ and r to $-\lambda$. When $s = 1$, the index r means the polarization of the lattice wave corresponding to one longitudinal and two transverse waves in an isotropic elastic con-

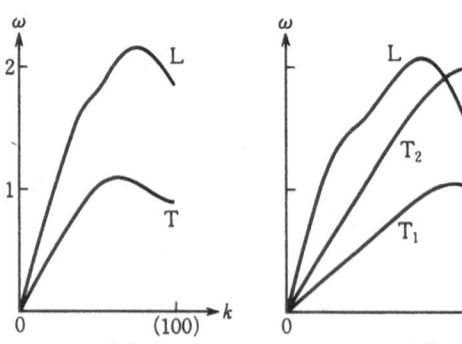

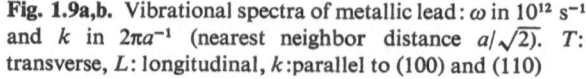

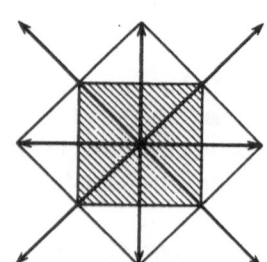

Fig. 1.9a,b. Vibrational spectra of metallic lead: ω in 10^{12} s^{-1} and k in $2\pi a^{-1}$ (nearest neighbor distance $a/\sqrt{2}$). T: transverse, L: longitudinal, k:parallel to (100) and (110)

Fig. 1.10. The first Brillouin zone of a square lattice (shaded)

tinuum. Figure 1.9 shows the dispersion relations of ω vs k in metallic lead observed by inelastic neutron scattering.

In general, among $3s$ branches, we always have three acoustic branches, for which $\omega \to 0$ as $k \to 0$. This derives from the translational invariance of our Hamiltonian, in which the interaction between atoms depends only on their relative positions. It is thus invariant under translations of the crystal as a whole along x, y, or z axes. Thus such translations are eigensolutions of (1.3.22) belonging to $\omega = 0$. This is the first example of the Goldstone theorem which we will discuss in detail in Sect. 4.7.

Finally three-dimensional Brillouin zone and reduced wave vectors may be defined in the following way. The reduced wave vector is the wave vector which can be drawn from the origin of the wave vector space (k space) without passing any one of the bisecting planes of reciprocal lattice vectors. End points of all reduced wave vectors constitute a polyhedron, which defines the (first) Brillouin zone. It is a unit cell of the reciprocal lattice. Figure 1.10 shows a two-dimensional example, which is easier to visualize.

1.4 Quantization of Lattice Vibrations

1.4.1 Phonons

In quantum mechanics, q_n, p_n in (1.3.9) are Hermitian operators operating on state functions. Similarly Q_λ^*, P_λ^* in (1.3.15) and the complex conjugate B_λ^* of (1.3.16) are replaced by Hermitian conjugate operators $Q_\lambda^\dagger$, $P_\lambda^\dagger$ and $B_\lambda^\dagger$, respectively. Algebra of these operators is defined by the well-known commutation relations $[q_n, p_{n'}] \equiv q_n p_{n'} - p_{n'} q_n = i\hbar \delta_{nn'}$, $[q_n, q_{n'}] = [p_n, p_{n'}] = 0$, where $2\pi\hbar$ is Planck's constant. With use of (1.3.12,14,15,18), we obtain $[q_\lambda, p_{\lambda'}] = i\hbar \delta_{\lambda\lambda'}$, etc. From the quantum mechanics of the harmonic oscillator we know that the eigenvalues of the Hermitian operator $(1/2)(p_\lambda^2 + \omega_\lambda^2 q_\lambda^2)$ are $\hbar\omega_\lambda(N_\lambda + 1/2)$ with $N_\lambda = 0, 1, 2, 3 \ldots$. The integer N_λ may be regarded as the number of phonons occupying the state λ of energy $\hbar\omega_\lambda$. Since N_λ can be arbitrarily large, phonons are Bose particles. The eigenvalues of (1.3.20) are given by

$$\sum_\lambda \frac{1}{2} \hbar\omega_\lambda + \sum_\lambda \hbar\omega_\lambda N_\lambda . \tag{1.4.1}$$

The second term means that the lattice may be regarced as a perfect gas of phonons as far as vibrational excitations are concerned. In contrast to ordinary gases, however, the total number of phonons is not fixed as is the number of photons enclosed in a cavity. In the harmonic approximation, the occupation number N_λ is constant in time since the energy of each normal mode is conserved. The mode–mode coupling arising from anharmonic terms means collisions between phonons in quantum mechanics, and thermal equilibrium is established in the phonon gas as well as in ordinary gases through collisions. The difference is that phonon–phonon collisions in general do not conserve the total number of phonons, as we will see explicitly in Sect. 1.7. In equilibrium, therefore, the phonon number should be such that against its variation the free energy of the phonon gas is at a minimum. According to statistical mechanics, it is equivalent to saying that the chemical potential of the phonon vanishes. Thus, for the phonon gas at the absolute temperature T, we have the Planck distribution, i.e., the Bose distribution with the vanishing chemical potential

$$\langle N_\lambda \rangle = \frac{1}{\exp(\hbar\omega_\lambda/k_\mathrm{B}T) - 1} \tag{1.4.2}$$

where k_B is Boltzmann's constant. At $T = 0$, it reduces to the phonon vacuum in which $N_\lambda = 0$ for all λ. Even then, we have the finite energy of zero-point vibration given by the first term of (1.4.1).

1.4.2 Specific Heat of the Phonon Gas

At high temperatures, where $k_\mathrm{B}T \gg \hbar\omega_\lambda$ for all λ, (1.4.2) reduces to the classical equipartition law $\hbar\omega_\lambda \langle N_\lambda \rangle \approx k_\mathrm{B}T$ and therefore gives the specific heat of the phonon gas as $\partial E_\mathrm{T}/\partial T \approx 3Nk_\mathrm{B}$, where N is the total number of atoms in the solid. In general, however, the specific heat depends on the distribution of frequencies ω_λ. Let $Vg(E)dE$ be the number of those normal modes for which $E < \hbar\omega_\lambda < E + dE$, where V is the total volume of the crystal. We call g the density of one-phonon states. With use of Dirac's delta, $\delta(x)$, we may write g as

$$g(E) = V^{-1} \sum_\lambda \delta(E - \hbar\omega_\lambda) \tag{1.4.3}$$

and the thermal energy of the phonon gas as

$$E_\mathrm{T} = V \int_0^\infty dE g(E) E \frac{1}{\exp(E/k_\mathrm{B}T) - 1}. \tag{1.4.4}$$

Phonons of long-wavelength acoustic branches, for which $\omega_\lambda = c_r k$, make a dominant contribution to E_T at low temperatures. The velocities of sound c_r depend also upon the direction of the wave vector $\boldsymbol{k}$. When the crystal is big enough, we may replace the sum over $\boldsymbol{k}$ in (1.4.3) by the integral in k space:

$$V^{-1} \sum_{\boldsymbol{k}} \ldots \to (2\pi)^{-3} \int d\boldsymbol{k} \tag{1.4.5}$$

where $d\boldsymbol{k} = dk_x dk_y dk_z$. We thus see that $g(E) = 12\pi(2\pi\hbar c)^{-3}E^2$ for small E, where c^{-3} is the average of c_r^{-3} over the direction of $\boldsymbol{k}$ and polarization r. Roughly speaking, a contribution of about $k_\mathrm{B}T$ is made to (1.4.4) by each of those modes which satisfy $\hbar\omega_\lambda \lesssim k_B T$. The number of these modes is proportional to T^3, and therefore E_T is proportional to T^4. In fact, taking $E/k_B T$ as the new variable and assuming $g \propto E^2$ up to $E = \infty$, we obtain the expression for the specific heat valid at low enough temperatures

$$C_v = \frac{2}{5}\pi^2 V\left(\frac{k_B T}{\hbar c}\right)^3 k_B = \frac{12}{5}\pi^4 N\left(\frac{T}{\theta}\right)^3 k_B . \tag{1.4.6}$$

The parameter θ defined by $k_B\theta = \hbar c k_m$, $k_m = (6\pi^2 N/V)^{1/3}$, is called the Debye temperature. Note that the number of wave numbers with $k < k_m$ is equal to N. Hence $\hbar c k_m$ is of the order of the maximum phonon energy. For ordinary solids, $k_m \approx 10^8$ cm^{-1}, $c \approx 10^5$ cm·s^{-1}, and therefore $\hbar c k_m \approx 10^{-14}$ erg, $\theta \approx 10^2$ K. The expression (1.4.6) is called the Debye T^3-law, which should hold for the vibrational specific heat of any solids as far as $T \ll \theta$. ·

1.4.3 Creation and Destruction Operations

As we have mentioned, we have to deal with creation and destruction of phonons once we introduce anharmonic terms which do not conserve the total number of phonons. The mathematical representation of creation and destruction processes is provided by the method of second quantization. Thus, instead of operators q_λ, p_λ defined by (1.3.18), we introduce

$$b_\lambda = (2\hbar\omega_\lambda)^{-1/2}(\omega_\lambda q_\lambda + ip_\lambda),\ b_\lambda^\dagger = (2\hbar\omega_\lambda)^{-1/2}(\omega_\lambda q_\lambda - ip_\lambda) \tag{1.4.7}$$

which are Hermitian conjugate with each other. They satisfy commutation relations

$$[b_\lambda, b_{\lambda'}^\dagger] = \delta_{\lambda\lambda'}, \quad [b_\lambda, b_{\lambda'}] = [b_\lambda^\dagger, b_{\lambda'}^\dagger] = 0 \tag{1.4.8}$$

where $[A,B] \equiv AB - BA$. From (1.4.7), we have

$$b_\lambda^\dagger b_\lambda = (2\hbar\omega_\lambda)^{-1}(p_\lambda^2 + \omega_\lambda^2 q_\lambda^2 - \hbar\omega_\lambda)$$

whose eigenvalues are thus phonon numbers $N_\lambda = 0, 1, 2, \ldots$. With use of Dirac's notation, we write the corresponding eigenfunction as $|N_\lambda\rangle$, so that $b_\lambda^\dagger b_\lambda |N_\lambda\rangle = N_\lambda |N_\lambda\rangle$. By the use of (1.4.8), we obtain

$$b_\lambda^\dagger b_\lambda b_\lambda^\dagger |N_\lambda\rangle = b_\lambda^\dagger(1 + b_\lambda^\dagger b_\lambda)|N_\lambda\rangle$$
$$= (N_\lambda + 1) b_\lambda^\dagger |N_\lambda\rangle$$

which shows that $b_\lambda^\dagger |N_\lambda\rangle$ is the eigenfunction belonging to the eigenvalue $N_\lambda + 1$. Thus $b_\lambda^\dagger$ is the creation operator which adds one phonon in the state λ. Similarly b_λ is the destruction operator which annihilates one phonon.

Let $|0\rangle$ be the state function which represents the phonon vacuum. This is the state from which we can annihilate no phonons, so that for all λ

$$b_\lambda |0\rangle = 0 . \tag{1.4.9}$$

This is the mathematical definition of $|0\rangle$. One advantage of the method of second quantization is that we have no need of knowing the explicit form

$$|0\rangle = \prod_\lambda \left[\frac{\omega_\lambda}{\pi\hbar}\right]^{1/4} \exp\left[-\frac{\omega_\lambda}{2\hbar} q_\lambda^2\right]$$

which satisfies (1.4.9) and is normalized.

The state in which we have N_λ phonons may be generated from $|0\rangle$ by means of the recursion formula

$$|N_\lambda\rangle = N_\lambda^{-1/2} b_\lambda^\dagger |N_\lambda - 1\rangle. \tag{1.4.10}$$

From this and (1.4.8), we see

$$b_\lambda |N_\lambda\rangle = N_\lambda^{1/2} |N_\lambda - 1\rangle, \quad b_\lambda^\dagger |N_\lambda\rangle = (N_\lambda + 1)^{1/2} |N_\lambda + 1\rangle. \tag{1.4.11}$$

From the algebraic point of view, all these properties of b operators may be derived solely from the commutation relations (1.4.8). Hence not only the system of phonons, but any system of identical Bose particles can be described by means of operators which satisfy the commutation relations (1.4.8).

1.4.4 Equations of Motion

In general the quantum-mechanical motion of an operator A is given by the Heisenberg representation $A(t) = \exp[(it/\hbar)\mathcal{H}] A \exp[-(it/\hbar)\mathcal{H}]$, which satisfies the equation of motion

$$\frac{dA(t)}{dt} = \frac{i}{\hbar} [\mathcal{H}, A(t)] . \tag{1.4.12}$$

Let us take the phonon operator b as A and the harmonic approximation (1.3.20) as $\mathcal{H}$, which is now written as

$$\mathcal{H} = \sum_\lambda \hbar\omega_\lambda \left(b_\lambda^\dagger b_\lambda + \frac{1}{2}\right). \tag{1.4.13}$$

We have $[\mathcal{H}, b_\lambda] = -\hbar\omega_\lambda b_\lambda$ from (1.4.8), so that

$$\frac{d}{dt} b_\lambda(t) = -i\omega_\lambda b_\lambda(t) . \tag{1.4.14}$$

Since $b_\lambda(0) = b_\lambda$ by definition, $b_\lambda(t) = b_\lambda \exp(-i\omega_\lambda t)$. This has the same form as the classical solution of (1.3.17); the well-known feature of harmonic oscillations.

Assuming that the unit cell contains one atom, we write (1.3.21) as $(NM)^{-1/2}$ $\times \varepsilon_\lambda \exp(i\mathbf{k}\cdot\mathbf{n})$, where $\varepsilon_\lambda = \varepsilon_r(\mathbf{k})$ (r = 1,2,3) are unit vectors of polarization and solutions of (1.3.22). From (1.3.13,16,18,1.4.7), we obtain the displacement vector of the atom as

$$\xi(\mathbf{n}, t) = \sum_\lambda \left(\frac{\hbar}{2NM\omega_\lambda}\right)^{1/2} \varepsilon_\lambda \varphi_\lambda(t) \exp(i\mathbf{k}\cdot\mathbf{n}) \tag{1.4.15}$$

$$\varphi_\lambda(t) = b_\lambda \exp(-i\omega_\lambda t) + b_{-\lambda}^\dagger \exp(i\omega_\lambda t) \tag{1.4.16}$$

where $-\lambda$ means replacing $\mathbf{k}$ by $-\mathbf{k}(\omega_{-\lambda} = \omega_\lambda)$.

1.5 Mössbauer Effect (Rigidity of Solids)

The concept of rigid body in classical mechanics is an idealization of solids with very large stiffness constants. In order to see how rigid a real solid is, however, we have no need of restricting ourselves to classical measurements of elastic moduli. We can measure the rigidity through a more microscopic phenomenon, e.g., the Mössbauer effect, as we will see below.

Suppose that our crystal contains radioactive isotopes whose nuclei undergo a transition from excited to ground levels by emitting a γ-ray. The well-known example is the case of ^{57}Fe, in which the excited level lies $E_0 = 14.4$ keV high above the ground level and has the mean lifetime of 1.45×10^{-7} s. The natural width of the excited level Γ is exceedingly small compared with E_0. For simplicity we neglect Γ.

Suppose first that an isolated nucleus emits a photon of energy E in the direction which makes the angle θ with respect to the translational momentum $\mathbf{P}$ of the nucleus. The nucleus receives recoil momentum and energy. Since the photon momentum is equal to E/c, the change in kinetic energy of the nucleus is given by $\Delta E = (EP/Mc)\cos\theta - (E^2/2Mc^2)$, where c is now the velocity of light and M is the nuclear mass. Since $E_0 = E + \Delta E$ and $\Delta E \ll E$ for $P \ll Mc$, we may replace E in the above expression for ΔE by E_0. Anyway the photon energy E differs from the level separation E_0 by ΔE and this depends on $\mathbf{P}$. For instance, if the nucleus is in a gas of temperature T, where the statistical distribution of $\mathbf{P}$ obeys Maxwell's law, the spectrum of the emitted γ ray is the Gaussian distribution centered at $E_0 - R$ with the width $R[k_B T/R]^{1/2}$. Here $R = E_0^2/2Mc^2$ is the recoil energy of the nucleus whose initial velocity vanishes. For $E_0 = 14.4$ keV of ^{57}Fe, we have $R = 1.95 \times 10^{-3}$ eV, which is much larger than the natural width.

When the nucleus in a crystal emits the photon, the recoil momentum is received by the motion ($k = 0$ acoustic modes) of the center of mass of the whole crystal. The change in its kinetic energy is given by the above expression for ΔE, in which the nuclear mass M is now replaced by the mass NM of the whole crystal, and is therefore negligible in practice. If the interatomic distances in the crystal were kept rigid, the emitted γ-ray would show a sharp spectrum centered at E_0 with the natural width only. This is the Mössbauer effect.

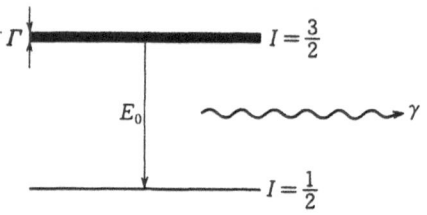

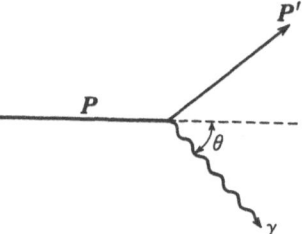

Fig. 1.11. γ-ray emission of ^{57}Fe. I: nuclear spin. $E_0 = 14.4$ eV. $\hbar/\Gamma = 1.45 \times 10^{-7}$ s

Fig. 1.12. Momentum conservation

The real crystal is not a rigid body, so that phonons may be emitted (or absorbed) simultaneously with the γ-ray emission. Let $P(\hbar\omega)d\hbar\omega$ be the probability of the processes in which the recoil energy between $\hbar\omega$ and $\hbar\omega + d\hbar\omega$ is given to the crystal in the form of phonon emission and absorption. Figure 1.13 shows the probability P at $T = 0$; the sharp peak at $\omega = 0$ represents the recoilless γ-ray emission, whereas the broad peak represents the γ-ray emission with the phonon emission. The intensity of the former peak measures the rigidity of the crystal.

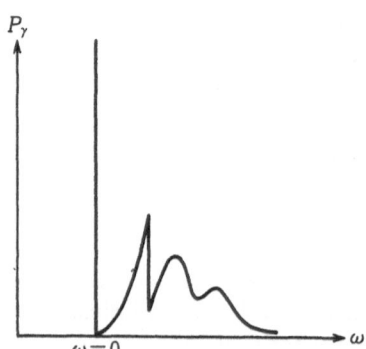

Fig. 1.13. The distribution of recoil energy

1.5.1 Spectrum of Recoil Energy

Let R be the center of mass of the radioactive nucleus and Q be the set of internal coordinates. The interaction with the electromagnetic field contains the term of the form $v(Q)\exp(i\boldsymbol{q}\cdot\boldsymbol{R})$, which gives rise to the emission of the photon of momentum $\hbar\boldsymbol{q}$. We apply the Born approximation to this. By emitting the photon, the nucleus undergoes the transition from excited to ground levels, the center of mass of the whole crystal changes its momentum from $\boldsymbol{P}$ to $\boldsymbol{P}'$, and the system of phonons also undergoes the transition from one eigenstate a of its Hamiltonian to another eigenstate b. We write $\boldsymbol{R} = \boldsymbol{X} + \boldsymbol{n} + \boldsymbol{\xi}$, where $\boldsymbol{X}$ is the center of mass of the crystal, $\boldsymbol{n}$ is a lattice vector, and $\boldsymbol{\xi}$ is the displacement of our radioactive nucleus from its lattice site. Without loss of generality, we take $\boldsymbol{n} = 0$.

The transition matrix element is factorized as

$$\langle f|v|i\rangle\langle b|\exp(i\boldsymbol{q}\cdot\boldsymbol{\xi})|a\rangle\int dX \exp\left[\frac{i}{\hbar}(\boldsymbol{P} - \boldsymbol{P}' - \hbar\boldsymbol{q})\cdot\boldsymbol{X}\right]. \tag{1.5.1}$$

We may ignore $\langle f|v|i\rangle$ to calculate the distribution $P(\hbar\omega)$ of recoil energy. The integral over $\boldsymbol{X}$ gives the momentum conservation $\boldsymbol{P} = \boldsymbol{P}' + \hbar\boldsymbol{q}$. As for the energy conservation, we may ignore the change of the kinetic energy of the crystal as a whole, so that $\hbar\omega = E_0 - \hbar cq$ should be equal to the change of the phonon energy, $E_b - E_a$. We further assume that the phonon system is initially in thermal equilibrium at temperature T. We then obtain $P(\hbar\omega)$ by taking the sum of $|\langle b|\exp(i\boldsymbol{q}\cdot\boldsymbol{\xi})|a\rangle|^2\delta(\hbar\omega + E_a - E_b)$ over final states b and averaging over initial states a with statistical weights $w_a \propto \exp(-E_a/k_B T)$. With use of the Fourier representation

$$2\pi\delta(x) = \int\limits_{-\infty}^{\infty} \exp(ixt)dt$$

and the Heisenberg representation (1.4.15), we can write the result as

$$P(\hbar\omega) = \int\limits_{-\infty}^{\infty} \frac{dt}{2\pi\hbar} K(t)\exp(i\omega t) \tag{1.5.2}$$

$$K(t) = \langle\exp[-i\eta(t)]\exp[i\eta(0)]\rangle . \tag{1.5.3}$$

Here $\eta = \boldsymbol{q}\cdot\boldsymbol{\xi}$, which is the nuclear displacement measured in units of the wavelength of the γ-ray. Note that $\langle A\rangle$ means the expectation value of A in thermal equilibrium, so that

$$\langle A(t)B(t')\rangle = \sum_a w_a \langle a|A(t)B(t')|a\rangle$$

$$= \sum_a\sum_b w_a \langle a|A|b\rangle\langle b|B|a\rangle \times \exp\left[\frac{i}{\hbar}(E_a - E_b)(t - t')\right]. \tag{1.5.4}$$

The total probability obtained by integrating (1.5.2) over $-\infty < \hbar\omega < +\infty$ is equal to $K(0) = \langle 1\rangle = 1$, as it should be.

1.5.2 Theorem of Bloch and DeDominicis

If we apply the harmonic approximation to the phonon system, we can calculate
(1.5.3) in a rather elementary way. Substituting (1.4.15), we write

$$\eta(t) = \sum_\lambda C_\lambda \varphi_\lambda(t) \tag{1.5.5}$$

where $C_\lambda = (\hbar/2NM\omega_\lambda)^{1/2}(\mathbf{q} \cdot \mathbf{s}_\lambda)$. As a brute-force method, we introduce the
expansion of exponentials into powers of η in (1.5.3), so that

$$K(t) = \sum_{p=0}^\infty \sum_{q=0}^\infty \frac{(-\mathrm{i})^p i^q}{p!q!} \langle \eta^p(t)\eta^q(0) \rangle . \tag{1.5.6}$$

Substituting (1.5.5), we see that our problem reduces to the calculation of expec-
tation values $\langle \varphi_1(t) \ldots \varphi_p(t)\varphi_{p+1}(0) \ldots \varphi_{p+q}(0) \rangle$. This type of expectation
value, with which we are frequently confronted in many-body problems, may
be factorized by applying a theorem due to Bloch and DeDominicis. It applies
to any product of operators, each of which is a linear combination of creation
and destruction operators of Bose (or Fermi) particles with the commutation
(or anticommutation) rules of the form (1.4.8) [or (2.3.22)], when the expectation
value of the product is taken with respect to the thermal equilibrium distribution
of noninteracting particles.

The last condition is satisfied in our case of phonons by assuming the har-
monic Hamiltonian (1.4.13), for which the orthonormalized functions $|a\rangle$ in
(1.5.4) can be taken as eigenfunctions of phonon number operators $b_\lambda^\dagger b_\lambda$. The
operator φ_λ emits or absorbs one phonon, so that $\varphi_\lambda|a\rangle$ and $|a\rangle$ are orthogonal
to each other since they belong to different eigenvalues of $b_\lambda^\dagger b_\lambda$. Thus $\langle a|\varphi_\lambda|a\rangle$
$= 0$, and, taking the average over the thermal distribution of $|a\rangle$, we have
$\langle \varphi_\lambda \rangle = 0$. Similarly, the expectation value vanishes for the product of an odd
number of φ operators since it does not conserve the total number of phonons.

Substituting (1.4.16), on the other hand, we obtain

$$\langle \varphi_\lambda(t)\varphi_{-\mu}(t') \rangle = \delta_{\lambda\mu}F_\lambda(t)$$
$$F_\lambda(t) = (1 + \langle N_\lambda \rangle) \exp(-\mathrm{i}\omega_\lambda t) + \langle N_\lambda \rangle \exp(\mathrm{i}\omega_\lambda t) \tag{1.5.7}$$

where $\langle N_\lambda \rangle$ is given by (1.4.2). For the product of an even number of φ operators,
the expectation value is obtained by first pairing each two of φ operators, then
replacing each pair by its expectation value, and finally taking the sum over all
possible ways of pairing. This set of elementary rules of calculation constitutes
the theorem of Bloch and DeDominicis. For example

$$\langle \varphi_1\varphi_2\varphi_3\varphi_4 \rangle = \langle \varphi_1\varphi_2 \rangle \langle \varphi_3\varphi_4 \rangle + \langle \varphi_1\varphi_3 \rangle \langle \varphi_2\varphi_4 \rangle + \langle \varphi_1\varphi_4 \rangle \langle \varphi_2\varphi_3 \rangle . \tag{1.5.8}$$

We will not enter into the proof, but only mention that in the case of Fermi
particles each term on the right should be multiplied by ± 1 depending on

whether the pairing needs an even or odd permutation of operators (see Sect. 3.4).

Applying the theorem to (1.5.6), let us first take the sum of those terms which nowhere contain factors of the form $\langle \varphi(t)\, \varphi(0) \rangle$ obtained by pairing φ operators at different times. Then both p and q are even, so that we write them $2r$ and $2s$, respectively. There are $(2r - 1)!$ ways of pairing $\varphi(t)$ operators and $(2s - 1)!$ ways of pairing $\varphi(0)$ operators, for given p and q. All of them give the same contribution to (1.5.6) and the sum we are interested in for the moment takes the form

$$K^{(0)}(t) = \sum_{r=0}^{\infty} \sum_{s=0}^{\infty} \frac{(-1)^{r+s}}{2^{r+s} r! s!} g^{r+s}(0)$$

$$= \exp[-g(0)] \tag{1.5.9}$$

with

$$g(t - t') = \sum_{\lambda} |C_{\lambda}|^2 F_{\lambda}(t - t')$$

$$= \langle \eta(t)\, \eta(t') \rangle. \tag{1.5.10}$$

We now turn to those terms which contain one $g(t)$. The simplest example is the term with $p = 1, q = 1$, which is identical with $g(t)$. Terms with $p = 3$, $q = 1$ and $p = 1, q = 3$ give $(i^2/2!)g^2(0)g(t)$ and so on. Thus the sum of terms containing one $g(t)$ is obtained as

$$K^{(1)}(t) = \sum_{r=0}^{\infty} \sum_{s=0}^{\infty} \frac{(-1)^{r+s}}{2^{r+s} r! s!} g^{r+s}(0) g(t)$$

$$= g(t) \exp[-g(0)]. \tag{1.5.11}$$

Similarly, the sum of terms containing $g^n(t)$ is equal to $(g^n(t)/n!)\exp[-g(0)]$, so that

$$K(t) = \exp[g(t) - g(0)]. \tag{1.5.12}$$

1.5.3 The Intensity of the Recoilless γ-Ray

The function (1.5.10) represents the correlation of the position of the same atom at different times and vanishes as $t - t' \to \infty$. Its Fourier transform may be written as

$$I(E) = \int_{-\infty}^{\infty} \frac{dt}{2\pi\hbar}\, g(t) \exp\left(\frac{i}{\hbar}\, Et\right)$$

$$= \sum_{\lambda} C_{\lambda}^2 [1 + N(E)]\, A_{\lambda}(E). \tag{1.5.13}$$

Here $N(E) = [\exp(E/k_{\rm B}T) - 1]^{-1}$ is Planck's distribution and

$$A_\lambda(E) = \delta(E - \hbar\omega_\lambda) - \delta(E + \hbar\omega_\lambda) . \tag{1.5.14}$$

For a three-dimensional crystal, $I(E)$ is a continuous function of E, and its Fourier transform $g(t)$ vanishes as $t \to \infty$.

When we are interested in the Mössbauer effect, we are concerned with (1.5.2) in the vicinity of $\omega = 0$ and therefore with $g(t)$ for large t. Hence we may expand (1.5.12) into powers of $g(t)$ and ignore terms of second and higher order. Thus

$$P(\hbar\omega) = \exp[-g(0)] \{\delta(\hbar\omega) + I(\hbar\omega) + \ldots\} . \tag{1.5.15}$$

The first term on the right is the probability density of the emission of the recoilless γ-ray. The second term is the probability density of γ-ray emission accompanied by emission or absorption of one phonon, as we see from the expression (1.5.13). Integrating the first term over $\hbar\omega$, we obtain the probability of the recoilless emission as $P_0 = \exp[-g(0)] = \exp[-\langle\eta^2\rangle]$, which is closer to unity for the smaller amplitude of the atomic vibration. The latter may be estimated by applying the simple Debye model $\omega_\lambda = c_s k$ to

$$\langle\eta^2\rangle = \sum_\lambda \frac{\hbar}{2NM\omega_\lambda} (\boldsymbol{q}\cdot\boldsymbol{s}_\lambda)^2(2\langle N_\lambda\rangle + 1) . \tag{1.5.16}$$

We replace the sum over wave vectors by the integral as (1.4.5) and cut it off at $k = k_m$ defined there. When the temperature T is much lower than the Debye temperature $\theta_D = \hbar c_s k_m/k_B$, (1.5.16) is mainly determined by the zero-point vibration and approaches the value

$$\langle\eta^2\rangle = \frac{\hbar q^2 V}{4\pi^2 NMc_s} \int_0^{k_m} k dk = \frac{3R}{2k_B\theta_D} . \tag{1.5.17}$$

Here $R = \hbar^2 q^2/2M$ is the recoil energy of the free atom. For ^{57}Fe, $R/k_B \approx 24$ K, $\theta_D \approx 420$ K, so that $P_0 \approx 0.9$ at low temperatures.

For $T \gg \theta_D$, (1.5.17) is multiplied by $(2T/\theta_D)$ and P_0 is thus much reduced by thermally excited phonons. In connection with this, it is to be noted that a rough estimate of the melting point of the crystal may be obtained as a temperature around which the average amplitude $\langle\xi^2\rangle^{1/2}$ calculated by the harmonic approximation is no longer negligible compared with the lattice constant a (being, say, 30% of the latter), though in actuality anharmonic terms should be quite important near the melting point. Such a type of criterion for melting is usually referred to as the Lindemann law. With use of our notation, it says that $\langle\eta^2\rangle \approx 0.1$ at the melting temperature for $q = k_m$. Provided that this temperature is higher than θ_D, we have $k_B T \approx 0.06Mc_s^2$.

Finally we notice that the integral (1.5.17) is divergent at $k = 0$ in the case of the one-dimensional crystal, since the volume element $\boldsymbol{dk}$ of k space is simply dk. The divergence implies that the atomic displacement due to long-wavelength zero-point vibrations would be divergent in the thermodynamic limit and therefore that the one-dimensional crystal would not be stable. In the case of the two-dimensional crystal, $dk = 2\pi k dk$, so that the zero-point amplitude is finite. But the integral is again divergent if we include $\langle N_\lambda \rangle$ in (1.5.16) at $T > 0$.

1.6 Inelastic Scattering of Neutrons and Phonon Spectrum

1.6.1 Van Hove's Formula

As we have mentioned in Sect.1.3, the inelastic scattering of neutrons will give us information about the dynamic behavior of atoms, in particular about phonons in our case of crystals. In Sect.1.3, we have discussed only the elastic scattering by a rigid lattice. If we take into account atomic vibrations, $R_j = n + \xi(n)$ in (1.3.3), so that ρ_q is now a dynamical variable, where we have denoted the mementum change $p - p'$ of the neutron by $\hbar q$. Note that the density of atoms at the point r inside the crystal is represented by the operator $\rho(r) = \sum_j \delta (r - R_j)$, whose Fourier coefficients are

$$\rho_q = \int dr\, \rho(r) \exp(iq \cdot r)\,.$$

Again we regard the interaction between the neutron and atoms, (1.3.2), as perturbation and apply the Born approximation to the scattering process, by which the neutron changes its momentum from p to p' and at the same time the crystal undergoes the transition from one eigenstate a to another eigenstate b of its unperturbed Hamiltonian. By thinking explicitly of the center of mass motion of the whole crystal, we see that the momentum change $p - p'$ of the neutron should be equal to the increase of momentum of the center of mass motion and also that the change of the kinetic energy of the crystal as a whole can be neglected. The energy loss $\hbar\omega = (2m_n)^{-1}(p^2 - p'^2)$, where m_n is the neutron mass, should be equal to the change of phonon energy, $E_b - E_a$. In the same way as in the preceding section, we take the sum of the transition probabilities over final states b and also the average over the thermal equilibrium distribution of initial states a. Thus, as the transition probability per unit time for the neutron to transfer momentum $\hbar q$ and energy $\hbar\omega$ to the crystal by scattering, we obtain the expression

$$W(q,\omega) = \frac{2\pi}{\hbar}\, |v_q|^2 S(q, \omega) \tag{1.6.1}$$

with

$$S(q, \omega) = \sum_a \sum_b w_a |\langle b|\rho_q|a\rangle|^2 \delta(\hbar\omega + E_a - E_b)$$

$$= \int_{-\infty}^{\infty} \frac{dt}{2\pi\hbar}\, S(q; t) \exp(i\omega t)\,. \tag{1.6.2}$$

We have made use of (1.5.4) and the Fourier representation of Dirac's delta, so that

$$S(q; t) = \langle \rho_{-q}(t)\, \rho_q(0) \rangle . \tag{1.6.3}$$

Note that the Fourier coefficient ρ_q represents the spatial fluctuation of the atomic density and that $S(q;t)$ describes its temporal correlation. The Fourier transform $S(q, \omega)$ is called the dynamical structure factor, or the spectral function, of the crystal. The expression (1.6.1), transcribed usually in the form of the differential cross section, is referred to as Van Hove's formula, which clearly shows that we observe the spectrum of the density fluctuation through the inelastic scattering. The formula is applicable to inelastic scatterings of particles in general as far as the Born approximation holds good.

If we do not select the energy of scattered particles, as is the case in x-ray scattering experiments, we should integrate (1.6.1) over ω. We write

$$\int\limits_{-\infty}^{\infty} W(q, \omega)\, d\hbar\omega = \frac{2\pi}{\hbar} |v_q|^2 N S(q) \tag{1.6.4}$$

where

$$S(q) = \frac{1}{N} \int\limits_{-\infty}^{\infty} S(q, \omega)\, d\omega = \frac{1}{N} \langle \rho_{-q}(0)\, \rho_q(0) \rangle \tag{1.6.5}$$

is the usual structure factor of the crystal.

1.6.2 Dynamical Structure Factors in the Harmonic Approximation

In the harmonic approximation we have (1.4.15) and

$$\rho_q(t) = \exp(iq \cdot n) \exp[iq \cdot \xi(n, t)] .$$

Substituting this in (1.6.3) we obtain

$$S(q; t) = \sum_{nm} \sum \exp[i(m - n) \cdot q] K_{nm}(t) \tag{1.6.6}$$

$$K_{nm}(t) = \langle \exp[-i\eta_n(t)]\, \exp[i\eta_m(0)] \rangle \tag{1.6.7}$$

$$\eta_n(t) = \sum_{\lambda} C_\lambda \exp(ik \cdot n)\, \varphi_\lambda(t) . \tag{1.6.8}$$

We expand exponentials in (1.6.7) into powers of η and apply the theorem of Bloch and DeDominicis. The sum of terms, in which two operators at different t values are never paired, is given again by (1.5.9). Substituting this in (1.6.6), taking the Fourier transform, and remembering (1.3.7), we obtain

$$S_0(q, \omega) = N^2 \delta_{q, K} \delta(\hbar\omega) \exp[-g(0)] , \tag{1.6.9}$$

where K is the reciprocal lattice vector defined by (1.3.5). Thus (1.6.9) is the contribution to the structure factor from elastic scatterings without phonon emission and absorption; the factor $\delta_{q,K}$ means the Bragg condition (1.3.8) and $\delta(\hbar\omega)$ means the vanishing energy loss of the neutron. It is not surprising, therefore, that (1.6.9) is proportional to the same factor $\exp[-g(0)]$ as the intensity of the Mössbauer γ-ray. This factor has been known for many decades in x-ray scattering theory by the name of Debye-Waller factor.

Now, for scatterings with one phonon emission or absoption, we have the correlation function

$$g_{nm}(t) = \langle \eta_n(t)\,\eta_m(0) \rangle$$
$$= \sum_\lambda C_\lambda^2 F_\lambda(t) \exp[i\mathbf{k}\cdot(\mathbf{n} - \mathbf{m})] \tag{1.6.10}$$

which corresponds to (1.5.11). Note that k in (1.6.10) is the wave vector to label normal modes and also that ω_λ, $\langle N_\lambda \rangle$ are invariant under $k \to -k$. The contribution of one-phonon processes to (1.6.7) is given by $g_{nm}(t)\exp[-g(0)]$, whose Fourier transform results in the structure factor

$$S_1(\mathbf{q},\omega) = N \exp[-g(0)] \sum_\lambda C_\lambda^2 \delta_{q,\,k+K}\,[1 + N(\hbar\omega_\lambda)]\,A_\lambda(\hbar\omega)\,. \tag{1.6.11}$$

Any wave vector q may be expressed uniquely as the sum of reduced wave vector and reciprocal lattice vector as

$$q = k + K\,. \tag{1.6.12}$$

We regard the sum over λ on the right of (1.6.11) as a function of ω for fixed q. We have a finite contribution only from those modes whose reduced wave vector k satisfies (1.6.12). We denote their indices of polarization by $r = 1,2,3$, supposing that the unit cell contains one atom. Then we expect three peaks at $\omega = \omega_{kr}$, corresponding to the phonon emission and three peaks at $\omega = -\omega_{kr}$ corresponding to the phonon absorption (Fig. 1.14). The latter peaks disappear as $T \to 0$, since the neutron cannot absorb energy from the crystal in its ground state. Anyway, by observing one-phonon peaks of the neutron inelastic scattering, we are able to know the energy spectrum of the phonon.

$S(\mathbf{q},\omega)$

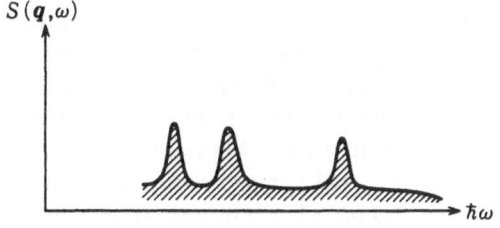

Fig. 1.14. One-phonon peaks of the cross section of inelastic neutron scattering (q = const.)

Integrating (1.6.11) over ω, we obtain the contribution of the one-phonon processes to the static structure factor (1.6.5). This part gives rise to the so-called diffuse scattering; it remains finite for any reduced wave vector k in (1.6.12), but the intensity is N^{-1} times as small as the so-called Laue spot which is obtained by integrating (1.6.9) over ω.

In contrast to one-phonon processes, we do not have isolated peaks in the case of multiphonon processes, in which two or more phonons are emitted or absorbed. In the case of the two phonon emission, for example, the conservation law of wave vectors takes the form $q = k_1 + k_2 + K$ and, for given q, determines only the sum $k_1 + k_2$ uniquely, so that the difference $k_1 - k_2$ may vary continuously in a certain range. Hence $\hbar\omega = \hbar\omega_1 + \hbar\omega_2$ may also form a continuous spectrum.

1.7 Anharmonic Terms

In classical mechanics, anharmonic terms mean the coupling between normal modes; even if only one mode is excited initially, it will decay by dissipating the vibrational energy to other modes. Since the number of other modes is macroscopic, it is most unlikely that we have the same concentration of energy in the particular mode as in the initial state. Thus the particular mode we are interested in may be regarded as an oscillator oscillating in a dissipative medium which represents the role of all other modes. The simplest model is the well-known damped oscillation described by $\ddot{x} - 2\gamma\dot{x} - \omega_0^2 x = 0$, whose solution $x \propto \exp(-\gamma t)\sin(\omega_0 - \Delta\omega)t$ is characterized by the decay time γ^{-1} and the frequency shift $\Delta\omega = \gamma^2/2\omega_0$ in the case of small damping ($\gamma \ll \omega_0$).

In quantum mechanics, anharmonic terms bring about the phonon–phonon collisions. For instance, terms cubic in atomic displacements give the interaction Hamiltonian

$$\mathscr{H}_1 = \sum B(\lambda_1, \lambda_2, \lambda_3)\, \varphi_{\lambda_1}\varphi_{\lambda_2}\varphi_{\lambda_3} \,. \tag{1.7.1}$$

The coefficient B is symmetric in $\lambda_1, \lambda_2, \lambda_3$, and reduced wave vectors of three modes should satisfy

$$k_1 + k_2 + k_3 = K \tag{1.7.2}$$

where K is the reciprocal lattice vector. When (1.7.1) is added to the harmonic Hamiltonian, it will give rise to collision processes involving three phonons; the term of the form $b_1^\dagger b_2 b_3$, for instance, annihilates two phonons and creates one phonon. Because of phonon–phonon collisions, the one-phonon state will also be subject to a certain energy shift as well as to a finite life time. The spectral function in (1.5.13,1.6.11) has infinitely sharp peaks in the harmonic approximation (1.5.14). These peaks will be shifted and broadened by anharmonic terms.

The width Γ is connected with the phonon life-time τ by the uncertainty relation $\Gamma\tau \approx \hbar$.

Anharmonic terms also provide the mechanism, through which the thermal equilibrium of the phonon gas is established. It is important that we have the reciprocal lattice vector $\boldsymbol{K}$ on the right of (1.7.2). Terms with $\boldsymbol{K} = 0$ in (1.7.1) are called normal processes and those with $\boldsymbol{K} \neq 0$, umklapp processes. If the latter processes are ignored, $\boldsymbol{P} = \sum_{\lambda} \hbar\boldsymbol{k}N_{\lambda}$ is conserved by collisions. In the presence of a macroscopic flow of energy in the phonon gas, N_{λ} is not invariant under $\boldsymbol{k} \to -\boldsymbol{k}$, so that $\boldsymbol{P} \neq 0$. Hence the energy flow cannot attain its equilibrium value $(= 0)$ through normal processes only, since $\boldsymbol{P} = 0$ for the equilibrium Planck distribution. That means that the thermal conductivity of the phonon gas is infinite; the temperature fluctuation in this case obeys a wave equation, instead of obeying the ordinary equation of heat conduction, and the wave of the temperature fluctuation is called the second sound in contradistinction to the ordinary elastic wave which is called the first sound. In order to obtain a finite thermal conductivity of the phonon gas, umklapp processes are indispensable.

1.7.1 General Definition of the Spectral Function

It is beyond the scope of the present volume to discuss every effect caused by anharmonic terms. We will restrict ourselves to energy shift and life time of the phonon, which can be observed as shift and width of the one-phonon peak of the neutron inelastic scattering. We will now show how to describe these concepts mathematically.

In this section, we denote the harmonic Hamiltonian (1.4.13) by $\mathscr{H}_0$ and the total Hamiltonian including the anharmonic part $\mathscr{H}_1$ by

$$\mathscr{H} = \mathscr{H}_0 + \mathscr{H}_1 \qquad (1.7.3)$$

In order to avoid confusion, we use capital letters to denote the Heisenberg operators whose time dependence is defined in terms of (1.7.3) and small letters to denote the Heisenberg operators whose time dependence is defined in terms of $\mathscr{H}_0$:

$$\left. \begin{aligned} \Phi_{\lambda}(t) &= \exp\!\left(\frac{i}{\hbar}\,\mathscr{H}t\right) \varphi_{\lambda} \exp\left(-\frac{i}{\hbar}\,\mathscr{H}t\right) \\ \varphi_{\lambda}(t) &= \exp\!\left(\frac{i}{\hbar}\,\mathscr{H}_0 t\right) \varphi_{\lambda} \exp\left(-\frac{i}{\hbar}\,\mathscr{H}_0 t\right) \end{aligned} \right\}. \qquad (1.7.4)$$

When anharmonic terms are taken into account, F_{λ} in the correlation function (1.6.10) to determine the one-phonon peak is no longer given by (1.5.7). This must be replaced by

$$F_{\lambda}(t - t') = \langle \Phi_{\lambda}(t)\,\Phi_{-\lambda}(t') \rangle \qquad (1.7.5)$$

where $\langle \ldots \rangle$ now means the expectation value with respect to the thermal equilibrium distribution defined in terms of *the total* Hamiltonian $\mathscr{H}$. Hence $|a\rangle$, $|b\rangle$ in (1.5.4) are now eigenstates of $\mathscr{H}$, so that

$$\int_{-\infty}^{\infty} \frac{dt}{2\pi\hbar} F_\lambda(t) \exp\left(\frac{i}{\hbar} Et\right) = [1 + N(E)] A_\lambda(E) . \tag{1.7.6}$$

The spectral function is defined by

$$\begin{aligned}
A_\lambda(E) &= \sum_a \sum_b w_a \langle a | \varphi_{-\lambda} | b \rangle \langle b | \varphi_\lambda | a \rangle \cdot \delta(E + E_b - E_a) \left[\exp\left(\frac{E}{k_B T}\right) - 1 \right] \\
&= \int_{-\infty}^{\infty} \frac{dt}{2\pi\hbar} \langle [\Phi_\lambda(t), \Phi_{-\lambda}(0)] \rangle \exp\left(\frac{i}{\hbar} Et\right) .
\end{aligned} \tag{1.7.7}$$

In terms of this, the one-phonon part of the dynamical structure factor keeps the same form as (1.6.11). Since $\varphi_\lambda^\dagger = \varphi_{-\lambda}$, $\langle a | \varphi_{-\lambda} | b \rangle = \langle b | \varphi_\lambda | a \rangle^*$, (1.7.7) is positive for $E > 0$ and negative for $E < 0$. With use of the time-reversal invariance of the Hamiltonian, we can prove that $A_\lambda(E) = -A_\lambda(-E)$. Finally we easily check that (1.7.7) reduces to (1.5.14) when $\mathscr{H}_1$ is ignored.

1.7.2 Retarded Green's Functions

As was mentioned in Sect. 1.1, information may be obtained about elementary excitations by observing the response of our macroscopic system against some external force. In our case of the crystal, we may apply an ultrasonic oscillator to produce forced oscillations of atoms near the surface and observe the propagation of the elastic wave through the crystal. It is well known in the theory of differential equations that forced solutions of classical wave equations with source terms, describing elastic body applied by external stress, electromagnetic field with external charge, and current, etc., can be expressed in the integral form with use of Green's functions. The Green's function we are going to introduce below may be regarded as the quantum mechanical version of the classical one.

Suppose that we expand the potential of the external force applied upon the crystal into powers of atomic displacements and retain linear terms only. The perturbing Hamiltonian added to (1.7.3) then has the form $\mathscr{H}_{ext}(t) = \sum_\lambda f_\lambda(t) \varphi_{-\lambda}$, where $f_\lambda(t)$ represents a weak external force coupled with the λ mode. We will restrict ourselves to linear effects caused by f_λ (linear response). Then we need to know only the response to the external force which acts only on the one particular mode. A linear combination of such responses represents the general linear response. Similarly we need only to consider the pulse of the form $f_\lambda(t) = f\delta(t)$, since the general force may be expressed as a linear combination of such pulses.

Suppose that the phonon system is in the eigenstate $|a\rangle$ of the Hamiltonian (1.7.3) before the pulse is applied ($t < 0$). The state function $|t\rangle$ for $t > 0$ is obtained as the solution of the Schrödinger equation

$$i\hbar \frac{\partial}{\partial t} |t\rangle = \mathcal{H} |t\rangle + \mathcal{H}_{ext}(t)|a\rangle \qquad (1.7.8)$$

where in the second term on the right we have replaced $|t\rangle$ by the initial state $|a\rangle$ since we need to know $|t\rangle - |a\rangle$ up to the first order of f. Remembering that $\mathcal{H}_{ext}(t) \propto \delta(t)$, we integrate (1.7.8) over t from $t = 0^-$ to $t = 0^+$ and obtain $|0^+\rangle = |0^-\rangle - (i/\hbar)f_{-\lambda}|0^-\rangle$, where $|0^-\rangle = |a\rangle$. For $t > 0$, where $\mathcal{H}_{ext}(t) = 0$, we have $|t\rangle = \exp[-(it/\hbar)\mathcal{H}] |0^+\rangle$.

With use of this solution, we now form the quantum-mechanical expectation value $\langle t| \varphi_\lambda |t\rangle$ of the operator φ_λ. Generalizing the initial condition slightly, we assume that the phonon system is in thermal equilibrium before the pulse is applied and therefore take the average over the equilibrium distribution of initial states $|a\rangle$. Then, to zeroth order, the expectation value of the phonon field is equal to the equilibrium expectation value $\langle \varphi_\lambda \rangle$ in the absence of the external force. We may assume $\langle \varphi_\lambda \rangle = 0$ since otherwise the crystal should undergo a *spontaneous* deformation and thus the original crystal structure is not stable (see Sect.4.4).

We can easily obtain the linear part of $\langle t| \varphi_\lambda |t\rangle$ in the form $(f/\hbar)D_r(\lambda;t)$, where

$$D_r(\lambda;t) = -i\theta(t) \langle [\Phi_\lambda(t), \Phi_{-\lambda}(0)]\rangle . \qquad (1.7.9)$$

We have introduced the step function $\theta(t)$, which is equal to unity for $t > 0$ and vanishes for $t < 0$, so that we can use the same expression for $t < 0$ as well as for $t > 0$. Since (1.7.9) describes the propagation for $t > 0$ of the disturbance produced in the crystal by the unit pulse applied at $t = 0$, it is called the retarded Green's function of the phonon field, or more simply the phonon propagator.

The phonon propagator in the harmonic approximation, denoted by $D_r^{(0)}$, is easily obtained by substituting (1.4.16) in (1.7.9). Thus $D_r^{(0)} = -2\theta(t)\sin \omega_\lambda t$, which satisfies the inhomogeneous wave equation

$$\left(\frac{\partial^2}{\partial t^2} - \omega_\lambda^2\right) D_r^{(0)}(\lambda; t) = -2\omega_\lambda \delta(t) \qquad (1.7.10)$$

and vanishes for $t < 0$.

From (1.5.4) and (1.7.7), we see

$$D_r(\lambda, E) = \lim_{\delta \to +0} \int_{-\infty}^{\infty} \frac{dt}{\hbar} D_r(\lambda; t) \exp\left[\frac{i}{\hbar} (E + i\delta)t\right]$$

$$= \int_{-\infty}^{\infty} \frac{A_\lambda(x)}{E - x + i0^+} dx . \qquad (1.7.11)$$

We have added $i\delta$ to E so that the integral is convergent at $t = +\infty$. Taking the imaginary part of (1.7.11) and remembering $\mathrm{Im}\,(x + i0^+)^{-1} = -\pi\delta(x)$, we have

$$A_\lambda(E) = -\frac{1}{\pi}\,\mathrm{Im}\,\{D_r(\lambda, E)\}. \tag{1.7.12}$$

This enables us to derive the spectral function from the Green's function once we know the latter.

From the form of $D_r^{(0)}$ in the harmonic approximation, we expect that D_r in the presence of anharmonic terms will be given approximately by the following damped oscillation ($\Gamma_\lambda > 0$):

$$D_r(\lambda; t) \approx -2\theta(t)\exp\left(-\frac{1}{\hbar}\,\Gamma_\lambda t\right)\sin\left(\omega_\lambda + \frac{1}{\hbar}\,\Delta_\lambda\right)t$$

$$D_r(\lambda; E) \approx \frac{1}{E - \hbar\omega_\lambda - \Delta_\lambda + i\Gamma_\lambda} - \frac{1}{E + \hbar\omega_\lambda + \Delta_\lambda + i\Gamma_\lambda}. \tag{1.7.13}$$

Substituting in (1.7.12), we obtain

$$A_\lambda(E) \approx \frac{1}{\pi}\left[\frac{\Gamma_\lambda}{(E - \hbar\omega_\lambda - \Delta_\lambda)^2 + \Gamma_\lambda^2} - \frac{\Gamma_\lambda}{(E + \hbar\omega_\lambda + \Delta_\lambda)^2 + \Gamma_\lambda^2}\right] \tag{1.7.14}$$

which shows two Lorentzian peaks at $E = \pm(\hbar\omega_\lambda + \Delta_\lambda)$ with the width Γ_λ. Thus Δ_λ and $\hbar/\Gamma_\lambda$ are energy shift and lifetime of the phonon, respectively.

1.8 Temperature Green's Function and Perturbation Expansion

One method to justify (1.7.13) and obtain expressions for Δ_λ, Γ_λ is to regard $\mathscr{H}_1$ as perturbation and apply the perturbation expansion to the Green's function. In order to carry out the expansion in a systematic way, however, it is much simpler to deal with the so-called temperature Green's function rather than the retarded Green's function. We will therefore explain the former concept first.

1.8.1 Thermal Green's Functions

Let us introduce the real parameter τ which runs from 0 to $\beta = (k_B T)^{-1}$, where T is the temperature of the system. Instead of the usual Heisenberg representation (1.7.4), we define

$$\Phi_\lambda(\tau) = \exp(\tau\mathscr{H})\,\varphi_\lambda\exp(-\tau\mathscr{H}) \tag{1.8.1}$$

and the thermal Green's function

$$\mathscr{D}(\lambda; \tau - \iota') = -\langle T_\tau \Phi_\lambda(\tau)\Phi_{-\lambda}(\tau')\rangle .$$ (1.8.2)

Here T_τ is called Wick's ordering operator and rearranges following quantum-mechanical operators in such an order that the operator with bigger τ stands to the left of the operator with smaller τ. Therefore, in the case of (1.8.2), the order of operators remains the same as it is written if $\tau > \tau'$, but should be rearranged as $\Phi_{-\lambda}(\tau')\Phi_\lambda(\tau)$ if $\tau < \tau'$. Replacing $it/\hbar$ by τ in (1.5.4), we can check that (1.8.2) is a function of $\tau - \tau'$ and has the periodicity $\mathscr{D}(\lambda; \tau - \tau' + \beta) = \mathscr{D}(\lambda; \tau - \tau')$. Hence we can expand it into the Fourier series as

$$\mathscr{D}(\lambda; \tau) = \beta^{-1} \sum_{l=-\infty}^{+\infty} \mathscr{D}(\lambda, i\nu_l) \exp(-i\nu_l\tau)$$

$$\nu_l = 2\pi l/\beta .$$ (1.8.3)

The Fourier coefficient may be expressed in terms of the spectral function as

$$\mathscr{D}(\lambda, i\nu_l) = \int_0^\beta d\tau\, \mathscr{D}(\lambda; \tau) \exp(i\nu_l\tau)$$

$$= \int_{-\infty}^\infty \frac{A_\lambda(E)}{i\nu_l - E} dE .$$ (1.8.4)

Comparing this with (1.7.11), we see the relation between thermal and retarded Green's functions. According to the theory of analytic functions, when $A_\lambda(E)$ is a continuous function of E,

$$\mathscr{D}(\lambda, z) = \int_{-\infty}^\infty \frac{A_\lambda(E)}{z - E} dE$$ (1.8.5)

defines two analytic functions on upper and lower half-planes of the complex z plane, respectively. These functions agree with (1.8.4) at $z = i\nu_l$ on the imaginary axis, and therefore provide analytic continuations of the latter. When z approaches a point E on the real axis from above, in particular, the limit $\mathscr{D}(\lambda, E + i0^+)$ gives the retarded Green's function (1.7.11).

1.8.2 Perturbation Expansions

The operator $\Phi(\tau)$ in (1.8.2) contains $\mathscr{H}_1$ through $\mathscr{H}$ in the definition (1.8.1) and also the thermal expectation value $\langle\ldots\rangle$ depends upon $\mathscr{H}_1$ through the Boltzmann factor $\exp(-\mathscr{H}/k_B T)$ in its definition. At first sight, therefore, it appears that we have multiple expansions into powers of $\mathscr{H}_1$. In actuality, however, we have the following simple expansion formula:

$$\mathscr{D}(\lambda; \tau - \tau') = - \sum_{n=0}^\infty \frac{(-1)^n}{n!} \int_0^\beta d\tau_1 \ldots \int_0^\beta d\tau_n \langle T_\tau \mathscr{H}_1(\tau_1) \ldots$$

$$\mathscr{H}_1(\tau_n)\varphi_\lambda(\tau)\varphi_{-\lambda}(\tau')\rangle_{0c} .$$ (1.8.6)

Operators on the right are so-called interaction representations defined as $A(\tau) = \exp(\tau \mathscr{H}_0) A \exp(-\tau \mathscr{H}_0)$. Similarly $\langle Q \rangle_0$ means the unperturbed expectation value defined by $\langle Q \rangle_0 = \text{tr} \{Q \exp(-\mathscr{H}_0/k_B T)\} \cdot [\text{tr} \{\exp(-\mathscr{H}_0/k_B T)\}]^{-1}$. The meaning of the symbol c will be explained below. That the perturbation expansion takes such a simple form is the greatest advantage of introducing thermal Green's functions, whose definition may look somewhat artificial at first sight. The expansion is not so simple if perturbation theory is applied directly to the retarded Green's function. We will not enter into the proof of the expansion (1.8.6).

In (1.8.6), the unperturbed expectation value $\langle \ldots \rangle_0$ can be factorized again with use of the Bloch-DeDominicis theorem. In doing so, it is convenient to depict the way of pairing operators with use of so-called Feynman diagrams. Thus, take $n + 2$ points $\tau_1, \ldots, \tau_n, \tau, \tau'$ on a sheet of paper. We call $\tau_1, \ldots, \tau_n$ internal vertices and τ, τ' external vertices. Corresponding to $3n + 2$ φ-operators appearing in the nth order $\langle \ldots \rangle_0$, draw three wavy lines from each one of the internal vertices and one wavy line from each of the external vertices, respectively. The pairing of two φ-operators is then represented by connecting two wavy lines together. For $n = 0$, for example, we have two φ operators only, so that we have only one way of pairing represented by the diagram

$$\tag{1.8.7}$$

We call it the phonon line. It is convenient analytically to suppose that the line represents $\langle T_\tau \varphi_\lambda(\tau) \varphi_{-\lambda}(\tau') \rangle_0 = - \mathscr{D}^{(0)}(\lambda; \tau - \tau')$.

We have no need to think of terms with odd n, since they contain odd numbers of φ-operators and therefore have vanishing expectation values. For $n = 2$, we have eight φ-operators and therefore obtain a number of terms when it is factorized. One of them has the form

$$- \frac{1}{2} \sum_{\lambda \mu \nu} |B(-\lambda, \mu, \nu)|^2 \int_0^\beta d\tau_1 \int_0^\beta d\tau_2 \mathscr{D}^{(0)}(\lambda; \tau - \tau_1) \, \mathscr{D}^{(0)}(\mu; \tau_1 - \tau_2)$$
$$\times \, \mathscr{D}^{(0)}(\nu; \tau_1 - \tau_2) \, \mathscr{D}^{(0)}(\lambda; \tau_2 - \tau')$$

which is represented by the diagram

$$\tag{1.8.8}$$

The phonon line connecting two internal vertices is called an internal line and the one connecting external and internal vertices is called an external line. We also have the term which is obtained by exchanging τ_1 with τ_2 in (1.8.8) and therefore makes the same contribution when integrated over τ_1, τ_2. Adding these two terms together, we can forget the factor 1/2. Thus the contribution of this type to $-\mathscr{D}$ is obtained by regarding each phonon line as the factor $-\mathscr{D}^{(0)}$ and each internal vertex as the interaction constant $-B$, then by taking the product of all these factors, and finally by integrating over τ_1, τ_2. Furthermore, corresponding to different combinations of λ indices, we have 18 terms which are represented by diagrams of the same form as (1.8.8) and make the same contribution because of the symmetry of the coefficient $B(\lambda,\mu,\nu)$.

For $n = 2$, we have one further way of pairing, which is represented by the topologically distinct diagram

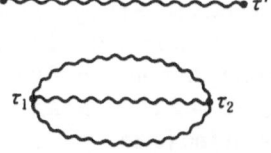

The symbol c in the expansion formula (1.8.6) means that we should disregard all such disconnected diagrams consisting of two or more subdiagrams which are not connected with each other by lines.

Integration over τ_1, τ_2 can be carried out conveniently by introducing the Fourier expansion (1.8.3). Substituting (1.5.13) in (1.8.4), we have the free-phonon propagator

$$\mathscr{D}^{(0)}(\lambda, i\nu_l) = \frac{-2\hbar\omega_\lambda}{\nu_l^2 + (\hbar\omega_\lambda)^2} \cdot \tag{1.8.9}$$

The second-order term of (1.8.6), when Fourier transformed, then takes the form

$$\mathscr{D}^{(2)}(\lambda, i\nu_l) = \mathscr{D}^{(0)}(\lambda, i\nu_l)\, \Pi^{(1)}(\lambda, i\nu_l)\, \mathscr{D}^{(0)}(\lambda, i\nu_l) \tag{1.8.10}$$

where

$$\Pi^{(1)}(\lambda, i\nu_l) = -\left(\frac{18}{\beta}\right) \sum_{n=-\infty}^{\infty} |B(-\lambda, \mu, \rho)|^2 \mathscr{D}^{(0)}(\mu, i\nu_n)\, \mathscr{D}^{(0)}(\rho, i\nu_l - i\nu_n)\,. \tag{1.8.11}$$

The second-order term (1.8.10) may be represented by the diagram of the same form as (1.8.8), in which we now regard each phonon line as the factor $-\mathscr{D}^{(0)}$ and take sums over ν variables associated with internal lines. The summation is restricted by the "energy" conservation law that the sum of ν variables associated with three phonon lines converging at each internal vertex should vanish.

Similarly, for $n = 4$, we have the following four connected diagrams:

The first diagram is a mere repetition of second-order diagrams and its contribution to $\mathscr{D}^{(4)}$ $(\lambda, i\nu_l)$ is given by $\mathscr{D}^{(0)} \Pi^{(1)} \mathscr{D}^{(0)} \Pi^{(1)} \mathscr{D}^{(0)}$. The sum of remaining three diagrams may be written as $\mathscr{D}^{(0)} \Pi^{(2)} \mathscr{D}^{(0)}$, which defines $\Pi^{(2)}$.

1.8.3 The Phonon Self-Energy

In general, the 2nth-order term of the perturbation expansion of $\mathscr{D}(\lambda, i\nu_l)$ has the form $\mathscr{D}^{(0)} S_n \mathscr{D}^{(0)}$, where S_n is the sum of so-called self-energy diagrams. Let $\Pi^{(n)}$ be the sum of irreducible self-energy diagrams of 2nth order, which does not decompose into disconnected pieces by cutting one phonon line. By taking the sum over n, we define the phonon self-energy $\Pi = \Sigma \Pi^{(n)}$. Rearranging terms in the perturbation expansion, we can write $\mathscr{D}$ in the form

$$\mathscr{D} = \mathscr{D}^{(0)} + \mathscr{D}^{(0)} \Pi \mathscr{D}^{(0)} + \mathscr{D}^{(0)} \Pi \mathscr{D}^{(0)} \Pi \mathscr{D}^{(0)} + \ldots$$
$$= \mathscr{D}^{(0)} + \mathscr{D}^{(0)} \Pi \mathscr{D}. \tag{1.8.12}$$

The last form is usually referred to as the *Dyson equation*. Substituting (1.8.9) and replacing $i\nu_l$ by the complex energy variable z, we write the solution as

$$D(\lambda, z) = \frac{2\hbar\omega_\lambda}{z^2 - (\hbar\omega_\lambda)^2 - 2\hbar\omega_\lambda \Pi(\lambda, z)}. \tag{1.8.13}$$

When the effect of anharmonicity is really small, we may replace Π by its first-order approximation $\Pi^{(1)}$ given by (1.8.11). This amounts to summing up an infinite number of specific terms in the original expansion (1.8.6), as we see by substituting $\Pi^{(1)}$ given by (1.8.11) into Π of the expanded form of (1.8.12).

The sum over ν_n in (1.8.11) may be taken with use of the formula

$$\frac{1}{\beta} \sum_{n=-\infty}^{+\infty} \frac{1}{(i\nu_n - a)(i\nu_n - b)} = -\frac{N(a) - N(b)}{a - b} \tag{1.8.14}$$

where $N(x)$ is Planck's distribution. We replace $i\nu_l$ by complex z after the summation and then let z approach a point E on the real axis from above. We write the result as

$$\Pi^{(1)}(\lambda, E + i0^+) = \Delta_\lambda(E) - i\Gamma_\lambda(E). \tag{1.8.15}$$

Real and imaginary parts of the self-energy are given respectively by

$$\Delta_\lambda(x) = 18 \sum_{1,2} |B|^2 \left[(1 + N_1 + N_2)\left(\frac{1}{x - x_1 - x_2} - \frac{1}{x + x_1 + x_2}\right) \right. $$
$$\left. + \frac{2(N_1 - N_2)}{x + x_1 - x_2} \right] \tag{1.8.16}$$

$$\Gamma_\lambda(x) = 18\pi \sum_{1,2} |B|^2 \{(1 + N_1 + N_2)[\delta(x - x_1 - x_2) - \delta(x + x_1 + x_2)]$$
$$+ 2(N_1 - N_2)\delta(x + x_1 - x_2)\}. \tag{1.8.17}$$

We have simplified notations as $\hbar\omega_1 = x_1$, $N(x_1) = N_1$. Note that $\Delta_\lambda(x) = \Delta_\lambda(-x)$, $\Gamma_\lambda(x) = -\Gamma_\lambda(-x)$, $\Gamma_\lambda(x) > 0$ for $x > 0$.

Now, taking the limit $z \to E + i0^+$ in (1.8.13) and substituting (1.8.15), we obtain the approximate expression for the retarded Green's function

$$D_r(\lambda, E) = \frac{2\hbar\omega_\lambda}{E^2 - (\hbar\omega_\lambda)^2 - 2\hbar\omega_\lambda\Delta_\lambda(E) + 2\hbar\omega_\lambda\Gamma_\lambda(\omega)\,i} \cdot \tag{1.8.18}$$

Since we assume that Δ, Γ are small, this function becomes appreciable in vicinities of $E = \pm\hbar\omega_\lambda$, where approximately $\Delta_\lambda(E) \approx \Delta_\lambda(\hbar\omega_\lambda) \equiv \Delta_\lambda$, $\Gamma_\lambda(E) \approx \Gamma_\lambda(\hbar\omega_\lambda) \equiv \Gamma_\lambda$ for $E \approx \hbar\omega_\lambda$ and $\Delta_\lambda(E) \approx \Delta_\lambda$, $\Gamma_\lambda(E) \approx -\Gamma_\lambda$ for $E \approx -\hbar\omega_\lambda$. Then D_r has simple poles at $E = \pm(\hbar\omega_\lambda + \Delta_\lambda) - i\Gamma_\lambda$, around which it takes the approximate form (1.7.13).

At $T = 0$, we have $N_1 = N_2 = 0$, but (1.8.17) may remain finite; the phonon, excited by the external force, can decay into two phonons and thus have a finite lifetime as far as wave vector conservation law (1.7.2) and energy conservation law represented by Dirac's delta in (1.8.17) are satisfied. This depends on the form of the phonon energy spectrum.

For $T \gg \theta_D$, N_1, N_2 are proportional to T and much bigger than unity. Hence Δ_λ, Γ_λ are also proportional to T. If Γ_λ becomes comparable with $\hbar\omega_\lambda$, the phonon concept defined with reference to the harmonic approximation will lose its significance of an elementary excitation. In this high-temperature region, the effect of anharmonicity is no longer small and there is no reason why we can stop at the first order in calculating the phonon self-energy. In inelastic scattering experiments of neutrons, however, we still recognize fairly well-defined one-phonon peaks even if the temperature is close to the melting point and therefore the effect of anharmonicity must be predominant. This suggests that we need to generalize the phonon concept beyond the harmonic approximation. We meet the same problem also in quantum solids.

1.9 Quantum Solids

We expect a large zero-point motion in bulk helium, since atomic mass is small and interatomic Van der Waals attraction is weak. In fact, we cannot solidify liquid helium by cooling under ordinary pressures. Solid ^{4}He and solid ^{3}He can exist only at pressures higher than about 25 atm and 30 atm, respectively. The zero-point vibration of atoms in solid helium is so violent that its amplitude may amount even to 30% of the atomic spacing. The solid thus certainly melts when the external pressure is removed even at $T = 0$. That is why liquid helium is called a *quantum liquid* and solid helium a *quantum solid*. The ordinary solid, in which the zero-point vibration is small, is often called the classical solid, but this does not mean that the atomic motion obeys classical mechanics.

1.9.1 Nuclear Magnetism of Solid ^{3}He

The nuclear magnetism of solid ^{3}He at ultralow temperatures shows clearly that it should be a quantum solid. The ^{3}He atom possesses the nuclear spin of the magnitude $\hbar/2$ and nuclear spins in the solid will be aligned spontaneously at low enough temperatures with the aid of their mutual interaction even in the absence of external magnetic field. In fact, it has been observed experimentally that solid ^{3}He undergoes the magnetic transition into a spin-ordered state (may be a non-cubic antiferromagnetic state with a big NMR line shift) at $T_N = 1.03$ mK $(= 1.03 \times 10^{-3}$K) under the melting pressure. Though this transition temperature might appear very low, it is much higher than the value we expect for a classical solid.

Thus, if solid ^{3}He were classical, atoms would be well localized around their lattice sites and therefore distinguishable in spite of being Fermi particles of the same kind. The transition temperature would then be determined by the dipole-dipole interaction between nuclear magnetic moments, so that $T_N \approx \mu_N^2/a^3 k_B \approx$ 0.1 μK $(= 10^{-7}$K) where $\mu_N(\approx 10^{-23}$ergG$^{-1})$ is the nuclear magnetic moment and a the lattice constant.

In actuality solid ^{3}He is a quantum solid; the amplitude of the zero-point vibration is so large that two atoms can approach close to each other and become no longer distinguishable. In other words, we need to take into account the condition that the state function of the whole crystal should be antisymmetric against the exchange of any two ^{3}He atoms. As is well known in quantum mechanics, we then have the exchange interaction acting between spins. Remember that we are concerned here with the exchange of atoms themselves, in contrast to the usual exchange interaction which arises from the exchange of magnetic electrons of fixed atoms.

The experiment shows that the exchange energy J in solid ^{3}He is much larger than the dipole-dipole interaction energy at least near the melting pressure and therefore determines T_N. The ratio of the exchange frequency $J/\hbar$ to the vibrational frequency ω is thus estimated as $J/\hbar\omega \approx T_N/\theta_D \sim 10^{-4}$, since the Debye temperature θ_D of solid ^{3}He is of the order of 10 K. We see that the exchange of atoms will occur once in 10^4 cycles of atomic vibration.

1.9.2 Defects in Quantum Solids

The periodic structure of the crystal is disturbed not only by atomic vibrations, but also by various defects; point defects such as an impurity atom, a vacancy (absence of atom), or an interstitial atom, one-dimensional defects such as dislocation lines, two-dimensional defects such as grain boundaries, and three-dimensional defects such as a cavity or precipitation. Here we will restrict ourselves to point defects.

The vacancy is produced in a perfect crystal by removing one atom from its regular site, and by placing this atom at a point in between regular lattice sites, the interstitial is produced. To produce such a point defect, we need an excitation energy of the same order of magnitude as the binding energy of the atom in the crystal. But the free energy of the crystal at $T > 0$ is lowered by producing such defects because of the entropy due to the random distribution of their positions. The concentration of defects will be proportional to $\exp[-\varepsilon_0/k_B T]$, where ε_0 is the excitation energy, and becomes appreciable at high temperatures near the melting point.

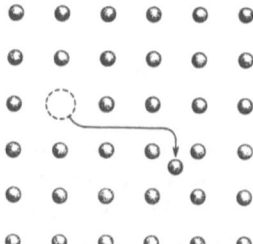

Fig. 1.15. Vacancy and interstitial atom

Now, if the atom next to the vacancy moves to occupy the vacant site, this is equivalent to the motion of the vacancy by one lattice constant in the opposite direction. In the course of this movement, however, a group of atoms surrounding the vacancy have to take a series of configurations of higher energy. Let $\Delta\varepsilon$ be the height in energy of this pass. The motion of the vacancy will be possible only when the thermal fluctuation allows the group of atoms to get excess energy higher than $\Delta\varepsilon$. Its probability is proportional to $\exp(-\Delta\varepsilon/k_B T)$ and therefore becomes exponentially small at lower temperatures. We have similar situations also for interstitials and impurities. When the crystal is cooled quickly (quenching), therefore, defects produced at high temperatures are frozen to survive even at low temperatures. This is not a state of true thermal equilibrium, but has an exceedingly long lifetime.

So far we have been concerned with defects in a classical solid. In solid helium, when a neighboring site happens to be vacant, the atom can move there without thermal excitation because of the large zero-point motion (tunneling). If the density is allowed to decrease by removing the external pressure, a large number of vacancies will be produced and move around through tunneling, so that the solid melts at $T = 0$. Even under the external pressure, the zero-point motion of the vacancy may be important. Then the vacancy is no longer a static structural defect, but must be regarded as a kind of elementary excitation. The same will also apply to interstitial and impurity atoms. We call them *defectons*.

For the sake of mathematical simplicity, let us consider the one-dimensional model consisting of N unit cells, each of which contains one atom in the state of a prefect crystal. Let $|n\rangle$ represent the state in which the nth unit cell is vacant. This is of course the state function of the whole crystal depending on coordinates of N-1 atoms and may include the effect of distortion produced around the vacancy. Let us assume that the expectation value of the Hamiltonian, $\langle n|\mathcal{H}|n\rangle$, is higher than the energy of the perfect crystal by ε_0, which is independent of n.

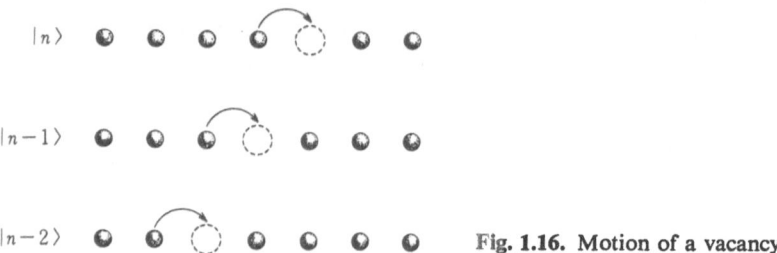

Fig. **1.16.** Motion of a vacancy

The quantum-mechanical motion of the vacancy means that we cannot neglect nondiagonal matrix elements $\langle n \pm 1|\mathcal{H}|n\rangle \equiv v$ and therefore should take a linear combination of $|n\rangle$ to represent the excited state. As is well known in perturbation theory of degenerate systems, the coefficients c_n of the linear combination should satisfy the secular equation

$$\eta c_n = v(c_{n+1} + c_{n-1}) \tag{1.9.1}$$

where η is the energy shift from ε_0 and we have ignored non-orthogonalities $\langle n \pm 1|1|n\rangle$. Since (1.9.1) has the same form as the secular equation to determine the frequencies of normal modes, we have the solution $c_n \propto \exp(ikna)$ under the periodic boundary condition, where a is the lattice constant and k is the reduced wave vector. The energy shift is given by

$$\eta(k) = 2v \cos ka \tag{1.9.2}$$

and the corresponding eigenfunction by

$$|k\rangle = N^{-1/2} \sum_n \exp(ikna)|n\rangle . \tag{1.9.3}$$

This represents the defecton of wave vector k and energy $\varepsilon(k) = \varepsilon_0 + \eta(k)$. Thus the energy level ε_0 splits into a band of N levels, which we call the defecton band.

It is known that the defecton plays an important role in nuclear spin relaxation phenomena of solid ^{3}He; for instance, the effect of ^{4}He added as dilute impurities to solid ^{3}He or vice versa.

We might add that we cannot a priori exclude the possibility $2|v| > \varepsilon_0$ in (1.9.2). This means that the minimum excitation energy of the defecton is nega-

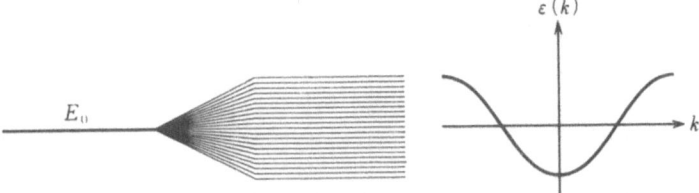

Fig. 1.17. Energy spectrum of the defecton

tive and therefore that the perfect crystal is actually not stable. It should be more favorable energetically to have a macroscopic number of defectons with wave vectors either around $k = 0$ or around $k = \pm\pi/a$. In other words, the crystal should contain a finite concentration of defects even at $T = 0$. They exist in the ground state of the crystal and are called *zero-point defectons* in contrast to defects in a quenched classical solid. The evidence for whether zero-point defects really exist or not in solid helium is not quite conclusive at present, but experimental results so far obtained are rather negative.

1.9.3 Self-Consistent Phonons

We cannot apply the usual harmonic approximation to atomic vibrations in solid helium. Solid ^{4}He under moderate pressure has the hexagonal close-packed structure, for instance, but the nearest neighbor distance is 1.5 times as large as the hard-core diameter of the He atom. Hence the potential energy of interatomic interaction is fairly larger than its minimum value. The harmonic approximation for small oscillations around the minimum potential cannot be applied. Nevertheless the inelastic scattering of neutrons shows "phonon" peaks, which are sharp enough to determine dispersion curves similar to Fig.1.9. Therefore we are forced again to generalize the phonon concept beyond the harmonic approximation. One possible answer is the method of self-consistent phonons.

For simplicity let us consider the one-dimensional Einstein model. Suppose that one atom is displaced along the x axis from its lattice site $x = 0$, all other atoms being fixed at their lattice sites. The potential $V(x)$ of the force acting upon our atom is qualitatively shown in Fig. 1.18. It is not a minimum, but a maximum at $x = 0$ because of the situation we have mentioned above. It is useless to expand $V(x)$ into powers of x. On the other hand, the energy of the zero-point vibration is much larger than the small bump of $V(x)$ around $x = 0$. We therefore expect that the state function must be rather insensitive to the bump and broadened smoothly. We may approximate it by the ground-state function

$$|0\rangle = (\kappa^2/2\pi)^{1/4} \exp(-\kappa^2 x^2/4) \tag{1.9.4}$$

for a smooth harmonic potential $V^*(x) = (f/2)x^2$ shown by the dotted line in Fig. 1.18. Note that $\hbar\kappa$ means the fluctuation of momentum and is related with the spring constant as $f = (\hbar^2 \kappa^4 / 2M)$. In fact the expectation value of the kinetic energy is given by

$$\langle 0 \left| - \frac{\hbar^2}{2M} \frac{d^2}{dx^2} \right| 0 \rangle = \frac{\hbar^2 \kappa^2}{8M} . \tag{1.9.5}$$

In order to calculate the expectation value of the potential energy, we introduce the Fourier expansion

$$V(x) = \int_{-\infty}^{\infty} \frac{dk}{2\pi} V_k \exp(ikx) . \tag{1.9.6}$$

From this and

$$\langle 0 | \exp(ikx) | 0 \rangle = \exp\left(-\frac{k^2}{2\kappa^2} \right) \tag{1.9.7}$$

we obtain

$$\langle 0 | V(x) | 0 \rangle = \int_{-\infty}^{\infty} \frac{dk}{2\pi} V_k \exp\left(-\frac{k^2}{2\kappa^2} \right) . \tag{1.9.8}$$

We regard, κ, f as variational parameters and demand that the sum of (1.9.5) and (1.9.8) be a minimum. With use of (1.9.6), we then have

$$f = \frac{\hbar^2 \kappa^2}{4M} = \langle 0 \left| \frac{d^2 V(x)}{dx^2} \right| 0 \rangle . \tag{1.9.9}$$

If we replace $V''(x)$ by $V''(0)$, we have the spring constant in the harmonic approximation. In (1.9.9), f is to be determined self-consistently because the state functions on the right depend on f through the parameter κ. Hence the name of the method of self-consistent phonons.

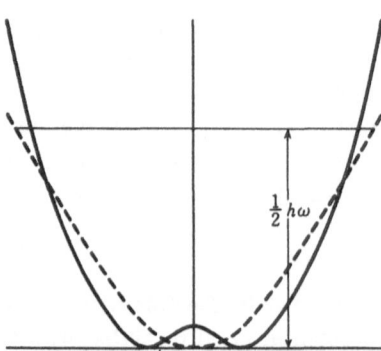

Fig. 1.18. Actual potential (full line) and harmonic potential of the self-consistent phonon

It is not difficult to extend the method to the three-dimensional Debye model. The phonon spectrum as well as the ground state energy are numerically computed. In doing so, it is essential to take account of the particle–particle correlation to avoid the direct contact of hard cores. The bare interaction potential, which has the form shown in Fig. 1.1, should be replaced by a certain pseudopotential, which has a soft core and may be calculated by means of methods known in the theory of nuclei (Jastrow function, K matrix, etc.).

Finally the method of self-consistent phonons can also be applied to classical solids when anharmonicity is essential.

2. Polarization Waves and Dielectric Dispersion

In this chapter, we will focus our attention on the dielectric aspect of various collective motions in condensed matter, and will study how they respond to oscillating electric fields or electromagnetic waves. The response coefficients, such as polarizability and dielectric constant, are basic quantities describing the electrical and optical properties of macroscopic system, with which we are so familiar in our everyday life. On the other hand, the frequency dependence of the response coefficients contains a variety of information on the microscopic motions in matter which can sometimes be described in terms of their energy quanta such as optical phonons, plasmons, and excitons.

We will start with an elementary description of dielectric dispersion associated with optical modes of lattice vibrations and then present a quantum theory of polarizability and dielectric function for more general systems. One of our subjects in this part is the old but yet unsettled problem in dielectrics: how to incorporate the long range part of the forces from other charged particles into the macroscopic electric field. The latter half of the chapter will be devoted to a somewhat detailed description of excitons, so as to clarify the conceptual structure of elementary excitations in general.

2.1 Optical Lattice Vibrations and Dielectric Dispersion

We consider the lattice vibration, a typical collective motion in solids, with particular attention on the optical modes which give rise to electric polarization As mentioned in Chap.1, the lattice vibrations of a crystal with σ ions in a unit cell consist of 3σ branches: three acoustic branches ($s = 1,2,3$) with (angular) frequencies $\omega_k^{(s)}$ which vanish for vanishing wave vector k and $3(\sigma - 1)$ optical branches ($s = 4,5, \ldots, 3\sigma$) with nonvanishing $\omega_0^{(s)}$ [with change of notation: (s,j) in Sect.1.3 into (σ,s) here]. The optical lattice vibration is accompanied by electric polarization even for $k \approx 0$ because of finite relative displacements of ions within each unit cell, thereby contributing to absorption and dispersion of electromagnetic waves. We will consider the interaction of electromagnetic wave and lattice vibration with wave vector k much smaller than the reciprocal lattice vector K_l (long-wavelength approximation), in view of the fact that electromagnetic waves with the frequencies resonating with those of lattice vibrations ($\omega \sim 10^{13}s^{-1}$) have much longer wavelengths ($\sim 10^{-2}$ cm) than the interatomic distance ($\sim 10^{-8}$ cm). We will take an anisotropic insulating crystal in order to clarify the relationship between the depolarizing field and the longitudinal component of polarization waves in this general situation of symmetry.

2.1.1 Incorporation of Long-Range Interionic Forces into the Macroscopic Electric Field

Consider the sum of the forces upon a particular ion exerted by all the other ions in the crystal. The forces would be unbalanced if the ions were slightly displaced from their respective equilibrium positions. The contribution of distant ions to this unbalance can be represented by the electrostatic dipolar field obtained by associating with each ion a dipole moment of effective ionic charge times displacement, although the short-range forces of quantum-mechanical origin such as Van der Waals attraction and hard-core repulsion should also be considered for nearby ions. Since the dipolar field decreases slowly as an inverse cube of the distance, the summation of the unbalanced force over distant ions must be performed with particular care for its convergence which generally depends on the macroscopic shape of the crystal.

This difficulty can be removed by appropriately incorporating the dipolar field from distant ions into the effective field E. E is nothing but the macroscopic electric field appearing in the Maxwell equations for dielectrics and is to be distinguished from the externally applied field $E^{(0)}$. The incorporation is therefore a concept essential in relating the microscopic field to the macroscopic field. We will show how to perform this incorporation in the simplest system of optical lattice vibrations, leaving the more general considerations to Sect.2.2.2.

Consider synchronizing lattice vibration and macroscopic electric field $E(r,t)$, both being proportional to $\exp(ikr - i\omega t)$. It is not specified for the moment whether E has external or internal (as in depolarizing field) origin, nor whether the lattice is undergoing forced or spontaneous vibration. Within a "macroscopically small" region, which by definition is much smaller than the wavelength $2\pi/k$, but much larger than interatomic distance, one can take $E(r,t)$ to be independent of r and the displacement $\xi_{n,\nu}$ of the νth ion in the nth unit cell to be independent of n. The local density of electric polarization is then given, in linear approximation, by

$$P(E;\xi_1, \ldots, \xi_\sigma) = \frac{\epsilon_\infty - 1}{4\pi} E + N_0 \sum_{\nu=1}^{\sigma} e_\nu \xi_\nu . \tag{2.1.1}$$

Here, N_0 is the number of unit cells within a unit volume, ϵ_∞ is the high frequency dielectric constant (well above the eigenfrequencies of lattice vibrations) contributed only from the polarization of bound electrons of the ions, and e_ν is the *effective charge* of the νth ion, satisfying the neutrality condition:

$$\sum_\nu e_\nu = 0 \tag{2.1.2}$$

as is evident from the invariance of P against uniform displacement of all the ions. ϵ_∞ and e_ν are assumed to be symmetric tensors.

The dynamical equation of motion for the (n,ν) ion with mass M_ν, in the absence of external field, is given by

$$M_\nu \ddot{\xi}_{n\nu} = -\sum_{m\mu} U_{n\nu, m\mu} \xi_{m\mu} \quad \begin{pmatrix} n = 1,2, \ldots, N \\ \nu = 1,2, \ldots, \sigma \end{pmatrix}, \tag{2.1.3}$$

where the force constants $U_{m\nu, m\mu}$ are symmetric tensors. Consider a macroscopically small sphere around the (n,ν) ion as its center, and divide the whole space into regions (I) and (II) according as the (m,μ) ion is within or outside the sphere as shown in Fig.2.1. One can then rewrite (2.1.3) as

$$M_\nu \ddot{\xi}_\nu = -\sum_{\mu=1}^{\sigma} U_{\nu,\mu}^{(I)} \xi_\mu - \sum_{m'\mu'}^{(II)} U_{n\nu, m'\mu'} \xi_{m'\mu'}, \tag{2.1.4}$$

$$U_{\nu,\mu}^{(I)} \equiv \sum_{m}^{(I)} U_{m\nu, m\mu}, \tag{2.1.5}$$

since $\xi_{m\mu}$ is nearly m independent ($\equiv \xi_\mu$) within region (I).

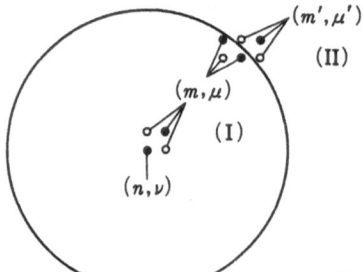

Fig. 2.1. Macroscopically small sphere

Suppose a spherical specimen extending over region (I) only, and let all the ions of respective species (ν) be subject to common displacement ξ_ν. As a result, the specimen is subject to uniform polarization P', and hence, to depolarizing field $E' = -(4\pi/3)P'$ characteristic of the spherical shape. Putting this E' for E in (2.1.1), and solving for P', one gets

$$P'(\xi_1, \ldots, \xi_\sigma) = N_0 \sum_{\nu=1}^{\sigma} e'_\nu \xi_\nu \tag{2.1.6}$$

$$e'_\nu = \frac{3}{\epsilon_\infty + 2} e_\nu. \tag{2.1.7}$$

It is to be noted that each ν ion in the spherical specimen is subject to the force $\tilde{e}'_\nu F$ under the externally applied electric field F, where $\tilde{e}'_\nu$ is the transposed matrix of e'_ν defined by (2.1.7). For, the work done by F on the spherical dielectric is given, per unit volume, by

$$F \cdot \delta P' = N_0 \sum_\nu F \cdot e'_\nu \delta \xi_\nu = N_0 \sum_\nu \delta \xi_\nu \cdot e'_\nu F \,. \tag{2.1.8}$$

Let us then suppose the dielectrics which fills up not only region (I) but extends well beyond the spherical boundary into region (II), and which is subject to the external electric field $E^{(0)}$. Noting that only the long range part of interionic forces remains in the second term of the right hand side of (2.1.4) because of the macroscopic size of the sphere, we can replace this dipolar field by the cavity field $+ (4/\pi 3)P$ where P is the uniform polarization outside the sphere. Then the forces upon the (n,ν) ion exerted by all the ions outside (I) and by the external electric field amount to

$$-\overset{(\mathrm{II})}{\sum} + \tilde{e}'_\nu E^{(0)} = \tilde{e}'_\nu \left(E + \frac{4\pi}{3} P \right), \tag{2.1.9}$$

since the dielectrics within the sphere is subject to the total cavity field $F = E + (4\pi/3)P$ [note that P is common to regions (I) and (II)].

Making use of (2.1.1,9) one can rewrite the equation of motion as

$$M_\nu \ddot{\xi}_\nu + \sum_{\mu=1}^{\sigma} U'_{\nu,\mu} \xi_\mu = e_\nu E \quad (\nu = 1,2, \ldots, \sigma) \tag{2.1.10}$$

$$U'_{\nu,\mu} \equiv U^{(\mathrm{I})}_{\nu,\mu} - \frac{4\pi}{3} N_0 e_\nu \frac{3}{\epsilon_\infty + 2} e_\mu \tag{2.1.11}$$

irrespective of the existence of the external field $E^{(0)}$. It is to be noted that e_ν (instead of e'_ν) appears as the effective charge for E, and that the contribution of distant ions to the force constant is explicitly given by the second term of (2.1.11). Both sides of (2.1.9) are independent of the radius of the sphere in so far as it is macroscopically small, and hence, the convergence of the summation in (2.1.5) is also assured.

It is pertinent here to consider the physical meaning of the two kinds of effective charge, e_ν and e'_ν. In introducing e_ν as a proportionality constant in (2.1.1), we have tacitly included in it the effect of the deformation of the electron cloud of the ion caused by the displacement ξ_ν. In fact, ξ_ν causes this deformation through two types of interactions. The first is the long-range dipolar field due to the relative displacement (by $- \xi_\nu$) of other ions ($\nu' \neq \nu$), and the second is the change of the short-range forces from nearby ions. While the same statement applies to the effective charge e'_ν except for different boundary conditions, the situation is greatly simplified in a special case that the ν ion site has cubic symmetry. For, the contribution of dipolar field to the summation on the right hand side of (2.1.5) vanishes due to the symmetry (the proof is left to the reader), only the short-range forces remaining to contribute to (2.1.5). Therefore, the effect of deformation of electron cloud which is included in the effective charge e'_ν is due solely to the short-range forces. This is the reason why Szigeti introduced the effective charge e'_ν as a physically more transparent quantity. In fact, he showed that in a variety of ionic crystals with cubic symmetry the effective charge e'_ν obtained from the observed dielectric dispersion is close to the value generally accepted for the ion concerned (see Table 2.1), and that the deformation of electronic cloud caused by the change in short-range forces has a relatively small effect.

To sum up, e_ν and e'_ν were defined respectively for a long (parallel to P) and spherical specimen. The depolarizing field is 0 and $-(4\pi/3)P$ in the respective cases. The latter is exactly cancelled by the Lorentz's molecular field $(4\pi/3)P$ which is valid for the cubic lattice as well as for the spherical cavity in the dielectric continuum.

Table 2.1. Static and high-frequency dielectric constants, transverse optical-phonon energies and Szigeti's effective charges of diatomic, cubic ionic crystals. [Fröhlich, H.: *Theory of Dielectrics*, Oxford University Press (1949)]

	$\epsilon(0)$	$\epsilon(\infty)$	$\hbar\omega_t$ [meV]	e'/e
LiF	9.3	1.92	38.0	0.83
NaF	6.0	1.74	30.5	0.94
NaCl	5.6	2.25	20.3	0.76
NaBr	6.0	2.62	16.6	0.85
NaI	6.6	2.91	14.5	0.71
KCl	4.7	2.13	17.5	0.80
KBr	4.8	2.33	14.0	0.76
KI	4.9	2.69	12.2	0.69
RbCl	5.0	2.19	14.6	0.86
RbBr	5.0	2.33	10.9	0.88
RbI	5.0	2.63	9.6	0.78
CsCl	7.2	2.60	12.2	0.88
CsBr	6.5	2.78	9.2	0.81
TlCl	32	5.10	10.6	1.11
CuCl	10	3.57	23.4	1.10
CuBr	8	4.08	21.8	1.0
MgO	10	2.95	71.6	2×0.88
CaO	12	3.28	45.2	2×0.76
SrO	13	3.31	26.4	2×0.60

2.1.2 Dielectric Dispersion

Coming back to the general case of anisotropy, let us solve (2.1.10) in the following three steps. First we solve the homogeneous equations with $E = 0$ to get the eigenvibrations of the form:

$$\xi_\nu(t) = \xi_\nu \exp(-i\Omega t). \tag{2.1.12}$$

The ith component of (2.1.10) is then given by

$$- M_\nu \Omega^2 \xi_{\nu i} + \sum_{\mu=1}^{\sigma} \sum_{j=1}^{3} U'_{\nu i,\mu j} \xi_{\mu j} = 0 \quad \left(\begin{matrix} \nu = 1,2,\ldots,\sigma \\ i = 1,2,3 \end{matrix} \right). \tag{2.1.13}$$

The eigenvalues Ω_s^2 and the eigenvectors $\xi_\nu^{(s)}(s = 1,2, \ldots, 3\sigma)$ of (2.1.13), if suitably normalized, satisfy the following relationships:

$$\sum_{\nu,i} M_\nu \xi_{\nu i}^{(s)} \xi_{\nu i}^{(s')} = \delta_{ss'} \qquad \text{(orthonormality)}, \tag{2.1.14}$$

$$\sum_{s} \sqrt{M_\nu M_\mu} \xi_{\nu i}^{(s)} \xi_{\mu j}^{(s)} = \delta_{\nu\mu} \delta_{ij} \quad \text{(completeness)}, \tag{2.1.15}$$

$$\sum_{\nu i, \mu j} \xi_{\nu i}^{(s)} U'_{\nu i, \mu j} \xi_{\mu j}^{(s')} = \delta_{ss'} \Omega_s^2 \quad \text{(diagonality)}. \tag{2.1.16}$$

Three of the eigensolutions ($s = 1,2,3$) represent the acoustic vibrations with $\xi_\nu^{(s)} = \xi^{(s)}$ and $\Omega_s = 0$ since $\sum_\mu U'_{\nu,\mu} = 0$. The last identity is obtained from (2.1.2,11) and $\sum_\mu U_{\nu,\mu}^{(1)} = 0$ (invariance of balance against uniform displacement).

For the second step, we specify the electric field as $E(t) = E\exp(-i\omega t)$ and study the "forced oscillation" of the form

$$\xi_\nu(t) = \xi_\nu^{(\omega)} \exp(-i\omega t). \tag{2.1.17}$$

Inserting this in (2.1.10), we get:

$$- M_\nu \omega^2 \xi_{\nu i}^{(\omega)} + \sum_{\mu j} U_{\nu i, \mu j}' \xi_{\mu j}^{(\omega)} = \sum_j e_{\nu i j} E_j. \tag{2.1.18}$$

$\xi_{\nu i}^{(\omega)}$, as 3σ-dimensional vector, can be expanded with the complete set of basis vectors: $\xi_{\nu i}^{(s')}(s' = 1,2, \ldots, 3\sigma)$ as

$$\xi_{\nu i}^{(\omega)} = \sum_{s'} a^{(s')}(\omega) \xi_{\nu i}^{(s')} \quad \left(\begin{matrix} \nu = 1,2, \ldots, \sigma \\ i = 1,2,3 \end{matrix} \right). \tag{2.1.19}$$

Putting this into (2.1.18), summing over (ν,i) after multiplying by $\xi_{\nu i}^{(s)}$, and making use of (2.1.14,16), one gets the equation for the expansion coefficients a:

$$(\Omega_s^2 - \omega^2) a^{(s)}(\omega) = p^{(s)} \cdot E, \tag{2.1.20}$$

where

$$p^{(s)} \equiv \sum_{\nu=1}^{\sigma} e_\nu \xi_\nu^{(s)} \tag{2.1.21}$$

is the polarization vector associated with the sth eigenvibration of (2.1.13)

Making use of (2.1.1,17,19, 20), we find that the dielectric function $\epsilon(\omega)$ relating $D = E + 4\pi P$ and P at frequency ω is given by

$$\epsilon(\omega) = \epsilon_\infty + 4\pi N_0 \sum_s \frac{p^{(s)} p^{(s)}}{\Omega_s^2 - \omega^2}. \tag{2.1.22}$$

Here, $p^{(s)}p^{(s)}$ represents a tensor whose (i,j) component is equal to $p_i^{(s)}p_j^{(s)}$, and satisfies the following *sum rule*:

$$\sum_s p_i^{(s)}p_j^{(s)} = \left[\sum_\nu \frac{e_\nu^2}{M_\nu}\right]_{ij}. \tag{2.1.23}$$

2.1.3 Lattice Vibrations in the Long-Wavelength Limit

The relation $D = \epsilon E$, derived in Sect.2.1.2, holds irrespective of the origin of E. Let us now study, as our final step, the eigenvibrations of the lattice in the absence of external field:

$$\xi_m(t) = \eta_\nu \exp(i\mathbf{k}\cdot\mathbf{r}_n - i\omega t) \tag{2.1.24}$$

in the long-wavelength limit ($k \ll K_t$). $\mathbf{r}_n$ represents the reference position of the nth unit cell. Although $E^{(0)} = 0$, we have to consider the depolarizing field E due to the polarization P accompanying the lattice vibration. So far as $k \ll K_t$, one can assume P and E to be proportional to $\exp(i\mathbf{k}\mathbf{r} - i\omega t)$ in conformity with (2.1.24). Neglecting retardation in the Maxwell equations, we get

$$0 = \text{rot } E = i\mathbf{k} \times E, \tag{2.1.25}$$

and further, in view of the absence of true charge,

$$0 = \text{div } D = i\mathbf{k}\cdot(E + 4\pi P). \tag{2.1.26}$$

From (2.1.25,26), the depolarizing field is given by

$$E = -4\pi\bar{\mathbf{k}}(\bar{\mathbf{k}} \cdot P), \quad \bar{\mathbf{k}} \equiv \frac{\mathbf{k}}{k}. \tag{2.1.27}$$

This means that only the longitudinal part of P is effective in producing a charge-density wave as the source of E. From (2.1.27) and the identiy: $4\pi P = [\epsilon(\omega) - 1]E$, we get the equation

$$[1 + \{[\epsilon(\omega) - 1]\bar{\mathbf{k}}\}\bar{\mathbf{k}} \cdot]P = 0 \tag{2.1.28}$$

for the polarization wave accompanying the lattice vibration (2.1.24). Taking the z' axis along the direction of $\mathbf{k}$, one can calculate the determinant of the tensor [...] in (2.1.28) as

$$\begin{vmatrix} 1 & 0 & \epsilon_{x'z'} \\ 0 & 1 & \epsilon_{y'z'} \\ 0 & 0 & \epsilon_{z'z'} \end{vmatrix} = \epsilon_{z'z'} \equiv \epsilon_{||}. \tag{2.1.29}$$

In order that (2.1.28) have a nontrivial solution, the determinant must vanish. In view of (2.1.22), this amounts to

$$\epsilon_{||}(\omega) = \epsilon_{||\infty} + 4\pi N_0 \sum_s \frac{(p_{||}^{(s)})^2}{\Omega_s^2 - \omega^2} = 0 \tag{2.1.30}$$

where $p_{||}^{(s)} \equiv \mathbf{k} \cdot \mathbf{p}^{(s)}$. The solutions ω of this equation represent the eigenfrequencies of lattice vibrations accompanied by polarization waves.

Let us assume for the moment that Ω_s are nondegenerate and $p_{||}^{(s)}$ are nonvanishing except for the three acoustic branches ($s = 1,2,3$). From (2.1.30), we find that the poles Ω_s and the zeros ω_s of $\epsilon_{||}(\omega)$ appear alternantly as shown in Fig.2.2a: it is convenient to number them in such a way that $\Omega_4 < \omega_4 < \ldots < \Omega_s < \omega_s < \Omega_{s+1} \ldots < \Omega_{3\sigma} < \omega_{3\sigma}$. (2.1.30) can then be factorized as

$$\frac{\epsilon_{||}(\omega)}{\epsilon_{||\infty}} = \sum_{s=4}^{3\sigma} \frac{\omega_s^2 - \omega^2}{\Omega_s^2 - \omega^2} . \tag{2.1.31}$$

Consequently, the dispersion of $-1/\epsilon_{||}(\omega)$ is similar to that of $\epsilon_{||}(\omega)$, as shown in Fig.2.2b. While the solution Ω_s for the homogeneous equation with $E = 0$ (fictitious sytem) is $\mathbf{k}$ independent, the eigenfrequency ω_s of lattice vibration with depolarizing field E being considered (real system) does depend on the direction $\mathbf{k}$ of wave propagation through $p_{||}^{(s)}$ and $\epsilon_{||\infty}$.

To clarify the physical meaning of Ω_s, let us take the x axis along $\mathbf{p}^{(s)}$ and consider the lattice vibration propagating with $\mathbf{k}$ almost perpendicular to $\mathbf{p}^{(s)}$ (Fig. 2.3). Take the z' axis to be parallel to $\mathbf{k}$ and (x',z) to be coplanar with $(x,$

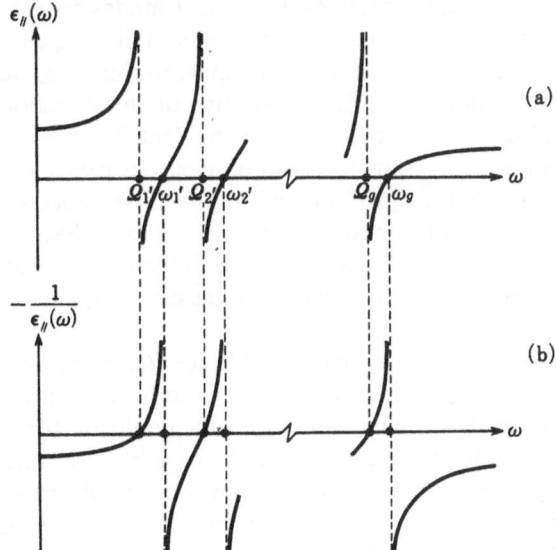

Fig. 2.2. Dielectric dispersion

z'). Since $p_{1}^{(s)} \equiv p_{x}^{(s)}$ is very small, (2.1.30) should have a solution in the neighborhood of Ω_s: $\omega_{s'} = \Omega_s + O[(p_{11}^{(s)})^2]$ ($s' = s$ or $s' = s - 1$ according as $\omega_{s'} \gtrless \Omega_s$). As for the components of dielectric tensor for this solution, $\epsilon_{x'z'}(\omega_{s'})$ is very large because the denominator $\Omega_s^2 - \omega_{s'}^2 = O[(p_{11}^{(s)})^2]$ and the numerator $p_{x'}^{(s)} p_{z'}^{(s)} = O(p_{11}^{(s)})$ while $\epsilon_{y'z'}(\omega_{s'})$ is of the order of unity. Because of (2.1.28,29), P turns out to be almost parallel to x'. Namely, there are a number of eigenvibrations approximately with frequency $\omega_{s'} = \Omega_s$ and $\bar{k} \perp P \parallel p^{(s)}$. Although they are transverse waves as regards the total electric polarization P, the motion of an individual ion ($\parallel \eta_\nu^{(s)}$) is not necessarily perpendicular to $\bar{k}$. As $\bar{k}$ becomes nonperpendicular to $p^{(s)}$, the eigenfrequency $\omega_{s'}$ deviates from Ω_s.

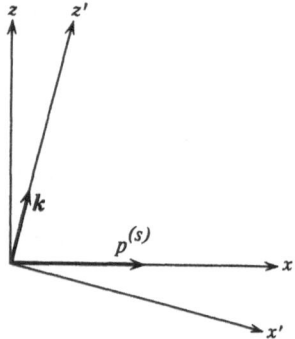

Fig. 2.3. Polarization $p^{(s)}$ and propagation vector k

While $p^{(s)}$ vanishes identically for the acoustic branches ($s = 1,2,3$) because of (2.1.2,21) and $\xi_\nu^{(s)} = \xi^{(s)}$, it can vanish for some of the optical modes because of the symmetry of the crystal. Such a situation can be reached as a limiting case ($p^{(s)} \to 0$) of the more general situation with lower symmetry. In this case, we have eigenvibrations with $\omega_{s'} = \Omega_s$ and $\eta_\nu^{(s')} = \xi_\nu^{(s)}$ irrespective of the direction of the propagation, $\bar{k}$, because of the absence of a depolarizing field. The branch s with nonvanishing $p^{(s)}$, or the corresponding eigenvibration $\eta_\nu^{(s')}$ is called the *infrared-active mode*, and that with vanishing $p^{(s)}$ the *infrared-inactive*. Since we are considering the long-wavelength limit, the above statement means that the eigenfrequency $\omega_{s'}(k)$ of the infrared-active branch tends to different values as k tends to zero from a different direction. Such a singularity at $k = 0$ originates in the long-range part of interionic force.

The singularity at $k = 0$ disappears in the cubic crystal in the following way. Because of the equivalence of three crystallographic axes x, y, and z, any eigensolution of (2.1.13) with a nonvanishing polarization vector should be triply degenerate: $\Omega_s = \Omega_{s+1} = \Omega_{s+2}$ with polarizations $p^{(s)} \parallel x$, $p^{(s+1)} \parallel y$ and $p^{(s+2)} \parallel z$. The corresponding eigenfrequencies of lattice vibration should be so located that $\Omega_s = \omega_s = \Omega_{s+1} = \omega_{s+1} = \Omega_{s+2} < \omega_{s+2}$, according to Fig.2.2. ω_{s+2} as well as ω_s and ω_{s+1}, should have no $\bar{k}$ dependence since (2.1.30) is now a scalar equation. ω_{s+2} corresponds to the longitudinal polarization wave as is easily seen by putting $\epsilon(\omega_{s+2}) = 0$ in (2.1.28), while ω_s and ω_{s+1} correspond to transverse polarization waves since the above equality between them and $\Omega's$ implies the absence of a depolarizing field. [It is only in cubic crystals that the infrared-active modes tend to purely longitudinal or purely transverse polarization waves (with $\mathbf{k}$-independent frequencies) as $\mathbf{k}$ tends to zero from any direction.] Denoting the transverse and longitudinal frequencies of each such set (α) by $\omega_{\alpha t}(=\Omega_{s+2})$ and $\omega_{\alpha l}(=\omega_{s+2})$, respectively, we find that (2.1.31) reduces to

$$\frac{\epsilon(\omega)}{\epsilon_\infty} = \prod_\alpha \frac{\omega_{\alpha l}^2 - \omega^2}{\omega_{\alpha t}^2 - \omega^2} \tag{2.1.32}$$

since the contributions from the modes s and $s + 1$ are unity.

In the particular case of diatomic cubic crystals such as NaCl, CsCl, and ZnS, there appears only one set of optical modes $(s = 4,5,6)$. Denoting effective charges of cation and anion respectively by e_+ and $e_-(= -e_+)$ and their reduced mass by M_r, one gets $|\mathbf{p}^{(s)}| = e_+/\sqrt{M_r}$ from the sum rule (2.1.23), and dielectric dispersion (2.1.22) in the form:

$$\epsilon(\omega) = \epsilon_\infty + \frac{4\pi N_0 e_+^2}{M_r(\omega_t^2 - \omega^2)} \tag{2.1.33}$$

$$= \epsilon_\infty + \frac{(\epsilon_0 - \epsilon_\infty)\,\omega_t^2}{\omega_t^2 - \omega^2}, \tag{2.1.33'}$$

where $\epsilon_0 \equiv \epsilon(0)$ is the static dielectric constant. From these equations, we get two relations between the longitudinal and transverse frequencies:

$$\omega_l^2 = \omega_t^2 + \frac{4\pi N_0 e_+^2}{\epsilon_\infty M_r}, \tag{2.1.34}$$

$$\frac{\omega_l^2}{\omega_t^2} = \frac{\epsilon_0}{\epsilon_\infty}, \tag{2.1.35}$$

The second term on the right hand side of (2.1.34) originates in the depolarizing field of the relative oscillation of cations and anions, being similar to the plasma oscillation of electrons in metals which will be described in Sect.2.2. Equation (2.1.35) is the *Lyddane–Sachs–Teller relation*, which was generalized to the multimode case by *Cochran* and *Cowley* and *Kurosawa* in the form of (2.1.32) (with $\omega = 0$).

2.2 Polarizability and Dielectric Constant

We have calculated, in the previous section, the dielectric dispersion associated with the optical modes of lattice vibration, with the use of classical mechanics. Within such a harmonic-oscillator model, quantum mechanics gives the same result for the dielectric constant as the classical one. This is no longer valid for more general systems, but some of the basic features of dielectric dispersion obtained there will be preserved. To show this, we will derive, in this section, the quantum-mechanical formulae for polarizability and dielectric constant, paying particular attention to the singular effect caused by the long-range part of interparticle interactions.

2.2.1 General Formula for Polarizability

Consider a system with Hamiltonian $\mathcal{H}$. If an external electric field $E^{(0)}(r,t)$, varying in space and time, is applied, the system is subject to the perturbation

$$\mathcal{H}_{ex}(t) = - \int P(r) E^{(0)}(r, t)\, dr \tag{2.2.1}$$

where $P(r)$ is an operator representing the density of electric polarization at the position r. In order to calculate the polarization density induced by the external field within the linear-response theory, we make use of the argument in Sect. 1.7.2. The linear-response coefficient connecting the external field at space–time point (r', t'), $E_j^{(0)}(r', t')$ ($\leftrightarrow f_\lambda$ of Sect.1.7.2), to the expectation value of the induced polarization density at space–time point (r, t), $\langle P_i(r) \rangle_t$, is given by replacing $\varphi_{-\lambda}$ by $-P_j(r')$ and φ_λ by $P_i(r)$ in (1.7.9):

$$\alpha_{ij}(r,r'; t - t') = \frac{i}{\hbar} \theta(t - t') \langle [P_i(r, t), P_j(r', t')] \rangle , \tag{2.2.2}$$

where

$$P_i(r,t) \equiv \exp\left(\frac{i}{\hbar} \mathcal{H}t\right) P_i(r) \exp\left(-\frac{i}{\hbar} \mathcal{H}t\right) , \tag{2.2.3}$$

and $\langle \, . \, . \, . \, \rangle$ denotes the expectation value in thermal equilibrium. It has been assumed that $\langle P(r) \rangle = 0$ as in Sect.1.7.2. Making use of the superposition principle, we have only to integrate the linear response over the space–time point (r', t') where the external force was applied, to get the total response:

$$\langle P(r) \rangle_t = \int dr' \int dt'\, \alpha(r, r', t - t')\, E^{(0)}(r', t') , \tag{2.2.4}$$

where α is the tensor whose (i, j) component is given by (2.2.2).

If the system is spacially homogeneous, namely, if it has translational symmetry, the response function $\alpha(r, r', t - t')$ should depend on r and r' only through $r - r'$:

$$\alpha(r, r', t - t') = \alpha(r - r', t - t') . \qquad (2.2.5)$$

Let us make use of the Fourier expansion, normalized to unit volume:

$$P(r) = \sum_k P_k \exp(ik \cdot r) , \qquad (2.2.6)$$

$$E^{(0)}(r, t) = \sum_k \int_{-\infty}^{+\infty} \frac{d\omega}{2\pi} E^{(0)}_{k\omega} \exp(ik \cdot r - i\omega t) . \qquad (2.2.7)$$

(2.2.4) is then written as

$$\langle P_k \rangle_t = \int_{-\infty}^{+\infty} \frac{d\omega}{2\pi} \alpha(k, \omega) E^{(0)}_{k\omega} \exp(-i\omega t) , \qquad (2.2.8)$$

where

$$\alpha(k, \omega) \equiv \int dr \int_{-\infty}^{+\infty} dt\, \alpha(r, t) \exp(-ik \cdot r + i\omega t) \qquad (2.2.9)$$

is the response coefficient for the component wave $E^{(0)}_{k\omega} \exp(-i\omega t)$.
Putting (2.2.6) into (2.2.2), one obtains

$$\alpha_{ij}(r, r', t) = \frac{i}{\hbar}\, \theta(t) \sum_k \sum_{k'} \langle [P_{ki}(t), P_{-k'j}] \rangle \exp[i(k \cdot r - k' \cdot r')] , \quad (2.2.2')$$

of which $\langle \ldots \rangle$ should vanish for $k' \neq k$ under the homogeneity assumption: (2.2.5). Inserting (2.2.2') into (2.2.9), one obtains

$$\alpha_{ij}(k, \omega) = \int_{-\infty}^{+\infty} dt \left(\frac{i}{\hbar}\right) \theta(t) \langle [P_{ki}(t), P_{-kj}] \rangle \exp(i\omega t) , \qquad (2.2.10)$$

Denoting the eigenvalue of $\mathcal{H}$ by $\varepsilon_m, \varepsilon_n \ldots$, and defining $\omega_{mn} \equiv (\varepsilon_m - \varepsilon_n)/\hbar$, we can rewrite $\langle \ldots \rangle$ in (2.2.10) as

$$\langle [P_{ki}(t), P_{-kj}] \rangle = \underset{m}{\mathrm{Av}} \sum_n [(P_{ki})_{mn} (P_{-kj})_{nm} \exp(-i\omega_{nm}t)$$
$$- (P_{ki})_{nm}(P_{-kj})_{mn} \exp(+i\omega_{nm}t)]$$

where use has been made of (2.2.3). While term-by-term integration of this as prescribed in (2.2.10) does not lead to proper convergence, the total integral should converge since $\langle \ldots \rangle$, being a kind of correlation function, should vanish at $t \to +\infty$ for a dissipative system. Therefore, by supplementing the integrand in (2.2.10) with a convergence factor $\exp(-\eta t)$ and taking the limit

$\eta \to +0$ after integration, one can assure the convergence of term-by-term integration as well without altering the final result. In this way one obtains

$$\alpha(\mathbf{k}, \omega) = \text{Av} \sum_m \sum_n \left[\frac{(\mathbf{P}_k)_{mn}(\mathbf{P}_{-k})_{nm}}{\hbar(\omega_{nm} - \omega - i\eta)} + \frac{(\mathbf{P}_k)_{nm}(\mathbf{P}_{-k})_{mn}}{\hbar(\omega_{nm} + \omega + i\eta)} \right]_{\eta \to +0}. \qquad (2.2.11)$$

Here the product $\mathbf{PP}'$ denotes the tensor whose (i,j) component is $P_i P'_j$, and Av/m means the statistical average with Boltzmann factor $\exp(-\beta\varepsilon_m)$.

For those systems in which free carriers, rather than bound charges, contribute to the electric polarization (e.g., for conduction electrons in metals), it is more convenient to express the polarizability in terms of charge density $\rho(\mathbf{r}) = -\text{div}\mathbf{P}(\mathbf{r})$. Making use of the Fourier component: $\rho_k = -i\mathbf{k} \cdot \mathbf{P}_k$, one can write the longitudinal component of (2.2.11) as

$$\alpha_{\parallel}(\mathbf{k}, \omega) = \frac{1}{k^2} \text{Av} \sum_m \sum_n \left[\frac{|(\rho_{-k})_{nm}|^2}{\hbar(\omega_{nm} - \omega - i\eta)} + \frac{|(\rho_k)_{nm}|^2}{\hbar(\omega_{nm} + \omega + i\eta)} \right]_{\eta \to +0}. \qquad (2.2.11')$$

This is related to the dynamical form factor introduced in Chap. 1, as will be described in more detail in Sect.3.3.4.

While the homogeneity assumption (2.2.5) is valid for uniform gases and liquids because of their translational symmetry, it would, at a first glance, not be the case for crystals whose translational symmetry is restricted to discrete lattice vectors. However, $\mathbf{P}(\mathbf{r})$ and $\alpha(\mathbf{r}, \mathbf{r}', t - t')$ can be well defined only for the macroscopically small region of $\mathbf{r}$ which contains a great number of unit cells, and on this macroscopic level the assumption (2.2.5) holds. Corresponding to this, we have to deal only with the long-wavelength components of external field and response such that $k \ll K_l$, and it is within this $\mathbf{k}$ region that we can put $\langle \ldots \rangle$ in (2.2.2') equal to zero for $\mathbf{k}' \neq \mathbf{k}$.

The step function $\theta(t - t')$ in (2.2.2) represents the *causality*: the result (response at t) never precedes the cause (external force at t'). This enabled us to introduce in (2.2.10) the convergence factor $\exp(-\eta t)$ ($\eta \to +0$), effective only for positive t, to get the result (2.2.11). By replacing the limit symbol $\eta \to +0$ with the inequality $\eta > 0$ in this expression one obtains the analytic continuation of $\alpha(\mathbf{k}, \omega)$ into the upper half of the complex ω plane. The causality is therefore equivalent to the analyticity in the upper half of the ω plne. Integrating $\alpha(\mathbf{k}, \omega')/(\omega' - \omega)$ along the ω' path shown in Fig.2.4 with the above mentioned analyticity in mind, one obtains the identity

$$\alpha(\mathbf{k}, \omega) = \frac{1}{i\pi} P \int_{-\infty}^{+\infty} \frac{d\omega'}{\omega' - \omega} \alpha(\mathbf{k}, \omega')$$

where P means the principal value. By writing the real and imaginary parts of this equation separately, one obtains the *dispersion formula*, or the so-called *Kramers-Kronig relations*:

$$\text{Re}\{\alpha(\boldsymbol{k}, \omega)\} = \frac{1}{\pi} P \int_{-\infty}^{+\infty} \frac{d\omega'}{\omega' - \omega} \text{Im}\{\alpha(\boldsymbol{k}, \omega')\} \;, \tag{2.2.12}$$

$$\text{Im}\{\alpha(\boldsymbol{k}, \omega)\} = -\frac{1}{\pi} P \int_{-\infty}^{+\infty} \frac{d\omega'}{\omega' - \omega} \text{Re}\{\alpha(\boldsymbol{k}, \omega')\} \tag{2.2.12'}$$

as another version of the causality.

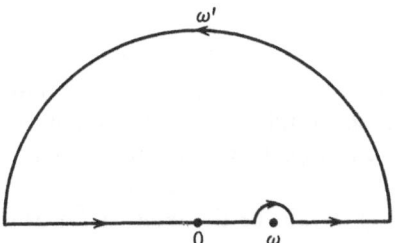

Fig. 2.4. Integration path on the complex ω' plane

2.2.2 The Relation Between Polarizability and Dielectric Constant

The externally applied electric field $E^{(0)}$ causes the redistribution of charged particles within the matter, which in turn gives rise to an induced electric field $E^{(1)}$. The total electric field $E = E^{(0)} + E^{(1)}$, which appears in the Maxwell equations for dielectric media, is not a purely external force. One must keep this in mind in relating dielectric constant ϵ to polarizability α: while the latter is usually considered to be a coefficient connecting response and external force, the former is not such since E is not purely external. As described below, there are two different ways of formulating this problem according as D or E is taken for (effective) external field. They should always give the same final answer for the dielectric constant if performed exactly. In practice, however, they often play complementary roles, leading to deeper understanding of the same physical entity, as will be shown later in a few examples.

(I) D **method.** Consider a longitudinal external field $E^{(0)} \propto \bar{\boldsymbol{k}} \exp(i\boldsymbol{k}\cdot\boldsymbol{r} - i\omega t)$, which is equivalent to assuming true charge $\rho^{(0)} = (4\pi)^{-1} \text{div}E^{(0)}$. Since $\text{div}D = 4\pi\rho^{(0)}$, the longitudinal component $D_{\parallel} \equiv (\bar{\boldsymbol{k}}\cdot\boldsymbol{D})\,\bar{\boldsymbol{k}}$ of the electric displacement is equal to $E^{(0)}$ and can be taken as an external force. The longitudinal component of polarization $\boldsymbol{P} = \alpha E^{(0)}$ is given by $P_{\parallel} = \alpha_{\parallel} E_{\parallel}^{(0)}$ where $\alpha_{\parallel} \equiv \bar{\boldsymbol{k}} \cdot \alpha\bar{\boldsymbol{k}}$. The ratio of response $4\pi P_{\parallel} = D_{\parallel} - E_{\parallel}$ to external field $D_{\parallel} = \epsilon_{\parallel} E_{\parallel}$ is then given by

$$1 - \frac{1}{\epsilon_{\parallel}(\boldsymbol{k}, \omega)} = 4\pi\alpha_{\parallel}(\boldsymbol{k}, \omega) \;. \tag{2.2.13}$$

Equation (2.2.13) gives only the longitudinal component of the dielectric tensor. If $\epsilon(\boldsymbol{k}, \omega)$ is independent of direction $\bar{\boldsymbol{k}}$, one can get the symmetric part of the ϵ tensor by calculating (2.2.13) for several directions $\vec{k}$, but nothing is obtained of the antisymmetric part which is nonvanishing in an optically active material or under a magnetic field. Only in the isotropic case does (2.2.13) give complete information on ϵ.

(II) E **method.** In order to treat E as an external field, one must assume that the long-range part of the interparticle interactions is already replaced phenomenologically by the induced field $E^{(i)}$ which constitutes a part of E. In order to calculate the effective polarizability which is defined by

$$\epsilon(\vec{k}, \omega) - 1 = 4\pi\alpha_{\text{eff}}(\vec{k}, \omega) \tag{2.2.14}$$

and which relates the polarization P to the macroscopic electric field E, one must make use of the effective Hamiltonian $\mathcal{H}_{\text{eff}}$ which would be obtained by suitably subtracting the long-range interactions from the true Hamiltonian $\mathcal{H}$. It is a sort of molecular field theory.

Dielectric constant ϵ appearing in (2.2.13,14) should be the same as far as it is definable, while the difference between α and α_{eff} comes from the use of different Hamiltonians in calculating them.

Each of the two methods mentioned above has its own merits and drawbacks. In the E method, there is no general prescription for deriving $\mathcal{H}_{\text{eff}}$ from $\mathcal{H}$, and case-by-case consideration is required in disposing the long-range part of the interactions. In this respect, the D method is rigorous and unambiguous. Its use, however, is restricted to the symmetric part of ϵ tensor. The disposal of the long-range interactions can be made with the use of a more or less sophisticated diagram technique of the many-body problem in the D method, while it is ingeneously concealed (renormalized) in the E method.

2.2.3 Applications to Optical Lattice Vibrations

It is instructive to study the relations between α_{eff} and α in the case of optical lattice vibrations described in Sect.2.1 as the simplest model system for dielectric dispersion. It was with the E method that we derived (2.1.22) [corresponding to (2.2.14)]; in fact, the argument leading from (2.1.3) (with force constant U) to (2.1.10) (with effective force constant U') amounts to deriving $\mathcal{H}_{\text{eff}}$ from $\mathcal{H}$. The pole Ω_s of $\alpha_{\text{eff}}(\omega)$, with corresponding polarization vector $p^{(s)}$, represents the eigenvibration of a fictitious system (2.1.13) in which the depolarizing field is equated to zero. The eigenvibration of the real system with the depolarizing field taken into account is given by a zero-point ω_s of $\epsilon_{||}(\omega)$ [see (2.1.30)], and hence, by a pole of $\alpha_{||}(\omega)$ due to (2.2.13). Ω_s and ω_s correspond to ω_{nm} of the systems with Hamiltonians $\mathcal{H}_{\text{eff}}$ and $\mathcal{H}$, respectively.

For diatomic crystals with cubic symmetry, one obtains

$$4\pi\alpha_{\text{eff}}(\omega) = (\epsilon_\infty - 1) + \frac{4\pi N_0 e_+^2}{M_r(\omega_t^2 - \omega^2)} \tag{2.2.15}$$

by comparing (2.1.33) and (2.2.14). Equation (2.1.33) can also be written in the form of (2.2.13), namely,

$$4\pi\alpha_{||}(\omega) = \left(1 - \frac{1}{\epsilon_\infty}\right) + \frac{4\pi N_0(e_+/\epsilon_\infty)^2}{M_r(\omega_l^2 - \omega^2)} . \tag{2.2.16}$$

Since $\alpha_{||}(\omega)$ [equal to $\alpha(\omega)$ in the present example] is longitudinal, only the longitudinal vibration ω_l with nonvanishing $\bar{k} \cdot P_{kmn}$ [see (2.2.11)] appears as a pole. Moreover, the effective charge e_+ of (2.2.15) is screened into e_+/ϵ_∞ in (2.2.16), because of the depolarizing field $E = -4\pi P$ for longitudinal wave.

Let us now apply the E method to the quantum-mechanical calculation of $\epsilon(\omega)$. The lattice vibrations of the fictitious system with $E = 0$ [as given in (2.1.13)] can be written, in quantum mechanics, as

$$\xi_{m\nu} = \sum_{s,k} \left(\frac{\hbar}{2N_0\Omega_{sk}}\right)^{1/2} \xi_\nu^{(s)}(k)[b_{sk} \exp(ik \cdot R_n) + b_{sk}^\dagger \exp(-ik \cdot R_n)] \tag{2.2.17}$$

with the use of (1.4.15) and the eigenvectors $\xi_\nu^{(s)}(k)$ normalized by (2.1.14). The Fourier component of polarization (as defined by (2.2.6)) for small k is then written as

$$P_k = \sum_s \left(\frac{\hbar N_0}{2\Omega_s}\right)^{1/2} p^{(s)}(b_{sk} + b_{s,-k}^\dagger) \tag{2.2.18}$$

with the use of (2.1.1,21). Since the phonon creation–annihilation operators $b^\dagger$ and b have matrix elements of only one quantum change as given in (1.4.11), ω_{nm} in (2.2.11) takes only the values $\pm \Omega_r$. The phonon numbers (m) disappear after the summation over n in (2.2.11), and one immediately gets (2.1.22), the same as in the classical case.

As an application of the D method, let us calculate the fluctuation of internal electric field caused by lattice vibrations. It should be possible to relate it with dielectric constant according to the fluctuation-dissipation theorem. Consider the eigenvibrations of real system with the depolarizing field being considered, impose on the eigenvectors $\eta_\nu^{(s)}(k)$ [see (2.1.24)] the same normalization conditions as were imposed on $\xi_\nu^{(s)}$ in (2.1.14), and define the polarization vector $q^{(s)}(k)$ as was done for $p^{(s)}$ in (2.1.21). Then, in place of (2.2.18), one obtains

$$P_k = \sum_s \left(\frac{\hbar N_0}{2\omega_s}\right)^{1/2} q^{(s)}(k)(b_{sk} + b_{s,-k}^\dagger) \tag{2.2.19}$$

[note that the meaning of b is also different in (2.2.18,19) although the same notation is used]. The electrostatic potential $\varphi(r)$ caused by this fluctuating polarization is given, by solving the Poisson equation: $\Delta\varphi = 4\pi\mathrm{div}\, P(r)$, as

$$\varphi(r) = \sum_k \left(-\frac{4\pi i}{k}\right) \bar{k} \cdot P_k \exp(ik \cdot r) . \tag{2.2.20}$$

The unknown coefficients, $\bar{k} \cdot q^{(s)}(k) \equiv q_{\parallel}^{(s)}(k)$, connecting $\varphi(k)$ and $(b, b^\dagger)$, can be related to the dielectric dispersion as follows. Putting (2.2.19) into (2.2.11) and making use of (2.2.13), one obtains.

$$\frac{1}{\epsilon_{\parallel}(k, \infty)} - \frac{1}{\epsilon_{\parallel}(k, \omega)} = \sum_s \frac{4\pi N_0 [q_{\parallel}^{(s)}(k)]^2}{\omega_s^2 - \omega^2} , \tag{2.2.21}$$

which corresponds to (2.1.22) of the E method. Since the s term is predominant at $\omega \approx \omega_s$ in (2.2.21), one gets immediately:

$$\left[\frac{\partial \epsilon_{\parallel}(k, \omega)}{\partial(\omega^2)} \right]_{\omega = \omega_s} = \frac{1}{4\pi N_0 q_{\parallel}^{(s)2}} . \tag{2.2.22}$$

Inserting (2.2.22) and (2.2.19) into (2.2.20), one obtains finally:

$$\varphi(r) = \sum_k \left[\frac{4\pi\hbar}{\partial \epsilon_{\parallel}(k, \omega_s)/\partial \omega_s} \right]^{1/2} \left(\frac{-i}{k} \right) (b_{sk} e^{ik \cdot r} - b_{sk}^\dagger e^{-ik \cdot r}) . \tag{2.2.23}$$

The transverse polarization waves $\omega_s = \Omega_s$, because of their divergent $\partial \epsilon_{\parallel}/\partial \omega_s$, do not contribute to the internal field, as is naturally expected.

An extra electron in the conduction band of an ionic crystal, with position coordinate r, is subject to fluctuating potential $(-e)\varphi(r)$ with $\varphi(r)$ given by (2.2.23). The electron–phonon interaction of this type plays an important role in the polaron effect to be discussed in Chap.6. For the diatomic crystal with cubic symmetry, one obtains, by putting (2.1.33′,35) into (2.2.23), the Fröhlich type interaction given in (6.1.15–17).

2.2.4 Plasma Oscillation and Screening Effect in Electron Gas

Consider an electron gas of average number density n put in the continuous medium of uniform positive charge $+ne$ and dielectric constant ϵ_∞. This is a model of favorite use for the conduction electrons in metals and highly doped semiconductors. In order to calculate the dielectric constant $\epsilon(k, \omega)$ of this system, we use the E method, on the assumption that all forces from other electrons and background positive charge can be effectively incorporated in the macroscopic field E. Namely, we calculate $\alpha_{\text{eff}}(\omega)$ of the noninteracting electron gas. (In a more advanced treatment, some short-range part of the Coulomb force remains unscreened in $\mathcal{H}_{\text{eff}}$, but it will be neglected here.) In thermal equilibrium, the electrons occupy various states, specified by momentum k and energy $\varepsilon(k) = \hbar^2 k^2 / 2m$, obeying the Fermi distribution: $f(k) = [\exp\{\beta[\varepsilon(k) - \varepsilon_F]\} + 1]^{-1}$ with Fermi energy ε_F.

The charge density operator of electrons

$$\rho(r) = (-e) \sum_i \delta(r - r_i) \tag{2.2.24}$$

has Fourier components given by

$$\rho_{-q} \equiv \int \rho(r) \exp(iq \cdot r) \, dr = (-e) \sum_i \exp(iq \cdot r_i) \ . \tag{2.2.25}$$

This operator has matrix elements removing an electron from occupied state k to unoccupied state $k + q$.

Therefore, the first half of the statistical average in (2.2.11′) (k is to be replaced by q) gives

$$\sum_k \frac{e^2 f(k)[1 - f(k + q)]}{\varepsilon(k + q) - \varepsilon(k) - \hbar\omega} \ ,$$

whereas the second half gives an expression with signs of q and ω reversed. Replacing k by $-k$ or k by $k + q$ in the latter, one obtains the following expressions for the effective polarizability of an ideal electron gas:

$$\alpha_{\mathrm{eff}}(q, \omega)_{\parallel} = \frac{e^2}{q^2} \sum_k \frac{2f(k)[\varepsilon(k) - \varepsilon(k + q)]}{(\hbar\omega)^2 - [\varepsilon(k) - \varepsilon(k + q)]^2} \tag{2.2.26}$$

$$= \frac{e^2}{q^2} \sum_k \frac{f(k) - f(k + q)}{\varepsilon(k + q) - \varepsilon(k) - \hbar\omega} \ . \tag{2.2.26′}$$

In the limit of $q \to 0$ (where α should be a scalar), (2.2.26) reduces to the well-known formula

$$\alpha_{\mathrm{eff}}(\omega) = - \frac{ne^2}{m\omega^2} \ . \tag{2.2.27}$$

By combining the relations: $\epsilon(\omega) = \epsilon_\infty + 4\pi\alpha_{\mathrm{eff}}(\omega)$ (E method) and $\epsilon(\omega)^{-1} = \epsilon_\infty^{-1} - 4\pi\alpha(\omega)$ (D method), the polarizability $\alpha(\omega)$ for interacting electron gas is obtained as

$$4\pi\alpha(\omega) = \frac{\omega_p^2/\epsilon_\infty}{\omega_p^2 - \omega^2} \tag{2.2.28}$$

where

$$\omega_p^2 = \frac{4\pi ne^2}{\epsilon_\infty m} \ . \tag{2.2.29}$$

(2.2.28) means that the interacting electron gas has eigenvibrations of frequency ω_p, which are evidently charge density waves (the fluctuation of ρ_q). Such a collective motion is called the *plasma oscillation* and its energy quantum $\hbar\omega_p$ the *plasmon*. The restoring force of (2.2.29) is nothing but the depolarizing field which also appeared in the longitudinal optical modes of lattice vibrations [compare (2.2.29) with the second term of (2.1.34)].

Let us next put $\omega = 0$ in (2.2.26'), keeping q small but finite. Since $f(k)$ depends on k through $\varepsilon(k)$, one can replace the summand by $-\partial f/\partial \varepsilon$. The static dielectric constant is then given by

$$\epsilon(q, 0) = \epsilon_\infty \left(1 + \frac{q_0^2}{q^2} \right) , \tag{2.2.30}$$

$$q_0^2 = \frac{4\pi e^2}{\epsilon_\infty} \sum_k \left(-\frac{\partial f}{\partial \varepsilon} \right) . \tag{2.2.31}$$

For the completely degenerate Fermi distribution with $-\partial f/\partial \varepsilon = \delta(\varepsilon - \varepsilon_F)$ (low temperature) and for the Boltzmann distribution with $-\partial f/\partial \varepsilon = \beta f$ (high temperature), one gets respectively

$$q_0^2 = \frac{4\pi n e^2}{\epsilon_\infty (2\varepsilon_F/3)} \quad \text{(Fermi)} , \tag{2.2.32}$$

$$= \frac{4\pi n e^2}{\epsilon_\infty k_B T} \quad \text{(Boltzmann)} . \tag{2.2.32'}$$

The physical meaning of q_0 can be seen as follows. Put a point charge Q at $r = 0$, which gives rise to the electric displacement $D_q = 4\pi Q q/i q^2$ [div$D(r) = 4\pi Q \delta(r)$] and in turn to the internal electric field $E_q = D_q/\epsilon(q,0)$. The electrostatic potential $\varphi(r) = \sum_q \varphi_q \exp(iq \cdot r)$ can then be calculated, with the use of $E_q = -iq\varphi_q$ and (2.2.30), as

$$\varphi(r) = \frac{Q}{\epsilon_\infty r} \exp(-q_0 r) . \tag{2.2.33}$$

This means that the long-range Coulomb potential due to a point charge is screened into a short-range potential with force range q_0^{-1} because of the redistribution of electron gas. q_0 is called the *screening constant*; in particular, the situations corresponding to (2.2.32,32') are called respectively the "Thomas-Fermi" case and the "Debye-Hückel" case. More systematic study of degenerate electron gas will be presented in Chap.6.

2.2.5 Absorption of Energy by Dielectrics

The most direct way of observing elementary excitations in dielectrics—generalized polarization waves—would be to put in a suitable electromagnetic probe and to measure the absorption spectra or the energy-loss spectra. For the probe, one may well think of charged particles and electromagnetic waves. In fact, these two types of probes give rise to dielectric response in two different ways mentioned in Sect.2.2.2 the effective external force being D in the former and E in the latter.

Electric conductivity $\sigma(\omega)$ and dielectric constant $\epsilon(\omega)$ are related by

$$\text{Re}\{\sigma(\omega)\} = \frac{\omega}{4\pi}\text{Im}\{\epsilon(\omega)\} \tag{2.2.34}$$

because of the identity $j = \partial P/\partial t$. The energy loss of electromagnetic waves with amplitude E_ω, per unit time and per unit volume of isotropic dielectrics, is given by

$$\text{Re}\{\sigma(\omega)\}\cdot\frac{E_\omega^2}{2} = \frac{\omega}{4\pi}\text{Im}\{\epsilon(\omega)\}\cdot\frac{E_\omega^2}{2} \tag{2.2.35}$$

where the k dependences of σ and ϵ are neglected because of the small k of electromagnetic waves.

Let us then consider a particle with charge Ze, mass M, and momentum $\hbar K$ coming into the same dielectrics. Because of the Coulomb potential $v(r - R) = Ze/|r - R|$ due to this particle, the dielectric is subject to the perturbation

$$\mathcal{H}_{ex} = \int \rho(r)v(r - R)\,dr = \sum_k v_k\rho_{-k}\exp(-ik\cdot R) \,, \tag{2.2.36}$$

where R is the position of the particle, $\rho(r)$ the charge density in dielectrics, ρ_k and $v_k = 4\pi Ze/k^2$ the Fourier components of $\rho(r)$ and $v(r)$, respectively. The scattering rate of this particle, with momentum loss $\hbar k$ and energy loss $\hbar\omega = \hbar^2 K^2/2M - \hbar^2(K - k)^2/2M$, is given, in the Born approximation, by

$$W(k, \omega) = \frac{2\pi}{\hbar}\text{Av}\sum_m\sum_n |v_k|^2|(\rho_{-k})_{nm}|^2\delta(\hbar\omega_{nm} - \hbar\omega) \,.$$

In order to relate this to (2.2.11') let us note that the imaginary part of the second term in (2.2.11') contributes $-\exp(-\beta\hbar\omega)$ times that of the first term. One can then rewrite W, with the use of (2.2.13), as

$$W(k, \omega) = \frac{2\pi}{\hbar}|ev_k|^2\left[\frac{k^2}{4\pi^2 e^2}\frac{1}{1 - \exp(-\beta\hbar\omega)}\text{Im}\left\{-\frac{1}{\epsilon(k, \omega)}\right\}\right] \tag{2.2.37}$$

where the expression in [...] corresponds to the dynamical structure factor introduced in Sect.1.6.

As shown in (2.2.35,37) the absorption spectrum of an electromagnetic wave is proportional to $\omega\,\text{Im}\{\epsilon(\omega)\}$, while the energy-loss spectrum of a high-energy charged particle is related to $\text{Im}\{-1/\epsilon(\omega)\}$. For the first example, let us consider the lattice vibrations. As is evident from (2.2.11) or (2.2.12,12') a pole of $\text{Re}\{\epsilon(\omega)\}$ corresponds to a δ singularity of $\text{Im}\{\epsilon(\omega)\}$. From (2.1.21,22), we have in fact

$$\omega\,\text{Im}\{\epsilon(\omega)\} = 2\pi^2 N_0\sum_s (p_x^{(s)})^2[\delta(\omega - \Omega_s) + \delta(\omega + \Omega_s)] \tag{2.2.38}$$

$$\omega \operatorname{Im} \left\{ -\frac{1}{\epsilon(\omega)} \right\} = 2\pi^2 N_0 \sum_s (q_x^{(\omega)})^2 [\delta(\omega - \omega_s) + \delta(\omega + \omega_s)] . \qquad (2.2.38')$$

The electromagnetic wave is absorbed at $\omega = \Omega_s$ while the charged particle loses energy by $\omega = \omega_s$. This is quite natural since in cubic crystals Ω_s and ω_s correspond respectively to transverse and longitudinal polarization waves for infrared active modes. Because of (2.2.34,38), the sum rule (2.1.23) can be brought into $\sigma(\omega)$ as follows:

$$\int_0^\infty \operatorname{Re} \{\sigma(\omega)\} \, d\omega = \frac{\pi}{2} N_0 \sum_\nu \frac{e_\nu^2}{M_\nu} . \qquad (2.2.39)$$

Let us consider, for the second example, the electron gas described in Sect. 2.2.4. As is evident from (2.2.28), the energy-loss spectrum consists of a line at $\omega = \omega_p$ corresponding to plasma excitation. The corresponding line in the absorption spectra should be at $\omega = 0$ because of the absence of restoring force $(\Omega_s = 0)$ for free electrons. In fact, we get $\operatorname{Re} \{\sigma(\omega)\} = \operatorname{Im} \{\omega \alpha_{eff}(\omega)\} = \operatorname{Im} \{-ne^2/m(\omega + i\eta)\}_{\eta \to +0} = (ne^2/m)\pi\delta(\omega)$. If the electrons were subject to scattering by other particles (e.g., phonons or impurities) with finite mean free time τ, this line spectrum would be broadened into a Lorentzian: $\operatorname{Re} \{\sigma(\omega)\} = (ne^2/m)\tau/(1 + \omega^2\tau^2)$ [without violating the sum rule: $\int_0^\infty \operatorname{Re} \{\sigma(\omega)\} \, d\omega = (\pi/2)$ (ne^2/m)]. This is the well-known *Drude formula* for the conductivity of nearly free electrons.

2.3 Exciton

In studying the dielectric dispersion associated with lattice vibrations of insulators (Sect.2.1), we have tacitly assumed that the electronic polarization—distortion of the electron cloud within each ion—follows the electromagnetic wave (ω being typically in the infrared region) without any delay, making an ω-independent contribution to $\epsilon_\infty - 1$. When ω approaches or gets into the region of eigenfrequencies of electronic polarization, the electronic polarizability shows its own dispersion accompanied by absorption. Because of the interatomic interactions, the electronic polarization as well makes a wave-like propagation through the crystal. In analogy to the phonons of lattice vibrations, one can conceive of the energy quanta of electronic polarization. We will introduce the concepts of Frenkel exciton and Wannier-Mott exciton representing for the two limiting situations of such energy quanta, and then present a theory unifying them.

2.3.1 Frenkel Exciton

Consider an isolated atom subject to an external electric field. The induced electronic polarization would be described, in quantum mechanics, with a slight mixing of, e.g., a p-like excited state $\phi_p(r)$ onto an s-like ground state $\phi_s(r)$. If the electric field were suddenly removed, the mixing coefficient, and hence the polarization, would thereafter vibrate with frequency $\omega = (\varepsilon_p - \varepsilon_s)/\hbar$. According to the correspondence principle, the quantum of electronic polarization of an atom is the energy of electronic excitation between the states of different parity.

In the case of a condensed system, the intraatomic excitation would propagate resonantly from atom to atom because of the interatomic interactions. The propagating energy of electronic excitation is called the *exciton*—in implication of its corpuscular aspect.

The wave function for the ground electronic state of an insulating crystal can be written as

$$\Psi^{(g)}(r_1, \ldots, r_N) = \prod_m \phi_s(r_m - R_m) , \tag{2.3.1}$$

with the neglect of the overlapping of atomic wave functions $\phi(r - R_m)$ between different sites ($R_m \neq R_{m'}$), and hence without the antisymmetrization of total wave function. Electronic spin will be neglected for the moment. Similarly, the wave function for the state with only the nth atom being excited can be written as (see Fig. 2.5.)

$$\Phi_n(r_1, \ldots, r_N) = \phi_p(r_n - R_n) \prod_{m(\neq n)} \phi_s(r_m - R_m) . \tag{2.3.2}$$

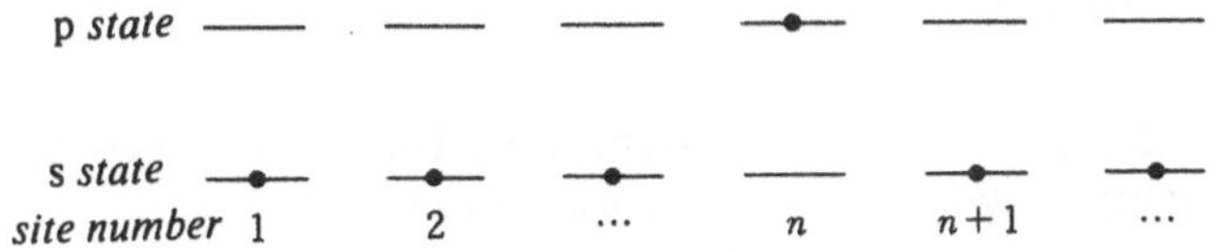

Fig. 2.5. The locally excited state [see (2.3.2)]

Note that the total Hamiltonian for this N-atom system consists of the atomic Hamiltonians h_n and the interatomic interactions v_{nm}:

$$\mathcal{H} = \sum_n h_n + \sum_{n<m} \sum v_{nm} . \tag{2.3.3}$$

The energy ε for the local excitation, namely, the energy-expectation value $\mathcal{H}_{nn} = (\Phi_n, \mathcal{H}\Phi_n)$ (which should be n independent) measured from that of the ground state $E^{(g)} = \mathcal{H}_{gg}$, would approximately be equal to the atomic excitation energy $\varepsilon_p - \varepsilon_s$. On the other hand, the off-diagonal matrix element is given by

$$\mathcal{H}_{nm} = (\Phi_n, \mathcal{H}\Phi_m) = \int dr_n \int dr_m \phi_p(r_n - R_n)\,\phi_s(r_m - R_m)$$
$$\cdot v_{nm}\phi_s(r_n - R_n)\,\phi_p(r_m - R_m)\ , \tag{2.3.4}$$

all other terms vanishing because of the orthogonalitity between ϕ_p and ϕ_s. Note that each atom consists of an atomic core and an electron. Among the various parts of v_{nm}, only the Coulomb interaction $e^2/|r_n - r_m|$ between the electrons remains to contribute to (2.3.4), since electron–core and core-core interactions contribute nothing to the integral, again because of the orthogonality between ϕ_s and ϕ_p. Assuming the spatial extensions of ϕ_s and ϕ_p to be smaller than the interatomic distance, one can make use of the multipole expansion of $e^2/|r_n - r_m|$ around $r_n = R_n$ and $r_m = R_m$. The first nonvanishing contribution to the integral is evidently the dipole–dipole interaction. Namely, with the transition dipole moment defined by

$$\mu \equiv \int \phi_p(r)\,(-er)\,\phi_s(r)dr\ , \tag{2.3.5}$$

one can put

$$\mathcal{H}_{nm} \approx \frac{\mu^2}{R_{nm}^3} - \frac{3(\mu \cdot R_{nm})^2}{R_{nm}^5} \equiv D(R_{nm})\ . \tag{2.3.6}$$

Noting that (2.3.4.), as well as (2.3.6.), is a function of only the distance $R_{nm} = R_n - R_m$, one can make use of the analogy to the tight-binding approximation for the one-electron Bloch state. In our case of an N electron system, the eigenstate with one-electron excitation which diagonalizes the translationally symmetric $\mathcal{H}$ should be the wave-like linear combination of locally excited states given by (2.3.2):

$$\Psi_K^{(e)} = \frac{1}{\sqrt{N}} \sum_n \exp(iK \cdot R_n)\,\Phi_n \tag{2.3.7}$$

where K can take any value within the first Brillouin zone. The excitation energy for this wave, $(\Psi_K^{(e)}, \mathcal{H}\Psi_K^{(e)}) - (\Psi^{(g)}, \mathcal{H}\Psi^{(g)})$, turns out to be

$$E_K = \varepsilon + \sum_{n(\neq 0)} D(R_n)\exp(-iK \cdot R_n) = \varepsilon + D_K\ , \tag{2.3.8}$$

which is a continuous function of K, forming as a whole the exciton band. In (2.3.7), it is not the electron but the excitation energy that is propagating from site to site; each electron stays within its own atom during excitation and de-excitation. Such an exciton, with only intra-atomic excitation (or intramolecular excitation in the case of molecular crystal), is called *Frenkel exciton*. The width of the Frenkel exciton band originates primarily from dipole–dipole interaction [see (2.3.8)] in so far as intra-atomic (intramolecular) excitation is dipole allowed and interatomic (intermolecular) overlap is negligibly small.

Since (2.3.6) is a long-range interaction decreasing only as R^{-3}, which moreover depends on direction, one must take care in the convergence of summation in (2.3.8), especially around $K = 0$. Let us consider a small wave vector K with $Kd \ll 1$ where $v_0 = d^3$ is the volume of a unit cell. Consider a macroscopic small sphere around a particular atom taken as the origin, with radius R_c such that $d \ll R_c \ll K^{-1}$. One can then put $\exp(-iK \cdot R_n) \approx 1$ for the atoms n within the sphere and replace the summation over atoms outside the sphere by integration: $N_0 \int dR$ where $N_0 = v_0^{-1}$ is the number of unit cells within a unit volume. Using the polar coordinate such that K is along the polar axis and μ lies within the reference longitudinal plane (see Fig. 2.6), one can write the integral as

$$I = N_0\mu^2 \int_{R_c}^{\infty} \frac{dR}{R} \int_0^{\pi} \exp(-iKP\cos\theta)\sin\theta\, d\theta \int_{-\pi}^{+\pi}(1 - 3\cos^2\theta')\, d\phi \ .$$

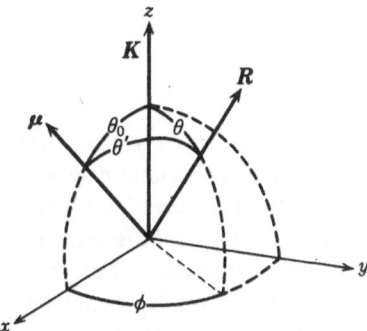

Fig. 2.6. Electric dipole moment μ, wave vector K, and integral variable R

Noting that $KR_c \ll 1$, one can calculate it as

$$I = -\frac{4\pi}{3}N_0\mu^2(1 - 3\cos^2\theta_0) + O(K^2R_c^2) \ .$$

Adding this to the intraspherical summation, one obtains finally

$$D_K = \sum_{R_n < R_c} D(R_n) - \frac{4\pi}{3}N_0[\mu^2 - 3(\mu \cdot \bar{K})^2] \quad (Kd \ll 1) \ , \qquad (2.3.9)$$

where $\bar{K}$ is the unit vector along K. Namely, D_K tends to different values depending on the direction along which K approaches zero. This signularity originates from the depolarizing field of the electronic polarization wave, the situation being the same as was encountered in the case of optical lattice vibrations. In fact, the energy of the longitudinal wave ($K \| \mu$) is higher than that of the transverse waves ($K \perp \mu$) by $4\pi N_0\mu^2$, according to (2.3.9).

In the cubic crystal, the p-like (T_{1u} under the crystalline field) excited states of an atom with cubic site symmetry are triply degenerate. Consequently, we have to consider, for each K, a set of three waves of the form (2.3.7) with dipole moments μ directed to the principal cubic axes x, y, and z (corresponding to p_x, p_y, and p_z excited states). In place of the second term of (2.3.9), we get a three-by-three energy matrix

$$- \frac{4\pi}{3} N_0 \mu^2 (\delta_{ij} - 3\vec{K}_i \vec{K}_j) \ . \tag{2.3.10}$$

This matrix can be diagonalized by taking the new z' axis along K; namely, the eigenstates of (2.3.10) consist of one longitudinal wave ($\mu \| K$) and two transverse waves, with eigenvalues $(8\pi/3) N_0\mu^2$ and $- (4\pi/3) N_0\mu^2$, respectively. It is interesting to note that the latter value for the transverse wave corresponds to the second term in (2.1.11), although comparison is made between the force constant and the energy.

2.3.2 Wannier–Mott exciton

In the Frenkel exciton described above, the excited electron was assumed to stay within the same atom. Besides such configurations of intra-atomic excitation: $\phi_s(r - R_m) \rightarrow \phi_p (r - R_m)$, we must take into account the "ionized" or "charge transfer" states in which the excited electron is removed to a different atom: $\phi_s(r - R_m) \rightarrow \phi_p(r - R_n)$ $(n \neq m)$ (since the spin is neglected, the excited electron cannot enter the s state of the nth atom which is already occupied), in order to extend our argument to the broader subspace of one-electron excitation. In these new configurations, we have a positively charged atom (m) and a negatively charged charged atom (n), namely, a pair with deficit electron and excess electron.

If the interatomic overlap of wave functions, neglected so far, is taken into account, the deficit electron as well as the excess electron can move from atom to atom. In terms of the band model, we have a *positive hole* in the *valence band* formed from the atomic state ϕ_s, and an electron in the *conduction band* formed from ϕ_p (see Fig. 2.7). The electron and the hole can move independently within their respective bands, except that they are subject to the attractive Coulomb force. In this way, we can describe the one-electron excited states of an insulating crystal in terms of a two-body problem of an electron in the conduction band and a hole in the valence band—analogous to that of the hydrogen atom. While the translational motion of the electron–hole pair, specified by wave vector K, forms a continuous energy band, the relative motion would have eigenenergies consisting of discrete spectra of bound states and continuous spectra of ionized states. The Frenkel exciton introduced before corresponds to the strong binding limit of the electron–hole pair such that the two particles are within the same atom (inter-atomic excitation).

In the other limit of the bound state such that the orbital radius a of electron–hole relative motion is much larger than the interatomic distance d, one can

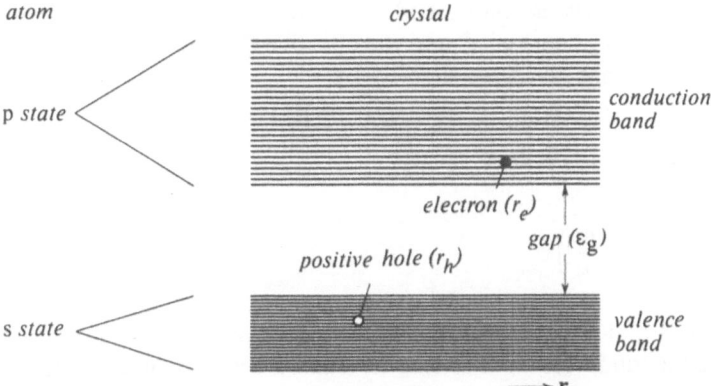

Fig. 2.7. The band model for an insulator. An electron–hole pair is excited across the gap

assume that the attractive Coulomb potential between the electron (at r_e) and the hole (at r_h) is screened into

$$V(r_e - r_h) = -\frac{e^2}{\epsilon|r_e - r_h|} \qquad (2.3.11)$$

by the macroscopic dielectric constant ϵ of the crystal. In the *effective mass approximation* which will be described in sect. 2.3.3, the electron (hole) can be considered as a free particle with mass m_e (m_h) characteristic of the conduction (valence) band. The total energy necessary to produce an electron–hole pair in the state specified by quantum number λ of relative motion and wave vector K of translational motion is then given by

$$E_{\lambda K} = \varepsilon_g + \frac{\hbar^2 K^2}{2M} + \bar{E}_\lambda . \qquad (2.3.12)$$

Here the energy gap ε_g between the valence and conduction bands (see Fig. 2.7) appears as the minimum energy necessary to produce an unbound electron–hole pair. The energy $\bar{E}_\lambda$ of relative motion takes discrete negative values (bound states)

$$\bar{E}_n = -\frac{R}{n^2} \quad (n = 1,2,3, \ldots) , \qquad (2.3.13)$$

and continuous positive values (ionized states)

$$\bar{E}_k = \frac{\hbar^2 k^2}{2\mu} , \qquad (2.3.14)$$

where n is the principal quantum number, k the wave vector for relative motion,

$M = m_e + m_h$ the *translational mass* and $\mu \equiv (m_e^{-1} + m_h^{-1})^{-1}$ the *reduced mass*. The effective Rydberg constant R and the effective Bohr radius a in the $1s$ state ($n = 1, l = m = 0$) are given respectively by

$$R = \frac{\mu e^4}{2\epsilon^2 \hbar^2} = \frac{e^2}{2\epsilon a} = \frac{1}{\epsilon^2} \left(\frac{\mu}{m_0} \right) R_H ,$$ (2.3.15)

$$a = \frac{\epsilon \hbar^2}{\mu e^2} = \epsilon \left(\frac{m_0}{\mu} \right) a_H .$$ (2.3.16)

In crystals with $\epsilon \gg 1$ and $m_e \lesssim m_0$ (true mass of the electron) as is usually the case for typical semiconductors with ε_g smaller than 2 eV, R is much smaller than the Rydberg constant $R_H = 13.6$ eV and a much larger than the radius $a_H = 0.53$ Å of the hydrogen atom. The bound pair of electron and hole with radius a large compared to the interatomic distance d is called the *Wannier–Mott exciton*.

In this way, the exciton can be visualized either as a propagation of excitation energy or as a composite particle consisting of an electron and a hole. They are nothing but two aspects of the same entity, the former (latter) predominating in the limiting situation of $a \ll d (a \gg d)$. A unified theory covering the both limits will be presented in the following.

2.3.3 Excited States of the Many-Electron System

Let us consider an insulating crystal to be composed of a rigid lattice of periodically arrayed nuclei and the electrons moving around them. Denote the one-electron Hamiltonian, consisting of kinetic energy and the potential due to the rigid lattice of the nuclei, by $h(p, r)$, and the repulsive Coulomb potential between electrons by $v(r - r') = e^2/|r - r'|$. Any wave function of this many-electron system can be represented by a linear combination of Slater determinants each of which corresponds to a particular choice (for occupation) out of a suitably determined orthogonal set of one-electron states. Wannier described the one-electron-excited states of an insulator by this representation, and derived a wave equation for an exciton. For the sake of compactness, we take here an equivalent representation with the second quantization.

With the use of electron annihilation–creation operators $\Psi(r)$ and $\Psi^\dagger(r)$, one can write the Hamiltonian of the many-electron system as

$$\mathscr{H} = \mathscr{H}^{(1)} + \mathscr{H}^{(2)} = \int dr\, \Psi^\dagger(r) h(p, r)\, \Psi(r)$$
$$+ \frac{1}{2} \iint dr dr'\, \Psi^\dagger(r) \Psi^\dagger(r') v(r - r')\, \Psi(r')\, \Psi(r) .$$ (2.3.17)

For the set of one-electron wave functions as an expansion basis of $\Psi(r)$, we take the *Bloch functions* $\phi_{\nu k}(r) = u_{\nu k}(r) \exp(ik \cdot p)$, or the *Wannier functions* obtained therefrom by orthogonal transformations

$$\phi_\nu(r - R_n) = \frac{1}{\sqrt{N}} \sum_k \exp(-ik \cdot R_n)\, \phi_{\nu k}(r) \ , \tag{2.3.18}$$

where ν denotes the band index, and the summation on the wave vector k extends over the first Brillouin zone. Leaving the Bloch functions unspecified for the moment, we assume only the orthonormality $(\phi^*_{\nu k},\ \phi_{\nu' k'}) = \delta_{\nu \nu'}\delta_{kk'}$, the completeness $\sum_{\nu k} \phi^*_{\nu k}(r)\, \phi_{\nu k}(r') = \delta(r - r')$, and the translational symmetry. From the last property, that $u_{\nu k}(r)$ be a periodic function with the periods of the lattice, one immediately finds (2.3.18) to be a function of $(r - R_n)$ only. As can be seen by comparing the inverse transform of (2.3.18) with the so-called tight-binding approximation for the Bloch states, these Wannier functions more or less resemble the atomic wave functions although only the former retain the orthonormality

$$\int dr \phi^*_\nu(r - R_n)\, \phi'_\nu(r - R_{n'}) = \delta_{\nu \nu'}\delta_{nn'} \tag{2.3.19}$$

and the completeness.

Expand $\Psi(r)$ in terms of Bloch or Wannier functions:

$$\Psi(r) = \sum_{\nu k} a_{\nu k}\phi_{\nu k}(r) = \sum_{\nu n} a_{\nu n}\phi_\nu(r - R_n) \ . \tag{2.3.20}$$

The two sets of expansion coefficients are related, because of (2.3.18), by

$$a_{\nu k} = \frac{1}{\sqrt{N}} \sum_n \exp(-ik \cdot R_n)\, a_{\nu n} \ . \tag{2.3.21}$$

$a_{\nu k}$ and $a^\dagger_{\nu k}$ (or $a_{\nu n}$ and $a^\dagger_{\nu n}$) represent the operators annihilating and creating an electron in the state $[\nu,\ k$ (or $n)]$, and satify the commutation relations:

$$a_{\nu k}a_{\mu k'} + a_{\mu k'}a_{\nu k} = 0, \quad a_{\nu k}a^\dagger_{\mu k'} + a^\dagger_{\mu k'}\, a_{\nu k} = \delta_{\nu \mu}\delta_{kk'} \tag{2.3.22}$$

(and those obtained by replacing k by n).

We will hereafter use the band indices μ for the filled bands, ν for the empty bands, and λ for any band without such specification. The ground state of this system within the one-electron approximation is given by

$$|g\rangle = \prod_{\mu' k} a^\dagger_{\mu' k}|0\rangle \ , \tag{2.3.23}$$

where $|0\rangle$ denotes the vacuum state and μ' runs over all the filled bands. Let us then consider the excited state $a^\dagger_{\nu k}a_{\mu k'}|g\rangle$ in which an electron has been raised from the $(\mu,\ k')$ state to the $(\nu,\ k)$ state. The total wave vector K of this state is given by $k - k'$. Since K is a quantum number of the translationally symmetric Hamiltonian (2.3.17), an eigenstate $|e\rangle$ within the subspace of one electron excitation can be written as a linear combination of the above mentioned excited states with all possible sets of (k, k') with a fixed value of $k - k' = K$. For the present, we confine ourselves to a single set of (ν, μ), leaving the more general treatments with linear combinations on (ν, μ) for later sections. Then we have

$$|e\rangle = B^\dagger|g\rangle \equiv \sum_k f(k) \, a^\dagger_{\nu k} a_{\mu, k-K}|g\rangle \tag{2.3.24}$$

$$= \frac{1}{\sqrt{N}} \sum_{m, l} \exp(i\mathbf{K} \cdot \mathbf{R}_m) \, F(\mathbf{R}_l) \, a^\dagger_{\nu, m+l} a_{\mu m}|g\rangle \; . \tag{2.3.24'}$$

The coefficients $f(k)$ and $F(\mathbf{R}_l)$, which are related through

$$f(k) = \frac{1}{\sqrt{N}} \sum_l \exp(-i\mathbf{k} \cdot \mathbf{R}_l) \, F(\mathbf{R}_l) \; , \tag{2.3.25}$$

will be normalized as

$$\sum_k |f(k)|^2 = \sum_l |F(\mathbf{R}_l)|^2 = 1 \; . \tag{2.3.26}$$

Equation (2.3.24') represents the state in which an electron in the ν band and a positive hole in the μ band make relative motion with wave function $F(\mathbf{R}_l)$ and translational motion with wave vector K—a reduction of the N-body problem to a two-body problem. What is then the Schrödinger equation for the wave function $F(\mathbf{R}_l)$ or $f(k)$ of relative motion?

For the given Hamiltonian $\mathscr{H}$ and the known ground state $|g\rangle$ of a many-body system, the following method for finding excited states $|e\rangle$ and excitation energies $E = E_e - E_g$ is in order: Putting $|e\rangle = B^\dagger|g\rangle$ with an appropriate operator $B^\dagger$, one can rewrite the eigenequation for $|e\rangle$ as

$$0 = (\mathscr{H} - E_e)|e\rangle = (\mathscr{H}B^\dagger - E_e B^\dagger)|g\rangle = ([\mathscr{H}, B^\dagger] + B^\dagger \mathscr{H} - E_e B^\dagger)|g\rangle \; .$$

with the use of $\mathscr{H}|g\rangle = E_g|g\rangle$, the equation for $B^\dagger$ is immediately obtained:

$$[\mathscr{H}, B^\dagger]|g\rangle = E B^\dagger|g\rangle \; . \tag{2.3.27}$$

Putting our $B^\dagger$ given by (2.3.24) into (2.3.27) and multiplying $\langle g| a^\dagger_{\mu k'-K} a_{\nu k'}$ from the left side, one gets the equation for $f(k)$

$$\sum_k H_{k'k} f(k) = E f(k') \tag{2.3.28}$$

with the effective Hamiltonian H defined by

$$H_{k'k} \equiv \langle g| a^\dagger_{\mu k'-K} a_{\nu k'}[\mathscr{H}, a^\dagger_{\nu k} a_{\mu k-K}]|g\rangle \; . \tag{2.3.29}$$

Let us now calculate (2.3.29). Making use of the commutation relations

$$[\Psi(r), a_{\nu k}^{\dagger} a_{\mu k - \kappa}] = \phi_{\nu k}(r)\, a_{\mu k - \kappa}$$
$$[\Psi\dagger(r), a_{\nu k}^{\dagger} a_{\mu k - \kappa}] = - \phi_{\mu k - \kappa}^{*}(r) a_{\nu k}^{\dagger} \qquad (2.3.30)$$

obtained from (2.3.20, 22) we can immediately calculate that part of the commutator in (2.3.29), which originates from the one-body Hamiltonian $\mathcal{H}^{(1)}$, as follows:

$$[\mathcal{H}^{(1)}, a_{\nu k}^{\dagger} a_{\mu k - \kappa}] = \sum_{\lambda} (h_{\lambda \nu k} a_{\lambda k}^{\dagger} a_{\mu k - \kappa} - h_{\mu \lambda k - \kappa} a_{\nu k}^{\dagger} a_{\lambda k - \kappa}) \ . \qquad (2.3.31)$$

In deriving (2.3.31), we have put the matrix elements of h in regard to the Bloch states as $h_{\lambda k', \nu k} = \delta_{kk'} h_{\lambda \nu k}$ since the translationally symmetric h should be diagonal in k. Considering the occupancy in the ground state $|g\rangle$, we see that only $\lambda = \nu$ and $\lambda = \mu$ respectively in the first and the second terms of (2.3.31) contribute to (2.3.29), with the result:

$$H_{k'k}^{(1)} = \delta_{k'k}(h_{\nu \nu k} - h_{\mu \mu k - \kappa}) \ . \qquad (2.3.32)$$

The contribution of the two-body Hamiltonian $\mathcal{H}^{(2)}$ to the commutator in (2.3.29) can be calculated, by repeated use of (2.3.30) (twice with Ψ and twice with $\Psi\dagger$), as

$$[\mathcal{H}^{(2)}, a_{\nu k}^{\dagger} a_{\mu k - \kappa}] = \iint dr dr' \Psi\dagger(r)\Psi\dagger(r') v \Psi(r')\phi_{\nu \kappa}(r) a_{\mu\ k - k}$$
$$- \iint dr dr' \Psi\dagger(r)\phi_{\mu\ k - \kappa}^{*}(r')\, a_{\nu k}^{\dagger} v \Psi(r')\, \Psi(r) \ , \qquad (2.3.33)$$

which, in turn, is to be put into the expression for the effective Hamiltonian

$$H_{k'k}^{(2)} = \langle g | a_{\mu\ k' - \kappa}^{\dagger} a_{\nu k'} [\mathcal{H}^{(2)}, a_{\nu k}^{\dagger} a_{\mu\ k - \kappa}] | g \rangle \ . \qquad (2.3.34)$$

With the use of the expansion (2.3.20), the contribution of the first term of (2.3.33) to (2.3.34) can be written as

$$\sum_{\lambda_1 k_1 \lambda_2 k_2 \lambda_3 k_3} v_{\lambda_1 k_1 \lambda_2 k_2 \lambda_3 k_3 \nu k} \langle g | a_{\mu\ k' - \kappa}^{\dagger} a_{\nu k'} a_{\lambda_1 k_1}^{\dagger} a_{\lambda_2 k_2}^{\dagger} a_{\lambda_3 k_3} a_{\mu\ k - \kappa} | g \rangle \qquad (2.3.35)$$

with

$$v_{\lambda_1 k_1 \lambda_2 k_2 \lambda_3 k_3 \lambda_4 k_4} \equiv \iint dr dr' \phi_{\lambda_1 k_1}^{*}(r) \phi_{\lambda_2 k_2}^{*}(r') v(r - r')\, \phi_{\lambda_3 k_3}(r') \phi_{\lambda_4 k_4}(r) \ . \qquad (2.3.36)$$

It is convenient to study the diagonal ($k' = k$) and nondiagonal ($k' = k$) matrix elements, separately.

With $k' = k$, the expectation value $\langle g | \ldots | g \rangle$ in (2.3.35) is nonvanishing only in the following two cases [the value being $+1$ in (i) and -1 in (ii)]:

(i) $(\lambda_1 k_1) = (\nu k)$, $(\lambda_2 k_2) = (\lambda_3 k_3) \equiv (\mu' k'')$,

(ii) $(\lambda_2 k_2) = (\nu k)$, $(\lambda_1 k_1) = (\lambda_3 k_3) \equiv (\mu' k'')$, where $(\mu'k'')$ runs over all the occupied (in the ground state $|g\rangle$) Bloch states except $(\mu, k - K)$. The corresponding contribution from the second term of (2.3.33) can be calculated in the same way. One gets altogether:

$$H_{kk}^{(2)} = \sum_{\mu'k''(\neq\mu,k-K)} [(v_{\nu k\mu'k''\mu'k''\nu k} - v_{\mu'k''\nu k\mu'k''\nu k})$$

$$- (v_{\mu'k''\mu k-K\mu k-K\mu'k''} - v_{\mu'k''\mu k-K\mu'k''\mu k-K})] . \qquad (2.3.37)$$

With $k' \neq k$, only the following two terms are nonvanishing in (2.3.35):

(i) $(\lambda_1 k_1) = (\nu k')$, $(\lambda_2 k_2) = (\mu k - K)$, $(\lambda_3 k_3) = (\mu k' - K)$,

(ii) $(\lambda_1 k_1) = (\mu k - K)$, $(\lambda_2 k_2) = (\nu k')$, $(\lambda_3 k_3) = (\mu k' - K)$.

Since the second term of (2.3.33) contributes nothing, one immediately obtains

$$H_{k'k}^{(2)} = -v_{\nu k'\mu k-K\mu k'-K\nu k} + v_{\mu k-K\nu k'\mu k'-K\nu k} \quad (k' \neq k) . \qquad (2.3.38)$$

Let us now choose the one-electron wave functions $\phi_{\nu k}(r)$, which have so far been unspecified, to be the solutions of

$$h_{\text{H.F.}}\phi_{\lambda k}(r) \equiv h(p, r) \phi_{\lambda k}(r) + \sum_{\mu'k''} \int dr' v(r - r') \phi_{\mu'k''}^*(r')$$

$$\cdot [\phi_{\mu'k''}(r') \phi_{\lambda k}(r) - \phi_{\mu'k''}(r)\phi_{\lambda k}(r')] = \varepsilon_\lambda(k)\phi_{\lambda k}(r) . \qquad (2.3.39)$$

The equations (2.3.39) provide not only the self-consistent Hartree-Fock solutions for the filled-band ($\lambda = \mu'$) Bloch states with the occupancy (2.3.23), but also the empty-band ($\lambda = \nu'$) Bloch functions under the "same" occupancy. Being eigenfunctions of the common Hermitian operator $h_{\text{H.F.}}$, they altogether form a complete orthonormal set. With the use of (2.3.36, 39), one can write the one electron energy as

$$\varepsilon_\lambda(k) = h_{\lambda\lambda k} + \sum_{\mu'k''} (v_{\lambda k\mu'k''\mu'k''\lambda k} - v_{\mu'k''\lambda k\mu'k''\lambda k}) . \qquad (2.3.40)$$

The results (2.3.32, 37, 38) can be combined, with the help of (2.3.40), to give the explicit form of the effective Hamiltonian $H = H^{(1)} + H^{(2)}$:

$$H_{k'k} = \delta_{k'k}[\varepsilon_\nu(k) - \varepsilon_\mu(k - K)]$$

$$- v_{\nu k'\mu k-K\mu k'-K\nu k} + v_{\mu k-K\nu k'\mu k'-K\nu k} . \qquad (2.3.41)$$

It implies the motion of an electron in the ν band and a positive hole in the μ band with attractive Coulomb and repulsive exchange interactions with the total wave vector being kept constant $[k - (k - K) = K]$. Note that the signs of the two types of interaction in this electron–hole system are reversed as compared to the two-electron system.

The equation of motion in the k representation, (2.3.28), can be transformed, with the use of (2.3.25), into the R representation

$$\sum_{l'} H_{ll'} F(\boldsymbol{R}_{l'}) = EF(\boldsymbol{R}_l) \tag{2.3.42}$$

with

$$H_{ll'} = \frac{1}{N} \sum_{\boldsymbol{k}\boldsymbol{k}'} \exp(i\boldsymbol{k}' \cdot \boldsymbol{R}_l) \, H_{\boldsymbol{k}'\boldsymbol{k}} \exp(-i\boldsymbol{k} \cdot \boldsymbol{R}_{l'}) \;, \tag{2.3.43}$$

where $\boldsymbol{R}_l$ denotes the electron-hole relative coordinate as is evident from (2.3. 24'). An explicit form of $H_{ll'}$, can be obtained as follows. Consider the matrix elements of $v(\boldsymbol{r} - \boldsymbol{r}')$ on the Wannier basis:

$$\begin{aligned} v_{\lambda_1 n_1 \lambda_2 n_2 \lambda_3 n_3 \lambda_4 n_4} &\equiv \iint d\boldsymbol{r} d\boldsymbol{r}' \phi_{\lambda_1}^* (\boldsymbol{r} - \boldsymbol{R}_{n_1}) \, \phi_{\lambda_2}^*(\boldsymbol{r}' - \boldsymbol{R}_{n_2}) \, v(\boldsymbol{r} - \boldsymbol{r}') \\ &\quad \cdot \phi_{\lambda_3}(\boldsymbol{r}' - \boldsymbol{R}_{n_3}) \, \phi_{\lambda_4}(\boldsymbol{r} - \boldsymbol{R}_{n_4}) \;. \end{aligned} \tag{2.3.44}$$

Let us neglect them except when $n_1 = n_4$ and $n_2 = n_3$, on the assumption that the Wannier functions $\phi_\lambda(\boldsymbol{r} - \boldsymbol{R}_n)$ are well localized around $\boldsymbol{R}_n$ like the atomic orbitals. Let us further approximate, in the case of $\lambda_1 = \lambda_4 = \nu$, $\lambda_2 = \lambda_3 = \mu$, as

$$\iint d\boldsymbol{r} d\boldsymbol{r}' |\phi_\nu(\boldsymbol{r} - \boldsymbol{R}_1)|^2 v(\boldsymbol{r} - \boldsymbol{r}') |\phi_\mu(\boldsymbol{r}' - \boldsymbol{R}_2)|^2 \approx v(\boldsymbol{R}_1 - \boldsymbol{R}_2) \tag{2.3.45}$$

and denote, in the case of $\lambda_1 = \lambda_3 = \nu$, $\lambda_2 = \lambda_4 = \mu$, as

$$\begin{aligned} &\iint d\boldsymbol{r} d\boldsymbol{r}' \phi_\nu^*(\boldsymbol{r} - \boldsymbol{R}_1) \, \phi_\mu(\boldsymbol{r} - \boldsymbol{R}_1) v(\boldsymbol{r} - \boldsymbol{r}') \, \phi_\mu^*(\boldsymbol{r}' - \boldsymbol{R}_2) \, \phi_\nu(\boldsymbol{r}' - \boldsymbol{R}_2) \\ &\equiv w(\boldsymbol{R}_1 - \boldsymbol{R}_2) \;. \end{aligned} \tag{2.3.46}$$

Then the last two terms of (2.3.41) contribute to (2.3.43)

$$-\delta_{ll'} v(\boldsymbol{R}_l) + \delta_{l0} \delta_{l'0} w_{\boldsymbol{K}} \;, \tag{2.3.47}$$

where

$$w_{\boldsymbol{K}} = \sum_m \exp(-i\boldsymbol{K} \cdot \boldsymbol{R}_m) w(\boldsymbol{R}_m) \;. \tag{2.3.48}$$

The first term of (2.3.41), $\delta_{\boldsymbol{k}'\boldsymbol{k}} \varepsilon_\nu(\boldsymbol{k})$, after the same transformation, gives

$$\frac{1}{N} \sum_{\boldsymbol{k}} \varepsilon_\nu(\boldsymbol{k}) \exp[i\boldsymbol{k} \cdot (\boldsymbol{R}_l - \boldsymbol{R}_{l'})] \equiv E_\nu(\boldsymbol{R}_l - \boldsymbol{R}_{l'}) \;. \tag{2.3.49}$$

Operating (2.3.49) on $F(\boldsymbol{R}_l)$ as prescribed in (2.3.42), one obtains the following:

$$\begin{aligned} \sum_{l'} E_\nu(\boldsymbol{R}_l - \boldsymbol{R}_{l'}) F(\boldsymbol{R}_{l'}) &= \sum_m E_\nu(\boldsymbol{R}_m) F(\boldsymbol{R}_l - \boldsymbol{R}_m) \\ &= \sum_m E_\nu(\boldsymbol{R}_m) \sum_{p=0}^\infty \frac{1}{p!} (-\boldsymbol{R}_m \cdot \boldsymbol{\nabla}_l)^p F(\boldsymbol{R}_l) \\ &= \sum_m E_\nu(\boldsymbol{R}_m) \exp[-\boldsymbol{R}_m \cdot \boldsymbol{\nabla}_l] F(\boldsymbol{R}_l) \\ &= \varepsilon_\nu(-i\boldsymbol{\nabla}_l) F(\boldsymbol{R}_l) \;. \end{aligned} \tag{2.3.50}$$

In the above we first extend the function $F(R_{l'})$ of discrete variable to a smooth function of continuous variable, then expand it around R_l in a Taylor series which is equivalent to applying the displacement operator $\exp(-R_m \cdot \nabla_l)$. Comparing the sum over m with the inverse transform of (2.3.49), we immediately get the final expression, in which $\varepsilon_\nu(-i\nabla_l)$ is an operator obtained by replacing k with $-i\nabla_l$ in the power series expansion of $\varepsilon_\nu(k)$. The effective Hamiltonian H for a pair excitation, as was defined in (2.3.43), can be written, because of (2.3.47, 50), as

$$H_{ll'} = [\varepsilon_\nu(-i\nabla_l) - \varepsilon_\mu(-i\nabla_l - K) - v(R_l)]\delta_{ll'} + \delta_{l0}\delta_{\nu 0}w_K . \qquad (2.3.51)$$

It is easy to see how (2.3.51) gives the Wannier–Mott type exciton and the Frenkel exciton in the two limits of $a \gtrless d$. But for the last term, (2.3.51) represents the two-particle picture—an electron in the conduction band with energy $\varepsilon_\nu(k)$ and a positive hole in the valence band with energy $-\varepsilon_\mu(k')$ are moving together under the attractive Coulomb potential $-v(R)$ and with a given total wave vector $K = k - k'$. To be more explicit, one expands the band energy of the pair with given K, $\varepsilon_\nu(k) - \varepsilon_\mu(k - K)$, in powers of the relative motion wave vector k around the minimum point $k_m(K)$:

$$\varepsilon_\nu(k) - \varepsilon_\mu(k - K) = \varepsilon^{(m)}(K) + (k - k_m)\frac{\hbar^2}{2\mu(K)}(k - k_m) + \cdots , \qquad (2.3.52)$$

the coefficient $\mu(K)$ being the effective reduced mass tensor. With the use of (2.3.51, 52), one can write the equation (2.3.42) for the relative motion as

$$\left[-\nabla \frac{\hbar^2}{2\mu(K)} \nabla - v(R)\right] \bar{F}(R) = \bar{E}\bar{F}(R) \qquad (2.3.53)$$

with

$$\bar{E} \equiv E - \varepsilon^{(m)}(K) , \qquad (2.3.54)$$

$$\bar{F}(R) \equiv F(R) \exp(-ik_m \cdot R) . \qquad (2.3.55)$$

The neglect of higher-order terms in (2.3.52), the so-called effective mass approximation, is justified when $\bar{F}(R)$ is a slowly varying function. The normalization (2.3.26) is to be written as

$$\frac{1}{v_0} \int |\bar{F}(R)|^2 dR = 1 \qquad (2.3.56)$$

where v_0 is the volume of a unit cell.

Consider the simplest model of direct band gap at $k = 0$ with isotropic effective masses:

$$\varepsilon_\mu(k) = \varepsilon_\mu(0) - \frac{\hbar^2 k^2}{2m_h}, \quad \varepsilon_\nu(k) = \varepsilon_\mu(0) + \varepsilon_g + \frac{\hbar^2 k^2}{2m_e} . \tag{2.3.57}$$

From (2.3.52), one immediately obtains $k_m(K) = (\mu/m_h) K$ and

$$\varepsilon^{(m)}(K) = \varepsilon_g + \frac{\hbar^2 K^2}{2M} \tag{2.3.58}$$

where $M = m_e + m_h$, $\mu(K)^{-1} = \mu^{-1} = m_e^{-1} + m_h^{-1}$. This is the isotropic Wannier–Mott exciton described in Sect. 2.3.2 [compare (2.3.53) to (2.3.13), and (2.3.54, 58) to (2.3.12)]. The expectation value of the last term of (2.3.51) in the hydrogenic 1s state, $F_{1s}(R) = (d^3/\pi a^3)^{1/2} \exp(- R/a)$, turns out to be $|\bar{F}_{1s}(0)|^2 w_K = \pi^{-1}(d/a)^3 w_K$. The neglect of this term in (2.3.51) and the neglect of higher-order terms in (2.3.52) (effective mass approximation) are both warranted under the same condition of "large exciton": $a \gg d$.

While the electron–hole interaction $-v(R)$ as obtained in (2.3.51) is the unscreened Coulomb potential $-e^2/R$, the actual interaction should be screened into $-e^2/\epsilon R$ with appropriate dielectric constant ϵ in the case of a large exciton since the electron and the hole are then considered to be embedded in a continuous dielectric medium. In order to derive this screened Coulomb potential *ab initio* in the context of our many-body problem, we must take into account that both the electron and the hole are dressed with dielectric polarization, which can be described by considering virtual excitation of further electron–hole pairs (excitons are quanta of electronic polarization as described in Sect. 2.3.1), and so on leading to an infinite chain of virtual excitations. Namely, the exact microscopic derivation of screening in a real exciton requires the subspaces of multiple excitation of virtual excitons to infinite order. All this complexity is in effect incorporated into a single screening constant ϵ^{-1} introduced phenomenologically in the above (see the arguments in Sect. 2.2.2, 4).

In the opposite limit of a "small" or "tight-binding" exciton such that $|F(R_l)|^2 = \delta_{l0}$, one can write the expectation value of the energy (2.3.51), with the help of (2.3.48, 50), as

$$H_{00} = [E_\nu(0) - E_\mu(0) - v(0) + w(0)] + \sum_{m(\neq 0)} w(R_m) \exp(-iK \cdot R_m) . \tag{2.3.59}$$

Such a strong binding is realized when the interatomic (intermolecular) interactions and hence the energy widths of ν and μ bands are sufficiently small (large effective masses m_e, m_h, and hence μ, result in the small radius of the exciton). It is evident in this situation that the first and the second terms in (2.3.59) correspond to those of (2.3.8), namely, to the intraatomic excitation energy ε and the interatomic dipole interaction D_K, respectively. Since $E_\nu(0)$ denotes the energy of the ν state when the μ state is occupied, the electronic excitation from μ to ν requires energy $E_\nu(0) - E_\mu(0)$ minus the sum of

Coulomb energy $v(0)$ and exchange energy $-w(0)$ between the ν and μ electrons. In this way, one regains the Frenkel exciton in the limit of tight binding.

2.4 Excitons in the Optical Spectra

While the exciton is a composite particle with internal (relative motion) as well as external (translation) degrees of freedom, it is a quantum of polarization wave which can be created or annihilated through the interaction with electromagnetic waves. This means that various energy states of the exciton can be observed more or less directly in the photoabsorption and emission spectra. How the internal (orbital and spin) and the translational motions are reflected in the optical spectra will be described in Sects. 2.4.1–3, respectively. Sections 2.4.4, 5 will be devoted respectively to the molecular bond and the fission–fusion processes, which represent two different types of conversions of composite particles.

2.4.1 Fundamental Absorption Spectra

The intrinsic photoabsorption spectra of an insulator corresponding to the electronic transition from the ground state to the excited states are called the *fundamental absorption spectra*. Considering that the absorption spectra are intimately related (proportional under certain conditions as stated in Sect. 2.2.5) to the imaginary part of the dielectric function $\epsilon(\omega)$, we will now calculate the latter spectra. Confining ourselves to single-pair excitations characterized by (λ, K) (see Sect. 2.3), we obtain

$$\text{Im}\{\epsilon(\boldsymbol{q}, \omega)\} = \text{Im}\{4\pi\alpha_{\text{eff}}(\boldsymbol{q}, \omega)\} = 4\pi^2 \sum_{\lambda K} (P_q)_{g,\lambda K}(P_{-q})_{\lambda K,g}\delta(E_{\lambda K} - \hbar\omega) ,$$

$$(2.4.1)$$

with the use of (2.2.11, 14) under the assumption that the $E_{\lambda K}$'s are much larger than the thermal energy.

Consider the current density operator

$$\dot{P}(\boldsymbol{r}) = j(\boldsymbol{r}) = \frac{-e}{m} \sum_i \frac{1}{2}[\delta(\boldsymbol{r} - \boldsymbol{r}_i)\boldsymbol{p}_i + \boldsymbol{p}_i\delta(\boldsymbol{r} - \boldsymbol{r}_i)] ,$$

whose Fourier component has the matrix element relevant to (2.4.1):

$$\frac{iE_{\lambda K}}{\hbar}(P_{-q})_{\lambda K, g} = \frac{-e}{m_0}\left\{\sum_i \frac{1}{2}[\exp(i\boldsymbol{q}\cdot\boldsymbol{r}_i)\boldsymbol{p}_i + \boldsymbol{p}_i\exp(i\boldsymbol{q}\cdot\boldsymbol{r}_i)]\right\}_{\lambda K, g} . \qquad (2.4.2)$$

The curly bracket on the right hand side, which is the sum of the one-body operator, can be written, in the second quantization, as

$$\{ \ \} \rightarrow \int dr \, \Psi\dagger(r) \frac{1}{2} \left[\exp(i\boldsymbol{q} \cdot \boldsymbol{r}) \, \boldsymbol{p} + \boldsymbol{p} \, \exp(i\boldsymbol{q} \cdot \boldsymbol{r}) \right] \Psi(r)$$

$$\approx \sum_{\lambda'\lambda''k} \boldsymbol{p}_{\lambda'\lambda''}(k) a^\dagger_{\lambda'k} a_{\lambda''k-q} \ , \tag{2.4.3}$$

$$\boldsymbol{p}_{\lambda'\lambda''}(k) \equiv \int \phi^*_{\lambda'k}(r)(-i\hbar\nabla) \, \phi_{\lambda''k}(r) \, dr \ . \tag{2.4.4}$$

Here the light-wave vector $\boldsymbol{q}$ has been assumed to be much smaller than the reciprocal lattice vector (note that the photon absorbed to create an exciton with energy ≤ 10 eV has the wave vector $\leq 10^5$ cm^{-1}). Making use of (2.4.3) and (2.3.24), one can write the matrix element of the curly bracket in (2.4.2) as

$$\{ \ \}_{\lambda K,s} = \delta_{K,q} \sum_k \boldsymbol{p}_{\nu\mu}(k) f^*(k) \ . \tag{2.4.5}$$

The Kronecker's δ represents the wave vector conservation between photon (inside the crystal) and exciton. Hereafter, $\boldsymbol{q} = \boldsymbol{K}$ will be neglected except when its finiteness or its direction plays an important role.

If the electron–hole interaction is neglected in the effective Hamiltonian (2.3.41) (the last two terms), (2.3.28) gives $E = E_{\lambda K} = \varepsilon_\nu(k) - \varepsilon_\mu(k)$, $f_{\lambda K}(k') = \delta_{k'k}(\boldsymbol{K} = \boldsymbol{q} \rightarrow 0, \, \lambda = k)$, and (2.4.1) leads to

$$\text{Im} \{\epsilon(\omega)\} = \frac{4\pi^2 e^2}{m_0^2 \omega^2} \sum_k \boldsymbol{p}_{\mu\nu}(k) \, \boldsymbol{p}_{\nu\mu}(k) \, \delta[\varepsilon_\nu(k) - \varepsilon_\mu(k) - \hbar\omega] \ , \tag{2.4.6}$$

which represents the photoabsorption spectrum for interband transition based on the one-electron picture. It is proportional except for the k dependence of $\boldsymbol{p}_{\mu\nu}(k)$ to the *joint density of states* of the two bands; in particular, it rises as $[\hbar\omega - \varepsilon^{(m)}(0)]^{1/2}$ at the absorption edge $\varepsilon^{(m)}(0)$ which corresponds to the smallest value of interband transition energy $\varepsilon_\nu(k) - \varepsilon_\mu(k)$ [see (2.3.52)]. It is in this spectral region that the electron–hole interaction manifests itself most clearly, as will be shown beow.

If the wave function $\bar{F}(R)$ for the relative motion [see (2.3.53)] is a slowly varying function, the Fourier transform $f(k)$ of $F(R) = \bar{F}(R)\exp[ik_m(0) \cdot R]$ [see (2.3.25, 55)] is expected to have a sharp peak around $k_m(0)$. It is then enough to make use of the expansion

$$\boldsymbol{p}_{\nu\mu}(k) = \boldsymbol{p}^{(m)}_{\nu\mu} + [k - k_m(0)] \cdot \nabla \boldsymbol{p}^{(m)}_{\nu\mu} + \cdots \tag{2.4.7}$$

in (2.4.5) and to consider only the leading term. There are two possibilities as described below.

(i) allowed-edge type: $\boldsymbol{p}^{(m)}_{\nu\mu} \neq 0$. Considering only the first term of (2.4.7), putting it into (2.4.5), and with the use of (2.4.1, 2, 2.3.25), one obtains

$$\text{Im} \{\epsilon(\omega)\} = \frac{4\pi^2 N_0 e^2}{m_0^2 \omega^2} \boldsymbol{p}^{(m)}_{\mu\nu} \boldsymbol{p}^{(m)}_{\nu\mu} \sum_\lambda |\bar{F}_\lambda(0)|^2 \delta(E_{\lambda 0} - \hbar\omega) \ . \tag{2.4.8}$$

It is to be noted that only the s states ($l = 0$) of the electron–hole relative motion contribute to the absorption spectra [$F_\lambda(0)$ vanishes for $l \neq 0$].

In the case of the isotropic Wannier–Mott exciton, the discrete spectral lines appear at

$$E_{n0} = \varepsilon^{(m)}(0) - \frac{R}{n^2} \quad (n = 1,2,3, \ldots) \tag{2.4.9}$$

with respective intensities

$$|\bar{F}_n(0)|^2 = \frac{v_0}{\pi a^3} \frac{1}{n^3} \, . \tag{2.4.10}$$

The continuous spectra, given by

$$E_{k0} = \varepsilon^{(m)}(0) + \frac{\hbar^2 k^2}{2\mu} \, , \tag{2.4.11}$$

$$|\bar{F}_k(0)|^2 = \frac{v_0 \pi \alpha \exp(\pi \alpha)}{\sinh(\pi \alpha)}, \quad \alpha \equiv \frac{1}{ak} \, , \tag{2.4.12}$$

tend to a finite value:

$$\sum_k |\bar{F}_k(0)|^2 \delta(E_{k0} - \hbar\omega) \to \frac{v_0}{2\pi R a^3} \tag{2.4.13}$$

as $\hbar\omega \to \varepsilon^{(m)}(0)$ ($k \to 0$). On the other hand, the discrete lines with large n are overlapping with and unresolvable from each other because of their finite widths (see Sect. 3). The smoothed-out spectra are then given by multiplying the intensity and the state density:

$$|\bar{F}_n(0)|^2 \left| \frac{dE_n}{dn} \right|^{-1} = \frac{v_0}{2\pi R a^3} \, . \tag{2.4.13'}$$

One finds that the quasicontinuum (2.4.13') connects smoothly to the true continuum (2.4.13) without any discontinuity (up to higher-order derivatives), as shown schematically in Fig. 2.8a.

(ii) forbidden-edge type: $p_{\nu\mu}^{(m)} = 0$. Considering only the second term of (2.4.7) and proceeding in the same way as before, one obtains the expression similar to (2.4.8) except that $p_{\nu\mu}^{(m)}$ and $\bar{F}_\lambda(0)$ are to be replaced by $\nabla_k p_{\nu\mu}^{(m)}$ and $\nabla_R \bar{F}_\lambda(0)$, respectively. Since $\nabla_R \bar{F}(0)$ is nonvanishing only for the p states ($l = 1$), the discrete spectra consist of the Bohr series (2.4.9) starting with $n = 2$ with intensities given by

$$|\nabla_x \bar{F}_{np_x}(0)|^2 = \frac{v_0}{\pi a^5} \left(\frac{1}{n^3} - \frac{1}{n^5} \right) , \tag{2.4.14}$$

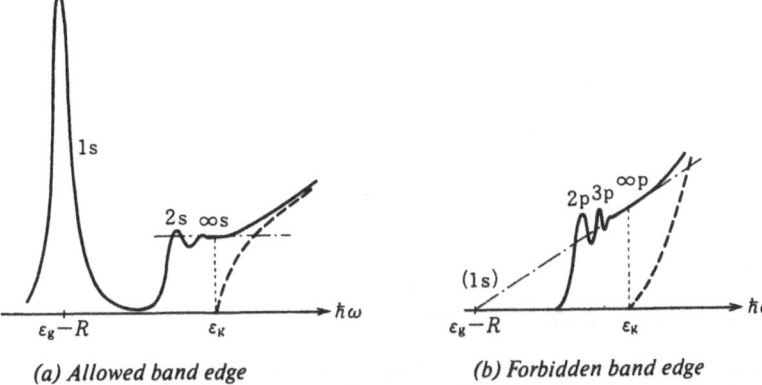

(a) Allowed band edge   (b) Forbidden band edge

Fig. 2.8a,b. The effect of electron–hole Coulomb interaction on the fundamental absorption edge. The dotted lines represent the interband absorption edge without the Coulomb effect

as shown schematically in Fig. 2.8b. The quasicontinuum for large n,

$$| \mathbf{V}_x \bar{F}_{np_x}(0) |^2 \left| \frac{dE_n}{dn} \right|^{-1} = \frac{v_0}{2\pi Ra^5} \frac{\hbar\omega - [\varepsilon^{(m)}(0) - R]}{R} , \qquad (2.4.15)$$

cuts the abscissa at the forbidden $n = 1$ line if extrapolated to lower energy, while on the high-energy side it connects smoothly to the true continuum as can be seen easily.

Considering that $\mathbf{V}_k p_{\nu\mu} : p_{\nu\mu}$ is of the order of lattice constant d and that $\mathbf{V}_R \bar{F}_n(0) : \bar{F}_n(0)$ is of the order of the reciprocal of the excitonic radius a, one finds that the absorption spectrum of type (ii) around $\varepsilon^{(m)}(0)$ is smaller than that of type (i) by a factor $\sim (d/a)^2$ if a is comparable in the two cases.

The fundamental absorption spectra near the edge region of KCl and Cu$_2$O crystals are shown in Figs. 2.9 and 2.10, representing type (i) and type (ii) spectra, respectively.

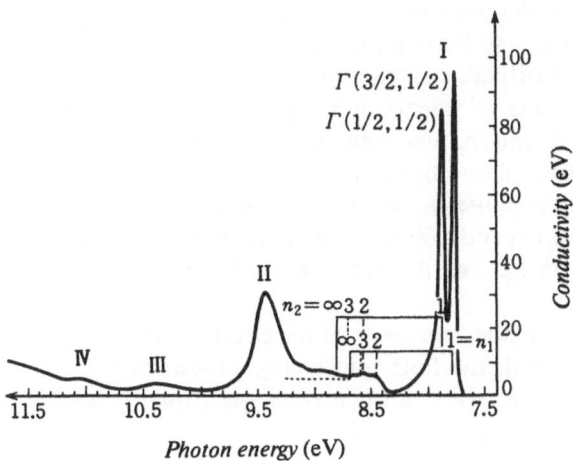

Fig. 2.9. The fundamental absorption spectra of a KCl crystal at 10 K. [Tomiki, T.: J. Phys. Soc. Jpn. **26**, 738 (1969)]

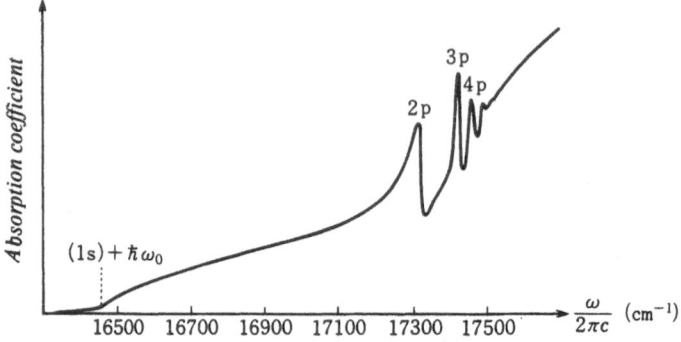

Fig. 2.10. The yellow series excitons at the fundamental absorption edge of a Cu_2O crystal at 4.2 K. [Nikitine, S., Grun, J. B., Sieskind, M: J. Phys. Chem. Solids **17**, 292 (1961)]

2.4.2 Spin–Orbit Versus Exchange Interactions

So far we have considered a single valence band (μ) and a single conduction band (ν). In real crystals with orbital as well as spin degeneracies, however, we have to consider sets of μ bands and ν bands. The wave function (2.3.24) for the one-electron excitation has to be generalized to $|e\rangle = \sum_{\nu, \mu, k} f_{\nu\mu}(k) a^{\dagger}_{\nu k} a_{\mu k-K} |g\rangle$ and the effective Hamiltonian (2.3.51) to

$$H_{\nu\mu l, \nu'\mu'l'} = [\delta_{\mu\mu'} \varepsilon_{\nu\nu'}(-i\nabla_l) - \delta_{\nu\nu'} \varepsilon_{\mu\mu'}(-i\nabla_l - K) \\ - \delta_{\nu\nu'} \delta_{\mu\mu'} v(R_l)] \delta_{ll'} + \delta_{l0} \delta_{l'0} w_{K, \nu\mu, \nu'\mu'} . \qquad (2.4.16)$$

Here $w_{K, \nu\mu, \nu'\mu'}$ is given by (2.3.48) after replacing the suffixes (ν, μ, μ, ν) by (ν, μ, μ', ν') in (2.3.46), and $\varepsilon_{\nu\nu'}(k)$ represents the (ν, ν') element of the energy matrix for the conduction band, the eigenvalues of which as functions of k give the branches of the conduction band.

Let us first consider the spin degeneracy only. The ground state of the crystal is assumed to be a singlet state due to the complete pairing of up and down spins within each atomic (or Bloch) orbital, as is the case in nonmagnetic insulators. Denote the spin angular momenta of valence and conduction electrons respectively by σ and τ in units of $\hbar$, and the eigenvalues of their z components by σ and τ. μ, ν in (2.4.16) are to be replaced by $\mu\sigma$, $\nu\tau$, with a single set of (μ, ν) but with each of σ and τ taking two values $\pm 1/2$ which are denoted by $\uparrow$ and $\downarrow$ for simplicity. The band energies $\varepsilon_{\mu}(k)$ and $\varepsilon_{\nu}(k)$ are spin independent in the absence of spin–orbit interaction, while $\delta_{\mu\mu'}$ and $\delta_{\nu\nu'}$ are to be written as $\delta_{\sigma\sigma'}$ and $\delta_{\tau\tau'}$, respectively.

The vector sum of the spin angular momenta of the electron and the positive hole is given by $S = \tau - \sigma$. With the four spin configurations of the excited states: $|\tau\sigma\rangle = \sum_k f(k) a^{\dagger}_{\nu k\tau} a_{\mu k-K\sigma} |g\rangle$, one can construct the triplet state $S = 1$ and the singlet state $S = 0$ as follows:

$$S = 1 \begin{cases} S_z = +1: & -|\uparrow\downarrow\rangle \\ S_z = 0: & (|\uparrow\uparrow\rangle - |\downarrow\downarrow\rangle)/\sqrt{2} \\ S_z = -1: & |\downarrow\uparrow\rangle \end{cases} \tag{2.4.17}$$

$$S = 0: \qquad (|\uparrow\uparrow\rangle + |\downarrow\downarrow\rangle)/\sqrt{2} .$$

Equation (2.4.17) could have been written in a more familiar form pertinent to the two-body problem if one had used the annihilation–creation operators of the positive hole, b_σ and $b_\sigma^\dagger$, defined by $a_\sigma \equiv 2\sigma b_{-\sigma}^\dagger$. The triplet (with $S_z = 0$) and singlet states of (2.4.17) can then be written as

$$\frac{1}{\sqrt{2}} (a_{\nu\uparrow}^\dagger a_{\mu\uparrow} \mp a_{\nu\downarrow}^\dagger a_{\mu\downarrow})|g\rangle = \frac{1}{\sqrt{2}} (a_{\nu\uparrow}^\dagger b_{\mu\downarrow}^\dagger \pm a_{\nu\downarrow}^\dagger b_{\mu\uparrow}^\dagger)|g\rangle .$$

Like a_σ and $a_\sigma^\dagger$, the above defined b_σ and $b_\sigma^\dagger$ satisfy not only the anti-commutation rules but also the identities: $\sigma_z = \sum_\sigma \sigma b_\sigma^\dagger b_\sigma$ and $\sigma_x \pm i\sigma_y = b_\sigma^\dagger b_{-\sigma} (\sigma = \pm 1/2$ on the right side).

Noting that the integrations in (2.3.46) should now include the summation over the spin variables, one can write the exchange integral in (2.4.16) as $w_{K,\nu\tau\mu\sigma,\nu\tau'\mu\sigma'} = \delta_{\tau\sigma}\delta_{\tau'\sigma'} w_K$, which vanishes for the triplet and equals $2w_k$ for the singlet. Hence the last term of (2.4.16) can be written as

$$2\delta_{s0}\delta_{l0}\delta_{l'0}w_K . \tag{2.4.18}$$

The dominant term in (2.3.48) is usually the intra-atomic (intramolecular) term $w(0)$, which is positive according to (2.3.46). Therefore, the singlet exciton has higher energy than the triplet exciton. In molecular crystals to which the Frenkel exciton model is applicable, they have good correspondence to the singlet and triplet excited states of an isolated molecule. While the singlet exciton moves rather rapidly mainly because of the K-dependent part of w_K originating from the intermolecular dipole interactions, the triplet exciton usually moves much more slowly through the k-dependent part of the band energies ε in (2.4.16) originating from the intermolecular overlap energies which are small in molecular crystals.

The importance of the spin–orbit interaction rises rapidly as the atomic numbers of constituent elements increase. Let us study the exciton with orbitally degenerate atomic (or Bloch) states, taking into account the spin–orbit interaction as well as the exchange interaction. As a typical example, we consider the alkali halide crystals, of which the bottom of the conduction band is at $k = 0$ with the orbitally nondegenerate Bloch function consisting mainly of the lowest unoccupied s state of the cation and the top of the valence band is at $k = 0$ with the triply degenerate (but for the spin–orbit interaction) Bloch function consisting of the outermost occupied p states of the anion.

The spin–orbit interaction $(\hbar/2m^2c^2) [\boldsymbol{\sigma} \times \nabla V(r)]\cdot\boldsymbol{p}$, can be approximated by $(\hbar^2/2m^2c^2)(dV/rdr)\boldsymbol{\sigma}\cdot\boldsymbol{l}$ inside each anion since the periodic crystalline potential $V(r)$ becomes almost spherically symmetric in the neighborhood of the nucleus where the gradient of $V(r)$ takes the greatest value. Here $\hbar l \equiv r \times p$ represents the orbital angular momentum of the electron around the anion nucleus. As in the case of an isolated atom, the triply degenerate p-like valence bands at $k = 0$ are split, due to the spin–orbit interaction, into $j_v = 3/2$ and $j_v = 1/2$ states where $j_v = l + \sigma$ is the total angular momentum of the electron. Since $2\sigma \cdot l = j_v^2 - l^2 - \sigma^2 = j_v(j_v + 1) - 2 - 3/4$, the spin–orbit energies in the two states are given by $+\lambda/3$ and $-2\lambda/3$, where the splitting energy $\lambda(> 0)$ is expected to be close to the atomic value. The bottom of the conduction band is not split since $l = 0$ gives only $j_c = 1/2$.

In the j–j coupling scheme, the excited states $|j_c, j_v\rangle$ consist of $|1/2, 3/2\rangle$ and $|1/2, 1/2\rangle$, for which the square bracket in (2.4.16), with addition of spin–orbit energy, can be written as

$$
\left[-\frac{\hbar^2}{2m_c}\Delta_1 - v(R_l) + \varepsilon_g\right] + \begin{cases} -\dfrac{1}{3}\lambda\left(j_v = \dfrac{3}{2}\right) \\[2mm] +\dfrac{2}{3}\lambda\left(j_v = \dfrac{1}{2}\right). \end{cases} \tag{2.4.19}
$$

Here ε_g denotes the band gap in the absence of λ and the dispersion of the valence band has been neglected. Since the sum of the total angular momenta of electron and hole is given by $J = j_c - j_v$, one obtains $J = 2$ and 1 from $|1/2, 3/2\rangle$, and $J = 1$ and 0 from $|1/2, 1/2\rangle$. Among them, the $J = 2$ and $J = 0$ states originate purely from spin triplet $S = 1$ since in the L–S coupling scheme $(J = L + S)$ one has $L = 1$ because $l_c = 0$ and $l_v = 1$. For these two states, one can use (2.4.19) as they stand, since the exchange interaction (2.4.18) vanishes for the spin triplet.

On the other hand, the two sets of $J = 1$ states should be a linear combination of spin triplet and singlet. According to the well-known formula relating j–j and L–S coupling schemes, the $J = 1$ states obtained from $|1/2, 3/2\rangle$ and $|1/2, 1/2\rangle$ are given respectively by $\sqrt{1/3}\,|\text{triplet}\rangle + \sqrt{2/3}\,|\text{singlet}\rangle$ and $-\sqrt{2/3}\,|\text{triplet}\rangle + \sqrt{1/3}\,|\text{singlet}\rangle$. Calculating the exchange energy (2.4.18) with the use of 1s-state wave function $F(R_l)$ of the purely orbital part (brace) of (2.4.19) and adding the spin–orbit part, one gets of the first-order energy matrix

$$
\begin{array}{c} \left\langle\frac{1}{2},\frac{3}{2}\right| \\[3mm] \left\langle\frac{1}{2},\frac{1}{2}\right| \end{array} \left(\begin{array}{cc} -\dfrac{1}{3}\lambda + \dfrac{2}{3}\Delta & \dfrac{\sqrt{2}}{3}\Delta \\[3mm] \dfrac{\sqrt{2}}{3}\Delta & +\dfrac{2}{3}\lambda + \dfrac{1}{3}\Delta \end{array} \right), \tag{2.4.20}
$$

where the exchange energy Δ is given by

$$
\Delta = 2|F(0)|^2 w_0 . \tag{2.4.21}
$$

To obtain w_0, one puts, in (2.3.46), the s-like Wannier function ϕ_s for ϕ_ν, and the p_x-like Wannier function ϕ_x for ϕ_μ, and takes the limit of $K \to 0(K \perp x)$ in (2.3.48) obtained therefrom, considering the optically active transverse exciton. The explicit form is given by

$$w_0 = \iint \phi_s(r)\phi_x(r)v(r - r')\,\phi_x(r')\,\phi_s(r')\,drdr' - \frac{4\pi}{3}N_0\mu^2 \ . \tag{2.4.22}$$

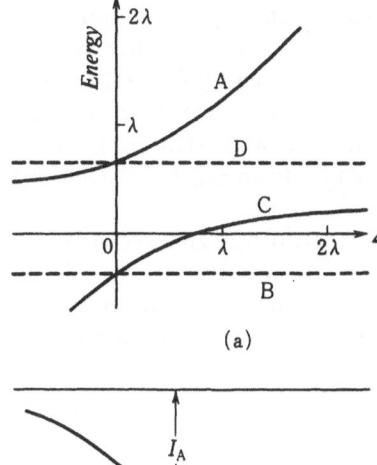

(a)

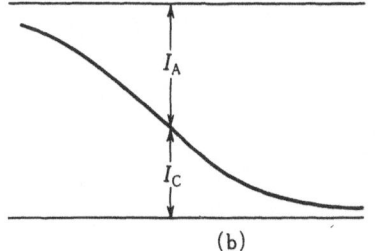

(b)

Fig. 2.11. (a) Schematic energy levels, with (b) their intensity ratio of absorption, of the 1s exciton at the Γ point in an alkali halide. λ and Δ represent the spin orbit splitting of the valence band at the Γ point and the electron–hole exchange energy, respectively. [Onodera, Y., Toyozawa, Y.: J. Phys. Soc. Jpn. **22**, 833 (1967)]

The eigenvalues of the energy matrix (2.4.20) for $J = 1$, as functions of Δ, are shown by solid lines C and A in Fig. 2.11a, together with the Δ-independent energies of the pure triplets $J = 2$ and 0 shown by broken lines B and D. In Fig. 2.11b is shown the partition of the oscillator strengths between C and A which is obtained from the weights of the optically allowed spin singlet state contained in the eigenstates of (2.4.20). With $\Delta = 0$ (j–j coupling), the intensity ratio is given by C($j_v = 3/2$): A($j_v = 1/2$) = 2:1 as a reflection of multiplicity $2j_v + 1$, while it tends to 0:1 in the L–S limit of $\Delta \to \infty$ (or $\lambda \to 0$) where C and A tend to pure triplet and singlet, respectively.

The deviation of the intensity ratio of the spin–orbit split components $\Gamma(1/2, 3/2)$ and $\Gamma(1/2, 1/2)$ from 2:1 as seen in Fig. 2.11 is due to the exchange interaction: conversely, one can estimate Δ and λ from this deviation and the splitting energy.

Continuous variation in the intensity ratio of the spin–orbit split lines as function of λ/Δ has been observed in a mixed-crystal system: $CuCl_{1-x}$ Br_x $(0 < x < 1)$. Both CuCl and CuBr crystals of the zinc-blende type have band structures similar to those of alkali halides in that the bottom of the conduction band and the top of the valence band are at $k = 0$, respectively with single and triple (except for the spin–orbit coupling) orbital degeneracies. However, the valence band consists not only of halide p orbitals (as in alkali halides) but also of Cu d orbitals. Depending sensitively upon this mixing ratio, the energy difference λ of the spin–orbit split bands $Z_{1,2}(j = 3/2)$ and $Z_3(j = 1/2)$ is negative in CuCl while positive in CuBr. Since the exchange energy Δ given by (2.4.21) is small compared to $|\lambda|$ in these crystals due to larger excitonic radii than in alkali halides, the situation is closer to the j–j coupling scheme so that the intensity ratio of the first and the second exciton peaks is not far from 1:2 in CuCl and 2:1 in CuBr.

It is known that the mixed crystals $CuCl_{1-x}$ Br_x are also of zinc-blende structure in the whole concentration range: $0 < x < 1$. The observed peak positions and intensity ratio of the first and the second exciton peaks are shown in Fig. 2.12a, b, respectively, as functions of the concentration x. Assuming that the

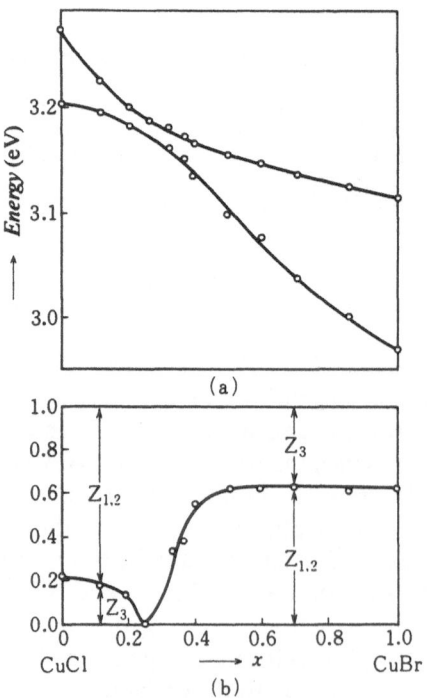

(a)

(b)

Fig. 2.12. The observed energies and intensity ratio of the $Z_{1,2}$ and Z_3 excitons in mixed crystals: $CuCl_{1-x}Br_x$ as functions of the concentration x. [Kato, Y., Yu, C. I., Goto, T.: J. Phys. Soc. Jpn. **28**, 104 (1970).

mixed crystals have band structures determined by the averaged periodic potential of the two components, one would expect that λ varies continuously from negative to positive values as x increases from 0 to 1, but that Δ, being always positive, does not vary significantly. Δ/λ would then vary in such a way that in Fig. 2.11 one proceeds to the left starting from a slightly negative value of the abscissa, turns from $-\infty(\lambda = -0)$ to $+\infty(\lambda = +0)$, proceeds again to the left down to a slightly positive value. For $\lambda < 0$, one must invert the ordinate of Fig. 2.11, and hence, transpose the first and the second peaks. In this way, one sees how the characteristic features of the observation shown in Fig. 2.12 can be explained in terms of the competition of spin–orbit and exchange interactions. In particular, at $x \doteq 0.23$ where λ vanishes, the first peak, with vanishing intensity because of its purely triplet nature (L–S coupling), is repelled by the second peak through the exchange energy alone.

Besides the spin and orbital degeneracies, there is another type of degeneracy originating from the fact that a unit cell contains two or more ($\sigma \geqq 2$) atoms or molecules of the same kind. One can then form as many as σ Frenkel excitons [for a given set of (λ, K)], depending on which atom or molecule in a unit cell is excited. The degeneracy is lifted by the dipole–dipole interactions, D_K[see (2.3.8)] which is now a σ-dimensional matrix; the energy splitting is larger for larger oscillator strength, both being proportional to μ^2. This type of splitting, the so-called *Davydov splitting*, has been observed in a number of molecular crystals of aromatic compounds such as anthracene.

2.4.3 The Observation of Translational Motion

As mentioned before, the translational wave vector K of an electron–hole pair is usually much smaller than the reciprocal lattice. This makes it difficult to observe the effect of translational motion in the fundamental absorption spectra which are mainly featured by relative motion. We will present here two examples of such observations, one relying on the high-resolution spectroscopy under external fields and another making use of phonon-assisted transition.

An exciton with translational mass M and momentum $\hbar K$ (transferred from photon) has velocity $v = \hbar K/M$. Application of magnetic field H gives rise to changes in the motions of the photoexcited electron and hole, namely, in the energies and wave functions of the excited states, and hence in the absorption spectra. This magneto-optical effect consists of three parts: the first is the spin and orbital (relative motion) Zeeman energies of the electron and the hole, the second is the orbital (relative motion) diamagnetic energy, while only the third is concerned with translational motion as described below. Since the electron and the hole in the exciton move with common average velocity $v = \hbar K/M$, they are subject to oppositely directed Lorentz forces: $\mp ev \times H/c$. In effect, a Lorentz electric field $E_L = v \times H/c$ is acting on the electron–hole relative motion. The effect of electrostatic field E (Stark effect) on the relative motion should start with the E^2 term because of the symmetry. The effect of simultaneous application of magnetic field H and electrostatic field E—magneto-Stark effect—should be such that each exciton line in the absorption spectra shifts as $(E + E_L)^2$. By varying E one can find $E = -E_L$, for which the energy shift takes the extremum value, and hence, the velocity v and the translational mass M as well.

The first successful observation as well as the underlying idea mentioned above were presented by Thomas and Hopfield, who chose the $n = 2$ lines of the A exciton series of wurtzite-type CdS crystal for the compatibility of sharpness and intensity. The translational mass M of exciton determined in this way is 0.92 times the true electronic mass, in approximate agreement with the sum of electron and hole effective masses determined with other methods.

If the phonon momentum, instead of very small photon momentum, can be utilized, one can extend the observable momentum of exciton to much wider region. As will be mentioned in Chap. 7, the exciton interacts with lattice vibrations so that it can be scattered from K' to $K' + K$, by emitting or absorbing a phonon with wave vector $\mp K$, though the K-dependent matrix element which extends up to the K values of the order of the reciprocal excitonic radius. This interaction gives rise to the second-order process in which a photon is absorbed to create an exciton ($K' \sim 0$) and then the exciton is scattered by phonons without energy conservation in the intermediate states. The optical spectra for this "indirect transition" extend over the energy region of $\varepsilon_K \pm \hbar\omega_{\mp K}$ for which the matrix element mentioned above has appreciable value. Since ε_K has much larger dispersion (K dependence) than $\hbar\omega_K$ except in molecular crystals, the spectral extension of indirect transition represents the recoil kinetic energy, $\varepsilon_K - \varepsilon_0$, in the final state.

Leaving further details of indirect photoabsorption spectra to Sect. 7.3.3, we consider now the reverse process, in which an exciton is annihilated to emit a photon with simultaneous emission or absorption of a phonon. In this second-order process, an exciton with initial wave vector K is converted to a photon with energy $\varepsilon_K \mp \hbar\omega_{\mp K}$ because of the energy momentum conservation rule. In Fig. 2.13 are shown the emission spectra of photoexcited CdS crystal, which consist of a sharp line (denoted by 0) for the direct transition and broad bands

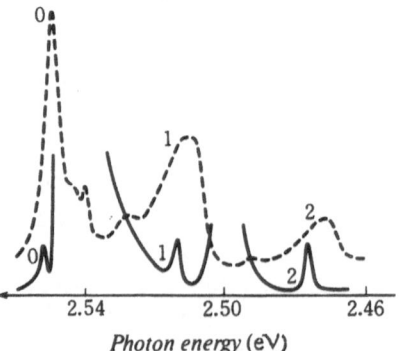

Photon energy (eV)

Fig. 2.13. The emission spectra due to exciton annihilation in a CdS crystal. The solid and broken lines represent the spectra at 4.2 K and 42 K, respectively. [Gross, E., Permogorov, S., Razbirin, B.: J. Phys. Chem. Solids **27**, 1647 (1966)]

(denoted by 1 and 2) for the indirect transitions in which one or two optical phonons are emitted simultaneously.

While only the excitons with $K \approx 0$ can be converted to a photon in the direct transition giving rise to a line spectrum at ε_0, the excitons with any K can be annihilated in the indirect transitions. If the dispersion of the optical phonon is negligible, the energy of the emitted photon accompanied by emission of one or two optical phonons is given by $\varepsilon_K - \hbar\omega_0$ and $\varepsilon_K - 2\hbar\omega_0$. The temperature-dependent asymmetric broadening of the bands 1 and 2 in Fig. 2.13 is the reflection of the Maxwell distribution of the excitons: $\exp[-(\varepsilon_K - \varepsilon_0)/k_B T]$, as is confirmed by the fact that the broadening on the high-energy side is of the order of $k_B T$. It is evidence that the excitons in this crystal are nearly free particles like atoms in gases or electrons and holes in semiconductors, even after they have been thermalized through the interaction with lattice vibrations.

2.4.4 Excitonic Molecule

It has been noted in a number of insulating crystals that two excitons can combine to form an *excitonic molecule* (*biexciton*) or vice versa (X + X $\rightleftarrows$ X$_2$ where X denotes a single exciton). According to the compound-particle model, the exciton molecule X$_2$ consists of two electrons (in the conduction band) and two positive holes (in the valence band), with respective effective masses m_e and m_h, bound together as a whole through the Coulomb interactions $\pm e^2/\epsilon r$. In either limit $m_e \lessgtr m_h$, X and X$_2$ have complete similarity to hydrogen atom and molecule, while the stability (positive binding energy) of X$_2$ for all mass ratio m_e/m_h has been established by a variational calculation of molecular binding energy B, the result being shown in the unit of excitonic binding energy R in Fig. 2.14.

According to the Wannier–Mott model, the energies of an exciton and an excitonic molecule with translational wave vector K are given respectively by

$$\left. \begin{array}{l} E_X(K) = (\varepsilon_g - R) + \dfrac{\hbar^2 K^2}{2M} \ , \\[3mm] E_{X_2}(K) = 2(\varepsilon_g - R) - B + \dfrac{\hbar^2 K^2}{4M} \ , \end{array} \right\} \tag{2.4.23}$$

because the translational mass of an excitonic molecule is twice that of an exciton.

The emission spectra corresponding to the optical conversion of an excitonic molecule to an exciton (X$_2$ → X + $h\nu$) have been observed in CuCl, as shown by the M band in Fig. 2.15. In contrast to the bands 1 and 2 in Fig. 2.13, the M band has asymmetry with tailing on the low-energy side, which has been explained as follows. The photon energy emitted in the optical conversion X$_2$ → X is given, with the use of (2.4.23), by

$$h\nu = E_{X_2}(K) - E_X(K) = (\varepsilon_g - R - B) - \frac{\hbar^2 K^2}{4M}$$

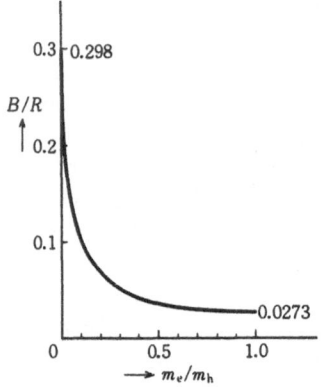

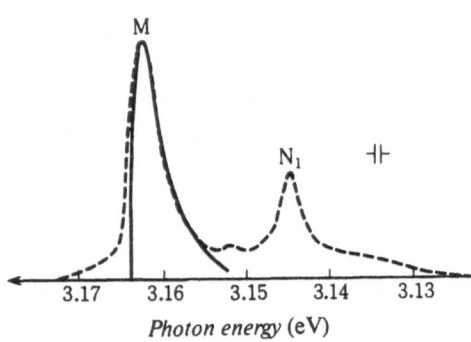

Fig. 2.14. The binding energy B of an excitonic molecule in units of the binding energy R of exciton, as a function of the electron–hole effective mass ratio m_e/m_h, calculated by a variation method. [Akimoto, O., Hanamiura E.: J. Phys. Soc. Jpn. **33**, 1537 (1972)]

Fig. 2.15. The emission spectra of CuCl crystal under intense laser light. The solid line represents the inverted Maxwellian distribution given by (2.4.24). [Souma, H., Goto, T., Ohta, T., Ueta, M.: J. Phys. Soc. Jpn. **29**, 697 (1970)]

since the photon momentum is negligibly small. Assuming the Maxwell distribution for the initial kinetic energy $\hbar^2 K^2/4M$ of the excitonic molecule, one expects the emission spectra with the inverted Maxwellian shape

$$E^{1/2}\exp\left(-\frac{E}{k_\mathrm{B}T}\right), \quad E \equiv (\varepsilon_g - R - B) - h\nu > 0 \ . \tag{2.4.24}$$

This is in accordance with the observed line shape of the M band except that one has to assume the effective temperature of the excitonic molecule to be several times higher than the lattice temperature.

The experiments mentioned in Sects. 2.4.3, 4 indicate that excitons and excitonic molecules are the particles moving nearly freely throughout the crystal. It should be noted, however, that they behave quite differently in other crystals, as will be described in Sect. 7.5.

2.4.5 Fission and Fusion of Excitons

In molecular crystals of aromatic compounds such as anthracene, the *singlet exciton* (abbreviated as S hereafter) has energy nearly twice as large as the *triplet exciton* (abbreviated as T) because of large intramolecular exchange energy. Since the spin–orbit interaction is usually very small in these compounds composed of lighter elements, the optical absorption for direct creation of T is extremely weak. However, this means that once created, T has a long lifetime. In contrast, S has strong optical absorption for its creation, and hence, decays rapidly by emitting blue light which is called *prompt luminescence*.

In the meanwhile, it was found in anthracene and tetracene crystals that luminescence spectrally identical to the above appears even when one creates only T by intense red (or near-infrared) light. It it called *delayed luminescence* because of its long duration. The dependence of the emission intensity upon the excitation intensity on the one hand, and the temporal behavior of emission intensity on the other hand, indicated that two T's can be united to form one S which in turn emits delayed luminescence. The existence of such a process and its reverse —*fusion* and *fission* of excitons ($T + T \rightleftarrows S +$ thermal energy)—was evidenced most clearly by the magnetic-field dependence of the intensities of delayed and prompt luminescences.

Figure 2.16 indicates how the magnitude (a) and the direction (b) of applied magnetic field H influences the intensities of prompt luminescence (broken line) obtained under blue-light illumination and of delayed luminescence (solid line) under near-infrared illumination on a tetracene crystal. The intensity is normalized so as to be unity for $H = 0$. Because of the principle of detailed balance, the rate constants of fission and fusion processes are related through a temperature dependent proportionality constant. The applied external field will not influence this constant in so far as the induced changes in energy of the relevant states of S and T are much smaller than the thermal energy. This means that the external field affects the rates of mutually reverse processes always in the same direction (through the common factor originating from transition matrix elements). The increase in the fission rate suppresses the prompt luminescence while the increase in the fusion rate enhances the delayed luminescence since both of them originate from S. This is the reason why the field-induced changes in prompt and delayed luminescences are in opposite directions as seen in Fig. 2.16. The magnetic-field dependence of the fusion rate is explained as follows.

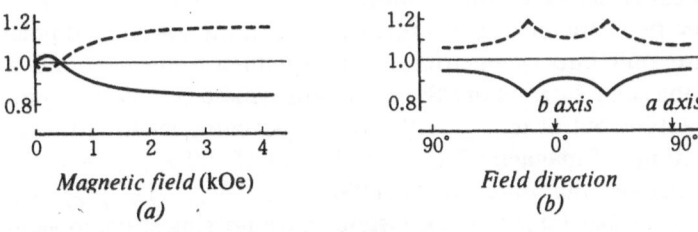

Fig. 2.16. The magnetic field dependences of the intensities of delayed (solid line) and prompt (broken lines) luminescences in a tetracene crystal at 238 K. [Groff, R. P., Avakian, P., Merrifield, R. E.: Phys. Rev. **B1**, 815 (1970)]

The binary collision rate of T per unit volume is given by $k_1 n^2$ where n denotes the number density of T. There are $3 \times 3 = 9$ possible spin states (ϕ_l, $l = 1,2, \ldots , 9$) of the collision complex (two T's located on two neighboring molecules), each of which will be formed with equal rate: $k_1 n^2/9$ at sufficiently high temperature. The collision complex will either be dissociated again into a pair of T's or be united into an S, while the former probability k_{-1} is spin independent, the latter probability is given by $k_2 S_l^2$ where $S_l \equiv |\langle S | \phi_l \rangle|$ in the amplitude of the singlet component contained in the spin state $|\phi_l\rangle$. The fusion rate is then given by

$$\gamma = \frac{k_1 n^2}{9} \sum_{l=1}^{9} \frac{k_2 S_l^2}{k_{-1} + k_2 S_l^2} . \tag{2.4.25}$$

In order to see how γ depends upon the structure of spin states ϕ_l, let us consider the two extreme cases: 1) all ϕ_l contain the singlet component with equal amplitude; 2) a particular ϕ_l is a pure singlet. Because of the completeness: $\sum_l S_l^2 = 1$, the ratio of γ between the two cases is given by $\gamma_1/\gamma_2 = (k_{-1} + k_2)/(k_{-1} + k_2/9)$: the more of the states l share the singlet component, the greater γ will be.

The spin Hamiltonian for T consists of Zeeman energy and magnetic dipolar interaction (of electron and hole within a molecule). The eigenstates of the latter alone are given by $|x\rangle$, $|y\rangle$, and $|z\rangle$ referred to the magnetic principal axes (the eigenstates of S_z with eigenvalues ± 1 are given by $|\pm\rangle = (|x\rangle \pm i |y\rangle)/\sqrt{2}$). The singlet state of the collision complex of two T's is then given by $(|xx\rangle + |yy\rangle + |zz\rangle)/\sqrt{3}$. In the absence of a magnetic field, one finds in this way that three states out of nine ϕ_l's share the singlet component with the same amplitude 1/3. As magnetic field is applied, ϕ_l will be reconstructed, generally in such a direction that the singlet component is shared by more of the states ϕ_l. This explains why the delayed luminescence is enhanced with application of a weak magnetic field as seen in Fig. 2.16a.

Under a strong magnetic field such that the Zeeman energy predominates over the dipolar interaction, the spin of T is quantized along the magnetic field, the eigenstates being given by $|0\rangle$ and $|\pm\rangle$. As for the collision complex, the singlet component is shared by three states: $|00\rangle$, $|+-\rangle$, and $|-+\rangle$, of which the latter two remain degenerate even with intramolecular dipolar interaction, but are split into symmetric and antisymmetric linear combinations under intermolecular interaction. Since the antisymmetric one is a pure triplet, only the two states out of nine ϕ_l's share the singlet component. However, for particular directions of magnetic field such that all of $|00\rangle$, $|+-\rangle$, and $|-+\rangle$ are degenerate, the pure singlet $(|00\rangle - |+-\rangle - |-+\rangle)/\sqrt{3}$ becomes eigenstates, so that remaining eigenstates have no singlet component. This corresponds to the extreme case 2) mentioned above in which γ takes the smallest value. In any case, one sees that γ under strong magnetic field is smaller than in the zero field and tends with $H \to \infty$ to a definite value (which depends upon the direction of H), in accordance with Fig. 2.16a. Sharp minima in delayed luminescence concurrent with sharp maxima in prompt luminescence in Fig. 2.16b correspond to the particular directions mentioned above.

The dipolar interaction and the Zeeman energy are given respectively by μ_B^2/r^3 and $\mu_B H$ with the Bohr magneton μ_B and the molecular radius $r(\sim 3 \times 10^{-8}$ cm$)$. They become comparable at $H \sim \mu_B/r^3 \sim 0.4$ kOe. This is in accordance with the border between weak and strong magnetic fields in Fig. 2.16a, where $\gamma(H) - \gamma(0)$ changes its sign. It is also to be noted that these energies of the spin system are much smaller than the thermal energy. In this way, all aspects of the magnetic field dependence shown in Fig. 2.16 can be explained in terms of the reconstruction of spin states of collision complex under the competing Zeeman and dipolar energies.

One should note an important difference between the fussion–fission processes $(T + T \rightleftharpoons S)$ on the one hand and the formation–dissociation processes of excitonic molecule $(X + X \rightleftharpoons X_2)$ on the other hand. The latter are, so to speak, chemical reactions conserving the numbers of constituents (two electrons and two holes), while the former are conversions into different particle(s) violating the conservation rule.

We have seen that the elementary excitations of the quasiparticles resemble in many respects the more basic particles such as atoms, nuclei, and elementary particles. While the former are the secondary concepts constructed upon the known background of static solids, and hence with known limitation of applicability (valid only for low energy excitations, low concentration, weak interactions, etc.), it is not clear how far this analogy can be applied to elementary particles whose background is yet unknown.

3. Fermi Liquids

In studying properties of matter, it is sometimes necessary to deal with a condensed system composed of Fermi particles, such as liquid ^{3}He or electrons in metals. An assembly of Fermi particles is often called a "Fermi liquid". The origin of this terminology stems from Landau's Fermi-liquid theory. In this chapter, we will discuss the problem of Fermi liquids, stressing in particular the importance of elementary excitations.

3.1 Models of Fermi Liquids

Before discussing the general features of a Fermi liquid, we will treat its simple model by means of an elementary method to see its characteristic aspects as a many-body system. In what follows, the method of second quantization is assumed to be known; the reader who is not familiar with it should refer to textbooks enumerated at the end of this book.

3.1.1 Hamiltonian of the System of Fermi Particles

Suppose N Fermi particles with mass m and spin 1/2 enclosed in a box of volume V. For simplicity, we assume that the box is a cube with the side of length L ($V = L^3$) and that the periodic boundary condition is imposed on the system. Let the interaction potential between the ith and the jth particles be $v(r_i - r_j)$. The Hamiltonian of the total system is expressed in the second quantized form as

$$\mathcal{H} = \int \psi^\dagger(x) \left(-\frac{\hbar^2 \Delta}{2m} \right) \psi(x) \, dx + \frac{1}{2} \int \psi^\dagger(x) \psi^\dagger(x') v(r - r') \psi(x') \psi(x) \, dx dx' \ .$$

$$(3.1.1)$$

Here $\psi(x)$ and $\psi^\dagger(x)$ are field operators, x represents both the spatial coordinate r and spin coordinate s, and the integration over x implies both the integration over r and the summation over s. Note that the integration over r is extended over the box of volume V.

In order to make (3.1.1) easier to see, we expand $\psi(x)$ and $\psi^\dagger(x)$ in terms of plane waves:

$$\psi(x) = \frac{1}{\sqrt{V}} \sum_{k\sigma} a_{k\sigma} \exp(i\mathbf{k} \cdot \mathbf{r}) \, \delta(s, \sigma) \ ,$$

$$(3.1.2a)$$

$$\psi^\dagger(x) = \frac{1}{\sqrt{V}} \sum_{k\sigma} a_{k\sigma}^\dagger \exp(-i k \cdot r) \, \delta(s, \sigma) \ . \tag{3.1.2b}$$

In the above equations, σ denotes the state of spin, e.g., the up spin is described by $\sigma = 1$ and the down spin by $\sigma = -1$. Furthermore, $a_{k\sigma}$ and $a_{k\sigma}^\dagger$ are annihilation and creation operators for the particle with wave number k and spin σ, respectively. The following anticommutation relations hold

$$[a_{k\sigma}, a_{k'\sigma'}]_+ = [a_{k\sigma}^\dagger, a_{k'\sigma'}^\dagger]_+ = 0, \quad [a_{k\sigma}, a_{k'\sigma'}^\dagger]_+ = \delta(k, k') \, \delta(\sigma, \sigma') \tag{3.1.3}$$

with $[A, B]_+ = AB + BA$. If we substitute (3.1.2a) and (3.1.2b) into (3.1.1), the first term in (3.1.1), i.e., the kinetic energy of the system is written as

$$\sum_{k\sigma} \varepsilon(k) \, a_{k\sigma}^\dagger \, a_{k\sigma}, \quad \varepsilon(k) = \frac{\hbar^2 k^2}{2m} \ . \tag{3.1.4}$$

In deriving this equation, we have used the following orthonormal relations:

$$\frac{1}{V} \int \exp[i(k - k') \cdot r] \, dr = \delta(k, k') \ .$$

In a similar way, the second term in (3.1.1) which represents the interaction part is expressed as

$$\frac{1}{2V^2} \sum_{k_1 k_2 k_3 k_4} \sum_{\sigma_1 \sigma_2 \sigma_3 \sigma_4} a_{k_1 \sigma_1}^\dagger a_{k_2 \sigma_2}^\dagger a_{k_4 \sigma_4} a_{k_3 \sigma_3} \sum_{ss'} \delta(s, \sigma_1) \, \delta(s', \sigma_2) \, \delta(s, \sigma_3) \, \delta(s', \sigma_4)$$
$$\cdot \int \exp(-i k_1 \cdot r - i k_2 \cdot r') \, v(r - r') \exp(i k_3 \cdot r + i k_4 \cdot r') \, dr dr' \ .$$

The sum over s, s' in the above equation is calculated to be

$$\sum_{ss'} \delta(s, \sigma_1) \delta(s', \sigma_2) \delta(s, \sigma_3) \delta(s', \sigma_4) = \delta(\sigma_1, \sigma_3) \delta(\sigma_2, \sigma_4) \ .$$

To carry out the integrations over r and r', we introduce center-of-mass coordinate r_1 and relative coordinate r_2 defined by (see Fig. 3.1)

$$r_1 = \frac{r + r'}{2}, \quad r_2 = r - r', \quad \therefore r = r_1 + \frac{r_2}{2}, \quad r' = r_1 - \frac{r_2}{2} \ .$$

Then the integration over r_1 leads to the term $V\delta(k_1 + k_2, k_3 + k_4)$. On the other hand, the integration over r_2 is expressed as

$$\int v(r_2) \exp\left[i \frac{r_2}{2} \cdot (k_2 - k_1 + k_3 - k_4)\right] dr_2 \ .$$

If we note the relation $k_2 - k_4 = k_3 - k_1$ and define the Fourier transform of $v(r)$ as

$$v(r) = \frac{1}{V} \sum_q v(q) \exp(iq \cdot r), \quad v(q) = \int v(r) \exp(-iq \cdot r) \, dr \tag{3.1.5}$$

we find that the integration over r_2 is equal to $v(k_1 - k_3)$. Furthermore, putting

$$k_1 - k_3 = q, \; k_1 = k_3 + q, \, k_2 = k_4 - q$$

and taking k_3, k_4 and q as independent variables, we have

$$\mathscr{H} = \sum_{k\sigma} \varepsilon(k) \, a_{k\sigma}^\dagger a_{k\sigma} + \frac{1}{2V} \sum_{qkk'\sigma\sigma'} v(q) \, a_{k+q,\sigma}^\dagger a_{k'-q,\sigma'}^\dagger a_{k'\sigma'} a_{k\sigma} . \tag{3.1.6}$$

The second term in the above equation represents a process in which particles with k, σ and k', σ' are annihilated and particles with $k + q$, σ and $k' - q$, σ' are created. Or, this process is described diagrammatically by Fig. 3.2 in which the particle with k' emits the wave vector q and the particle with k absorbs it and their spins σ and σ' do not change. Note that the conservation of wave number is satisfied in a sense that the wave vector of the particle with k changes to $k + q$ when it receives the wave vector q.

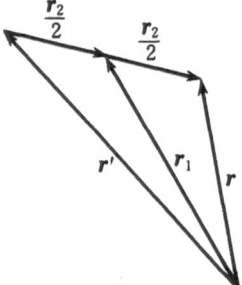

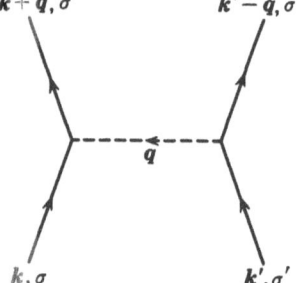

Fig. 3.1. Center of mass position r_1 and relative position r_2

Fig. 3.2. Mutual scattering of two particles described by the second term of (3.1.6)

3.1.2 The Electron-Gas Model

In dealing with electrons in metals, to simplify the problem, one sometimes introduces a model in which the metallic ions are smeared out to a uniform background and electrons exert on each other the Coulomb interaction. The electrical neutrality, however, is assumed for the total system. This is called the electron-gas model and is one of the most typical examples of a Fermi liquid. In this section, we will study some problems associated with the electron gas on the basis of simple perturbation procedure.

If we put the electronic charge to be $-e$, the potential $v(r)$ is given by

$$v(r) = \frac{e^2}{r} , \quad r \equiv |r| .$$

Taking into account charge neutrality, we have

$$\Delta v(r) = -4\pi e^2 \left[\delta(r) - \frac{1}{V} \right]$$

as a relation between the potential and the charge density (Poisson's equation). The δ function on the right-hand side represents the point charge at the origin and $1/V$ the uniform positive charge. Noting that

$$\delta(r) = \frac{1}{V} \sum_q \exp(iq \cdot r)$$

and using (3.1.5), we find

$$-\frac{1}{V} \sum_q v(q) q^2 \exp(iq \cdot r) = -4\pi e^2 \cdot \frac{1}{V} \sum_{q \neq 0} \exp(iq \cdot r) .$$

Therefore, comparing the coefficients of $\exp(iq \cdot r)$ on both sides, we are led to

$$v(q) = \frac{4\pi e^2}{q^2} \quad (q \neq 0) . \tag{3.1.7}$$

However, the $v(0)$ remains indefinite in the above argument. For convenience we put it to be zero; the reason for this choice is as follows.

If we assume that $v(0)$ is not zero, the summand with $q = 0$ in the second term in (3.1.6) becomes

$$\frac{1}{2V} \sum_{kk'\sigma\sigma'} v(0) a^\dagger_{k\sigma} a^\dagger_{k'\sigma'} a_{k'\sigma'} a_{k\sigma}$$

$$= \frac{1}{2V} \sum_{kk'\sigma\sigma'} v(0) [a^\dagger_{k\sigma} a_{k\sigma} a^\dagger_{k'\sigma'} a_{k'\sigma'} - \delta(k, k') \delta(\sigma, \sigma') a^\dagger_{k\sigma} a_{k'\sigma'}]$$

$$= \frac{1}{2V} v(0) (N^2 - N) .$$

In deriving this relation, we have used the fact that the total number N of electrons is given by

$$N = \sum_{k\sigma} a^\dagger_{k\sigma} a_{k\sigma} . \tag{3.1.8}$$

As is seen from the above, the term with $q = 0$ is just a constant. A constant added to the Hamiltonian simply shifts the origin of energy, so that it can be put to zero without violating essential physics. This is the reason why we put $v(0) = 0$.

3.1.3 The Exchange Energy of the Electron Gas

As was explained in Sect. 3.1.2, the Hamiltonian of the electron gas is expressed as

$$\mathcal{H} = \mathcal{H}_0 + \mathcal{H}' \tag{3.1.9a}$$

$$\mathcal{H}_0 = \sum_{k\sigma} \frac{\hbar^2 k^2}{2m} a_{k\sigma}^\dagger a_{k\sigma} \tag{3.1.9b}$$

$$\mathcal{H}' = \frac{1}{2V} \sum_{q \neq 0} \sum_{kk'\sigma\sigma'} \frac{4\pi e^2}{q^2} a_{k+q,\sigma}^\dagger a_{k'-q,\sigma'}^\dagger a_{k'\sigma'} a_{k\sigma} . \tag{3.1.9c}$$

In the following, we will make the first-order perturbation calculation, regarding $\mathcal{H}_0$ as the unperturbed system and $\mathcal{H}'$ as the perturbation. This kind of perturbation procedure is not always applicable to a real metal, but it is expected that some of essential features of Fermi liquid may be understood to some extent by such a treatment.

In the case without the Coulomb interaction, i.e., in the case of free electrons, the inside of a Fermi sphere with the radius k_F is occupied by electrons and the outside unoccupied at absolute zero (Fig. 3.3). Thus, if we denote the ground state of $\mathcal{H}_0$ by $|0\rangle$, the relations

$$n_{k\sigma}|0\rangle = \begin{cases} |0\rangle & (k < k_F) \\ 0 & (k > k_F) \end{cases}$$

hold, where $n_{k\sigma} \equiv a_{k\sigma}^\dagger a_{k\sigma}$ is the operator for the number of electrons occupying the state k and σ. The Fermi wave number k_F is determined from the condition that the the number of electrons inside the Fermi sphere should be equal to N. Noting that the sum over k is expressed as the integral (1.4.5) in k space in the limit $V \to \infty$ and taking account of the factor 2 coming from the spin degeneracy, we find

$$N = \frac{2V}{(2\pi)^3} \frac{4\pi k_F^3}{3}$$

whence it follows that

$$k_F = (3\pi^2 n)^{1/3}, \quad n \equiv \frac{N}{V} . \tag{3.1.10}$$

In the same way, the energy eigenvalue E_0 of $\mathcal{H}_0$ for $|0\rangle$ is calculated to be

$$E_0 = \sum_{k<k_F,\sigma} \frac{\hbar^2 k^2}{2m} = \frac{2V}{(2\pi)^3} \frac{\hbar^2}{2m} \int_{k<k_F} k^2 dk = \frac{\hbar^2}{2m} \frac{V k_F^5}{5\pi^2} .$$

Substituting the relation $k_F^3 = 3\pi^2 n$ into the above equation, we obtain

$$E_0 = \frac{3}{5} N \frac{\hbar^2 k_F^2}{2m} \ . \tag{3.1.11}$$

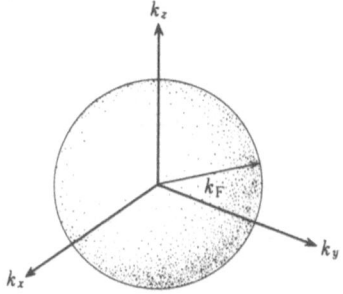

Fig. 3.3. In the case of the free electron, the inside of the Fermi sphere (radius k_F) is occupied by electrons and the outside unoccupied at absolute zero

As we have explained so far, the state $|0\rangle$ is an eigenstate of $\mathcal{H}_0$ and the relation $\mathcal{H}_0|0\rangle = E_0|0\rangle$ holds. However, if we consider the Coulomb interaction $\mathcal{H}'$, the correction term appears in the energy eigenvalue. According to perturbation theory in quantum mechanics, the correction term E_1 of the first order is given by

$$E_1 = \langle 0| \mathcal{H}' |0\rangle$$

$$= \frac{1}{2V} \sum_{q \neq 0} \sum_{kk'\sigma\sigma'} \frac{4\pi e^2}{q^2} \langle 0| a_{k+q,\sigma}^\dagger a_{k'-q,\sigma'}^\dagger a_{k'\sigma'} a_{k\sigma} |0\rangle \ .$$

As seen from the above equation, electrons with k and k' are annihilated, so that one of the following two conditions should be satisfied in order to have the initial state after the operation by $a^\dagger$:

(I) $k + q = k, \quad k' - q = k'$
(II) $k + q = k', \quad k' - q = k, \sigma = \sigma'$.

Because of the restriction $q \neq 0$, the condition (I) is not allowed. Therefore, if we omit the subscript σ, corresponding to the condition (II) we obtain

$$\langle 0| a_{k+q}^\dagger a_k^\dagger a_{k+q} a_k |0\rangle = - n_{k+q} n_k \ .$$

In this way, E_1 is expressed as

$$E_1 = -\frac{1}{2V} \sum_{q \neq 0} \frac{4\pi e^2}{q^2} \sum_{k\sigma} n_{k+q,\sigma} n_{k\sigma}$$

$$= -\sum_{q \neq 0} \frac{4\pi e^2}{q^2} \frac{1}{(2\pi)^3} \int_{|k+q|<k_F, \, k<k_F} dk \ .$$

If we denote the integral over $\mathbf{k}$ by $I(q)$, as is shown in Fig. 3.4, it is equal to the volume of overlapping region (shaded part in the figure) of two identical spheres of radius k_F when their centers are separated by the distance q. To calculate $I(q)$, suppose that we cut the sphere with radius k_F by a plane whose distance from the center is h, and denote the volume of the region (shaded part in Fig. 3.5) not including the center by $\Omega(h)$. Then, it is easily seen that

$$I(q) = 2\Omega\left(\frac{q}{2}\right) .$$

Since $\Omega(h)$ is a decreasing function of h, if we put the volume corresponding to the interval $h \sim h + dh$ to be $d\Omega$, we find

$$d\Omega = -\pi(k_F^2 - h^2)\, dh .$$

Solving this differential equation, we have

$$\Omega(h) = -\pi k_F^2 h + \frac{\pi h^3}{3} + A$$

and a constant A is determined from the condition at $h = 0$ to yield $(2\pi/3)k_F^3 = A$. As a result, $I(q)$ is calculated to be

$$I(q) = 2\pi \left(\frac{2k_F^3}{3} - \frac{k_F^2 q}{2} + \frac{q^3}{24}\right) .$$

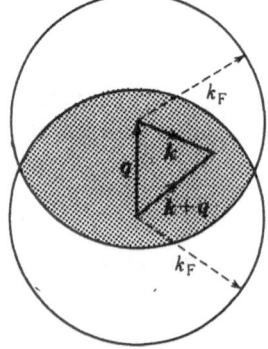

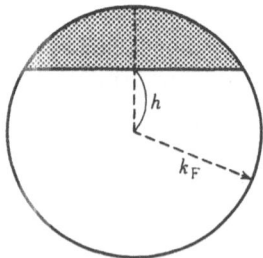

Fig. 3.4. The integral $I(q)$ is the volume shared by two spheres of radius k_F

Fig. 3.5. $\Omega(h)$ is the volume of the shaded region

It is obvious that $I(q) = 0$ for $q \geq 2k_F$.

In this way, the integral over $\mathbf{k}$ appearing in the expression for E_1 is calculated. Accordingly, expressing the sum over $\mathbf{q}$ by the corresponding integral, we see that E_1 is written as

$$E_1 = -\frac{Ve^2}{\pi}\frac{4\pi}{(2\pi)^3}\int_0^{2k_F}\left(\frac{2k_F^3}{3} - \frac{k_F^2 q}{2} + \frac{q^3}{24}\right)dq$$

$$= -\frac{Ve^2 k_F^4}{4\pi^3}\ .$$

Furthermore, by the use of $k_F^3 = 3\pi^2 n$, E_1 is expressed as

$$E_1 = -\frac{3e^2 N k_F}{4\pi}\ . \tag{3.1.12}$$

As is clear from the above-mentioned calculation, the term E_1 arises from a process in which the electron with k in the Fermi sphere absorbs q to go to the state with k' and conversely the electron with k' loses q to go to the state with k. That is, the exchange of wave number takes place between two electrons. For this reason, E_1 is called the "exchange" energy.

3.1.4 r_s Expansion

The result for the electron gas obtained above can be converted to a more transparent form if dimensionless quantities are introduced. In particular, what the expansion parameter is will be clarified. For this purpose, let us define r_0 by $(4\pi/3)r_0^3 = 1/n$; r_0 is the radius of sphere which one electron occupies. Since r_0 has the dimension of length, in order to introduce a dimensionless quantity, we divide it by the Bohr radius a and define r_s by

$$r_s = \frac{r_0}{a}\ , \quad a = \frac{\hbar^2}{me^2} \quad (= \text{Bohr radius}).$$

Also, let us measure the energy per electron in units of Rydberg which is given by

$$\frac{me^4}{2\hbar^2} = 13.6\text{eV}\ .$$

Then, the exchange energy is expressed as

$$\varepsilon_x = \frac{E_1/N}{me^4/2\hbar^2} = -\frac{3}{2\pi}\left(\frac{9\pi}{4}\right)^{1/3}\frac{1}{r_s} \approx -\frac{0.916}{r_s}\ . \tag{3.1.13}$$

In deriving this equation, it is convenient to use the unit system: $m = 1/2$, $\hbar = 1$, $e^2 = 2$. The Bohr radius or Rydberg is unity in this system, so that the length and the energy are measured in units of Bohr radius and Rydberg, respectively. If we use this unit system, from the relation $k_F^3 = 3\pi^2 n = 3\pi^2 (3/4\pi r_0^3)$, it follows that

$$k_{\mathrm{F}} = \left(\frac{9\pi}{4}\right)^{1/3} \frac{1}{r_s} \ .$$
(3.1.14)

In the same way, (3.1.11) is expressed as

$$\varepsilon_0 = \frac{3}{5} \left(\frac{9\pi}{4}\right)^{2/3} \frac{1}{r_s^2} \approx \frac{2.21}{r_s^2}$$
(3.1.15)

so that the energy per electron in units of Rydberg is expanded as

$$\varepsilon = \frac{2.21}{r_s^2} - \frac{0.916}{r_s} + \dots \ .$$
(3.1.16)

This expansion is called the r_s expansion and yields a good approximation for $r_s \ll 1$, i.e., when the number density of electrons is high. The sum of terms not explicitly written in (3.1.16) is called the "correlation energy" and is denoted usually by ε_c. It is impossible to calculate ε_c by means of conventional perturbation theory[1]; we will discuss the problem in Sect. 3.3.

3.1.5 Systems with Short-Range Force

In the case of liquid ^{3}He which is another typical example of a Fermi liquid, the interaction between particles has a property entirely different from that in the electron gas. Namely, whereas the Coulomb interaction in the electron gas is of long range, taking a form e^2/r, $v(r)$ in liquid ^{3}He is, for example, given approximately by the Lennard-Jones potential

$$v(r) = v_0 \left[\left(\frac{\sigma}{r}\right)^{12} - 2\left(\frac{\sigma}{r}\right)^6 \right] ,$$

so that the force between particles is of short range in contrast to the Coulomb force. Or, we sometimes simplify the above potential and use the hard-sphere potential:

$$v(r) = \begin{cases} 0 & (r > a) \\ \infty & (r < a) \ . \end{cases}$$

The simplest way to express the characteristic of short-range force is to employ the potential of δ function type given by

$$v(r) = v\delta(r) \ .$$
(3.1.17)

As a matter of fact, it is possible to express the hard-sphere potential in the above form under a certain condition. However, we will postpone this discussion and

---

[1] For example, the perturbation energy of the second order diverges to infinity.

for the time being assuming (3.1.17) we will calculate the ground-state energy by means of perturbation theory. From (3.1.5) it follows that $v(q) = v$ for $v(r)$ given by (3.1.17). In other words, in the present case the Fourier transform of the potential is a constant independent of q. Therefore, the correction term for energy of the first order is expressed as

$$E_1 = \frac{v}{2V} \sum_{qkk'\sigma\sigma'} \langle 0 | a_{k+q,\sigma}^\dagger a_{k'-q,\sigma'}^\dagger a_{k'\sigma'} a_{k\sigma} | 0 \rangle .$$

In the case of the electron gas, due to the rectriction $q \neq 0$ only the condition (II) was allowed, but in the above expression both (I) and (II) are possible. Taking into account these two possibilities, after a straightforward calculation, we have

$$E_1 = \frac{v}{2V} [\sum_{kk'\sigma\sigma'} n_{k\sigma} n_{k'\sigma'} - \sum_{kk'\sigma} n_{k\sigma} n_{k'\sigma}] . \tag{3.1.18}$$

In calculating (3.1.18) we assume that the number of up spins is equal to that of down spins; both are equal to $N/2$. Then, the second term in the brackets in the above equation yields $N^2/4$ for up spins and $N^2/4$ for down spins, leading to $N^2/2$ in total. Also, it is clear that the first term is equal to N^2. In this way, E_1 is calculated to be

$$E_1 = \frac{vN^2}{4V} = \frac{Nvn}{4} . \tag{3.1.19}$$

In the present case, since the correction term is proportional to n, the perturbational calculation gives rise to the better result, the lower the number density is. Thus, the situation becomes completely opposite to that in the electron gas. This distinction has its origin in the difference in whether the potential is of long range or of short range.

So far we have been assuming that the potential is of the δ function type. In what follows it will be shown that the hard-sphere potential can be written in such a form under a suitable condition. If we consider two hard spheres of the diameter a, $v(r)$ becomes infinity for $r < a$ since they cannot overlap with each other (Fig. 3.6). From the viewpoint of quantum mechanics, this implies that the wave function vanishes for $r \leq a$. If we substitute the hard-sphere potential itself into (3.1.5), $v(q)$ diverges so that we cannot get a result of physical significance. One way to overcome this difficulty is to replace the hard-sphere potential by something like a potential which is mathematically equivalent to the hard-sphere potential. It is called the "pseudopotential". Let us discuss this method in the following.

The Schrödinger equation for the relative motion of two hard spheres of the mass m is expressed as

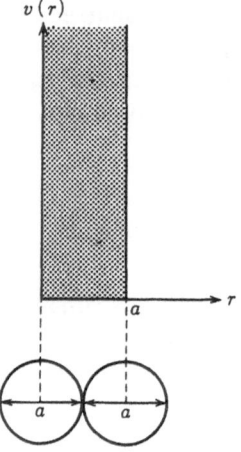

$v(r)$

r

Fig. 3.6. Hard-sphere potential

$$-\frac{\hbar^2}{m}\Delta\varphi = E\varphi \ , \quad \varphi = 0 \quad (r \leq a) \tag{3.1.20}$$

where we have used the fact that the reduced mass is $m/2$. Putting $k^2 = Em/\hbar^2$, we have

$$(\Delta + k^2)\,\varphi = 0, \quad \varphi = 0 \quad (r \leq a).$$

For simplicity consider the case where the wave function is spherically symmetric. For the S wave, the Schrödinger equation is written as

$$\frac{1}{r}\frac{d^2}{dr^2}(r\varphi) + k^2\varphi = 0 \ .$$

If we solve this equation under the boundary condition that $\varphi = 0$ at $r = a$, we find

$$r\varphi = A \sin k(r-a), \quad (r \geq a)$$

where A is an arbitrary constant. Since $\varphi = 0$ for $r \leq a$, the solution is represented as shown in Fig. 3.7a.

Now, suppose that we extend the solution for $r \geq a$ to the inside of the hard sphere keeping its original form, as is shown in Fig. 3.7b. Then, by the use of the relation $\Delta(1/r) = -4\pi\delta(r)$, the following equation

$$(\Delta + k^2)\,\varphi = 4\pi A\delta(r) \sin ka$$

holds for the extended φ. This implies that there appears a singularity of the δ function type at the origin, if we extend the solution to the inside of the hard

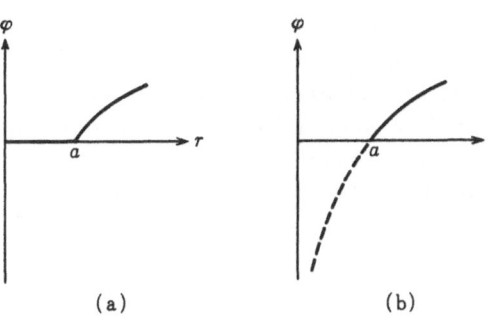

Fig. 3.7a,b. If we extend the wave function φ in the outside of the hard sphere to its inside as it is, the extended φ takes a form represented by the dotted line in (b)

sphere. The physical reason for this singularity may be understood in the following way. If we charge a spherical perfect conductor with the charge Q, the charge will be distributed uniformly over the spherical surface and the electrostatic potential φ outside the sphere is expressed as Q/r. On the other hand, since the electric field inside the sphere is zero, φ takes a constant value there. However, as long as one observes the electrostatic potential outside the sphere, one may treat it as if the charge Q were concentrated at the center of the sphere. In other words, the electrostatic potential outside the sphere can be regarded as arising from the charge distribution $Q\delta(r)$. The $\delta(r)$ in the right-hand side in the above equation corresponds to this charge distribution.

In order to eliminate the arbitrary constant A in the above equation, we use the following relation

$$\delta(r)\frac{\partial}{\partial r}(r\varphi) = \delta(r)\,Ak\cos ka\ .$$

Then, if a is sufficiently small, or more precisely for $ka \ll 1$, we obtain

$$(\Delta + k^2)\,\varphi = 4\pi\,\frac{\tan ka}{k}\,\delta(r)\frac{\partial}{\partial r}(r\varphi) \approx 4\pi a\delta(r)\frac{\partial}{\partial r}(r\varphi)\ .$$

Writing the above equation in a form of Schrödinger equation as given by (3.1.20), we find

$$-\frac{\hbar^2}{m}\,\Delta\varphi + \frac{4\pi a\hbar^2}{m}\,\delta(r)\frac{\partial}{\partial r}(r\varphi) = E\varphi\ .$$

This equation means that the potential $v(r)$ between particles is replaced by an operator

$$v(r) = \frac{4\pi a\hbar^2}{m}\,\delta(r)\frac{\partial}{\partial r}\,r$$

and this is the pseudopotential for the hard sphere.

In the foregoing discussion, we have considered only the S wave $(l = 0)$ for simplicity. However, a generalization to the P wave $(l = 1)$ or D wave $(l = 2)$ is straightforward. It turns out that in general the pseudopotential coming from the lth partial wave is proportional to a^{2l+1} for small a. Therefore, it is sufficient to take into account the S wave alone when a is small.

Returning to the above pseudopotential, we notice that the following relation is valid

$$\delta(r) \frac{\partial}{\partial r} (r\varphi) = \delta(r) \varphi$$

provided that the wave function is differentiable at $r = 0$. This implies that the operator $(\partial/\partial r)r$ can be put to 1. When we calculate the first-order term of perturbed energy, the wave function has just the property mentioned above. Therefore, the pseudopotential takes the form (3.1.17) and ν is given by

$$\nu = \frac{4\pi a \hbar^2}{m} . \tag{3.1.21}$$

If we proceed to the second-order term of energy, the operator $(\partial/\partial r)r$ cannot be put simply to 1, since the perturbed wave function of the first order involves a part proportional to $1/r$. If we take into account this situation, after a complicated calculation the ground-state energy per particle is shown to be

$$\frac{E}{N} = \frac{3\hbar^2 k_F^2}{10m} + \frac{\pi \hbar^2 n}{m} a + \frac{6\hbar^2 n k_F}{35m} (11 - 2 \ln 2) a^2 + \dots .$$

This equation is exact up to the order of a^2 and was derived by T. D. Lee and C. N. Yang [Phys. Rev. **105**, 1119 (1957)]. Of course, the term of the order of a is obtained by substituting (3.1.21) into (3.1.19). It is easily seen that the expansion parameter in this case is $(na^3)^{1/3}$. Therefore, the above formula leads to an asymptotic expansion which is valid at the low-density limit $(na^3 \ll 1)$.

Unfortunately the results obtained so far for models of a Fermi liquid are not applicable to real physical systems. We have neither $r_s \ll 1$ for electrons in metals, nor $na^3 \ll 1$ in liquid ^{3}He. Nevertheless, the results derived here are not always meaningless. In dealing with an actual physical system, we have to develop a theory which involves, more or less, approximations. The results obtained in this section are then useful to judge their validity. That is, by applying a given theory to an extreme case, either a high density limit or a low density limit, and by comparing the result so obtained with the rigorous one, we may infer the validity of the approximation to some extent.

3.2 Stimulus to a Many-Body System and Its Response

In studying the properties of a Fermi liquid, it is important to see what kind of elementary excitations exist in the system and what properties they have. In general, if one wants to know the properties of a given physical system, it is necessary to stimulate the system in one way or another. For example, as mentioned in Chap. 1, neutron scattering or the Mössbauer effect is used to observe the phonon experimentally. In this case, the bombardment of the system by a neutron or γ-ray is employed as the means to stimulate the system. To the stimulus given to the system, it shows some response. In this section, we will discuss the relationship between the stimulus and response and see that through it useful information can be obtained with respect to elementary excitations in the system. The following formulation is an extension of the theory of dielectric constant mentioned in Sect. 2.2 and is applicable not only to a Fermi liquid but also to any physical system.

3.2.1 Schrödinger Equation in the Presence of External Field

Let the Hamiltonian of the system under consideration be $\mathscr{H}$ and let its nth energy eigenvalue and eigenfunction be ε_n and ϕ_n, respectively. We assume that we apply a suitable external field to the system in order to stimulate it and denote the Hamiltonian[2] describing the external field by $\mathscr{H}'(t)$. As in Sect. 2.2, we consider that $\mathscr{H}'(t) = 0$ at $t = -\infty$ and assume that $\mathscr{H}'(t)$ is proportional to the factor $\exp(\delta t)$ ($\delta > 0$). After the calculation is finished, we take the limit $\delta \to +0$. Furthermore, we assume that the system is in the state ϕ_m at $t = -\infty$.

Since the Hamiltonian $\mathscr{H}'(t)$ is added to the original Hamiltonian, the temporal development of the wave function is described by the time-dependent Schrödinger equation

$$i\hbar \frac{\partial \psi}{\partial t} = [\mathscr{H} + \mathscr{H}'(t)] \psi \ . \tag{3.2.1}$$

Here, we put

$$\psi(t) = \exp\left(-\frac{i}{\hbar} \mathscr{H} t\right) \varphi(t) \tag{3.2.2}$$

and make a transformation from ψ to φ. Then, by the use of the relation $i\hbar\dot{\psi} = \mathscr{H}\psi + \exp(-i\mathscr{H}t/\hbar) i\hbar\dot{\varphi}$, we obtain

$$\frac{\partial \varphi}{\partial t} = \frac{1}{i\hbar} \mathscr{H}''(t) \varphi, \quad \mathscr{H}''(t) = \exp\left(\frac{i}{\hbar} \mathscr{H} t\right) \mathscr{H}'(t) \exp\left(-\frac{i}{\hbar} \mathscr{H} t\right) \ . \tag{3.2.3}$$

---

[2] Explicit examples of $\mathscr{H}'(t)$ will be discussed in the next section.

If we note that[3] $\varphi(t) = \phi_m$ at $t = -\infty$, formal integration of the above equation leads to

$$\varphi(t) = \phi_m + \frac{1}{i\hbar} \int_{-\infty}^{t} \mathscr{H}''(t') \, \varphi(t') \, dt' \ . \tag{3.2.4}$$

Suppose, in order to study a response of the system to the external field, we observe a physical quantity A associated with the system, e.g., electric current, electric polarization, number density, etc. According to the general principle of quantum mechanics, the expectation value $\bar{A}$ of A at time t is expressed as

$$\bar{A} = \int \psi^* A \psi \, d\tau = \int \left[\exp\left(-\frac{i}{\hbar} \mathscr{H} t\right) \varphi \right]^* A \left[\exp\left(-\frac{i}{\hbar} \mathscr{H} t\right) \varphi \right] d\tau \ .$$

Here, the integration over τ implies both the spatial integrations over the coordinates of all particles and the summations over their spin coordinates. Also, we have used (3.2.2). In order to simplify the above expression, we notice that for arbitrary two functions f and g, and for an arbitrary operator P the following identity holds:

$$\int (Pf)^* \, g \, d\tau = [\int g^*(Pf) \, d\tau]^* = \int f^* P^\dagger g \, d\tau \ .$$

In the above, $P^\dagger$ is the operator Hermitian conjugate to P. If we put $f \to \varphi$, $P \to \exp(-i\mathscr{H} t/\hbar)$, $g \to A \exp(-i\mathscr{H} t/\hbar)\varphi$, since we have $P^\dagger = \exp(i\mathscr{H} t/\hbar)$ it follows that

$$\bar{A} = \int \varphi^* \exp\left(\frac{i}{\hbar} \mathscr{H} t\right) A \, \exp\left(-\frac{i}{\hbar} \mathscr{H} t\right) \varphi \, d\tau = \int \varphi^* A(t) \varphi \, d\tau \tag{3.2.5}$$

where $A(t)$ is the operator in the Heisenberg representation defined by

$$A(t) = \exp\left(\frac{i}{\hbar} \mathscr{H} t\right) A \, \exp\left(-\frac{i}{\hbar} \mathscr{H} t\right) \ . \tag{3.2.6}$$

3.2.2 Linear Responses

When we consider the response of the system to the external field, the term linear with respect to $\mathscr{H}'(t)$ has a particularly important meaning. This is called the "linear response". From (3.2.4) we see that $\varphi(t) = \phi_m + O(\mathscr{H}'')$, so that as far as the term linear in $\mathscr{H}'(t)$ is concerned we can replace $\varphi(t')$ in the integral in this equation by ϕ_m. That is, we find

---

[3] Strictly speaking, from (3.2.2) we have at $t = -\infty$

$$\varphi(t) = \exp\left(\frac{i}{\hbar} \varepsilon_m t\right) \phi_m \ .$$

However, the exponential factor here is physically irrelevant so that we can assume $\varphi(t) = \phi_m$.

$$\varphi(t) = \phi_m + \frac{1}{i\hbar} \int_{-\infty}^{t} \mathcal{H}''(t') \phi_m \, dt' + O(\mathcal{H}''^2) . \qquad (3.2.7)$$

Therefore, in linear response theory, we have from (3.2.5)

$$\bar{A} = \int \phi_m^* A(t) \phi_m d\tau + \frac{1}{i\hbar} \int_{-\infty}^{t} dt' \int \phi_m^* A(t) \, \mathcal{H}''(t') \phi_m d\tau$$

$$- \frac{1}{i\hbar} \int_{-\infty}^{t} dt' \int [\mathcal{H}''(t')\phi_m]^* A(t) \phi_m d\tau .$$

By means of the relation $\exp(-i\mathcal{H}t/\hbar)\phi_m = \exp(-i\varepsilon_m t/\hbar)\phi_m$, the first term in the above equation is equal to

$$\int \phi_m^* \exp\left(\frac{i}{\hbar} \mathcal{H}t\right) A \exp\left(-\frac{i}{\hbar} \mathcal{H}t\right) \phi_m d\tau$$

$$= \int \left[\exp\left(-\frac{i}{\hbar} \mathcal{H}t\right) \phi_m\right]^* A \left[\exp\left(-\frac{i}{\hbar} \mathcal{H}t\right) \phi_m\right] d\tau$$

$$= \int \phi_m^* A \phi_m d\tau .$$

Also, by noting that $\mathcal{H}''(t')$ is the Hermitian operator, the τ integral in the third term is written as

$$\int [\mathcal{H}''(t')\phi_m]^* A(t) \phi_m d\tau = \int \phi_m^* \mathcal{H}''(t') A(t) \phi_m d\tau .$$

In this way, by using Dirac's notation, if we write for an arbitrary operator Q as

$$\int \phi_m^* Q \phi_m d\tau \equiv \langle m|Q|m \rangle ,$$

we obtain

$$\bar{A} = \langle m|A|m \rangle + \frac{1}{i\hbar} \int_{-\infty}^{t} dt' \langle m|A(t)\mathcal{H}''(t') - \mathcal{H}''(t')A(t)|m \rangle .$$

We have been assuming so far that the state is in a quantum-mechanical pure state at $t = -\infty$. However, from the viewpoint of statistical mechanics, the state is subject to a certain probability distribution. Keeping this in mind, we assume that the system is in thermal equilibrium at $t = -\infty$ and let the probability that the state ϕ_m is realized be w_m. In order to calculate statistical-mechanical average in thermal equilibrium, we multiply the above equation by w_m and sum over all m. To simplify the notation, we denote this average value by $\langle \ldots \rangle$, i.e.,

$$\sum_m w_m \langle m| \ldots |m \rangle = \langle \ldots \rangle .$$

Furthermore, we assume that the average of the physical quantity A is zero in thermal equilibrium. Then, the statistical mechanical-average of A at time t is expressed as

$$\langle A \rangle_t = - \frac{i}{\hbar} \int_{-\infty}^{t} dt' \, \langle [A(t), \mathscr{H}''(t')] \rangle \tag{3.2.8}$$

where $[A,B] \equiv AB - BA$.

Here we assume that the Hamiltonian $\mathscr{H}'(t)$ describing the external field has a form

$$\mathscr{H}'(t) = B \exp(-i\omega t + \delta t) \tag{3.2.9}$$

with an appropriate operator B. The above equation implies that the external field oscillates in time with the angular frequency ω. If we remember (3.2.3), by using the Heisenberg representation for B, (3.2.8) is written as

$$\langle A \rangle_t = - \frac{i}{\hbar} \int_{-\infty}^{t} dt' \exp(-i\omega t' + \delta t') \, \langle [A(t), B(t')] \rangle \ . \tag{3.2.10}$$

This is a fundamental formula in the theory of linear responses and is applicable to a large number of problems. For example, if we consider the electric field applied to a metal as the external field and take the electric current density as the physical quantity A, we can derive from the above formula the coefficient describing the linear relation between electric current density and electric field, i.e., electrical conductivity. We will return to this problem in Chap. 6.

3.2.3 Retarded and Temperature Green's Functions

It is convenient to express the results obtained above in terms of Green's functions for both formal and practical purposes. Although we have mentioned Green's functions in connection with the problem of phonons in Chap. 1, we would like to repeat some explanation of Green's functions to the extent which will be necessary in the following discussion. First, let us introduce the retarded Green's function

$$G_r[t, t'] = - \frac{i}{\hbar} \theta(t - t') \, \langle [A(t), B(t')] \rangle \tag{3.2.11}$$

for A and B. Here, $\theta(t - t')$ is a step function which implies that $\theta(t) = 1$ $(t > 0)$ and $\theta(t) = 0$ $(t < 0)$. By the use of (3.2.11), (3.2.10) is written as

$$\langle A \rangle_t = \int_{-\infty}^{\infty} \exp(-i\omega t' + \delta t') \, G_r[t, t'] \, dt' \ .$$

As will be proved later on, $G_r[t, t']$ is a function of $t - t'$, so that we will express $G_r[t, t']$ as $G_r[t - t']$. Also, we define the Fourier transform of $G_r[t]$ by

$$G_r(\omega) = \int_{-\infty}^{\infty} G_r[t] \exp(i\omega t - \delta t)\, dt \ . \tag{3.2.12}$$

Taking $t - t'$ as a new integral variable in the equation for $\langle A \rangle_t$, we obtain

$$\langle A \rangle_t = G_r(\omega) \exp(-i\omega t + \delta t) \ . \tag{3.2.13}$$

This equation means that $\langle A \rangle_t$ has the same time dependence as $\mathcal{H}'(t)$.

From (3.2.13) it turns out that $G_r(\omega)$ is the amplitude with which $\langle A \rangle_t$ oscillates in time. At the same time, important information about elementary excitations in the system is included in the behavior of $G_r(\omega)$. To see this, let us consider the case at absolute zero. Noting that $G_r[t, t']$ is a function of $t - t'$, we put $t' = 0$ in (3.2.11) and denote the ground state of the system by $|0\rangle$. Then, at absolute zero we have

$$
\begin{aligned}
\langle [A(t), B] \rangle &= \langle 0 | A(t)B - BA(t) | 0 \rangle \\
&= \sum_n \left\{ \langle 0|A|n\rangle\langle n|B|0\rangle \exp\left[-\frac{i}{\hbar}(\varepsilon_n - \varepsilon_0)\, t \right] \right. \\
&\quad \left. - \langle 0|B|n\rangle\langle n|A|0\rangle \exp\left[\frac{i}{\hbar}(\varepsilon_n - \varepsilon_0)\, t \right] \right\} \ .
\end{aligned}
$$

Therefore, writing $\omega_{n0} = (\varepsilon_n - \varepsilon_0)/\hbar$ for simplicity, we find

$$
\begin{aligned}
G_r[t] = &- \frac{i}{\hbar} \theta(t) \sum_n [\langle 0|A|n\rangle\langle n|B|0\rangle \exp(-i\omega_{n0}t) \\
&- \langle 0|B|n\rangle\langle n|A|0\rangle \exp(i\omega_{n0}t)] \ .
\end{aligned}
$$

If we substitute the above equation into (3.2.12) and carry out the integration over t, after a simple calculation we obtain

$$G_r(\omega) = \frac{1}{\hbar} \sum_n \left[\frac{\langle 0|A|n\rangle\langle n|B|0\rangle}{\omega - \omega_{n0} + i\delta} - \frac{\langle 0|B|n\rangle\langle n|A|0\rangle}{\omega + \omega_{n0} + i\delta} \right] \ . \tag{3.2.14}$$

Taking the limit $\delta \to 0$ in (3.2.14), we see that $\omega = \omega_{n0}$ or $\omega = -\omega_{n0}$ is the pole of $G_r(\omega)$. Conversely, regarding $G_r(\omega)$ as a function of ω, if we can find the point where it diverges, we can calculate the value of ω_{n0}. Since ω_{n0} is proportional to the energy difference between excited and ground states of the system, the energy of elementary excitations is calculated in principle from the pole of $G_r(\omega)$. Explicit examples will be discussed in Sect. 3.4. From the physical point of view, what is stated above represents the fact that if the external frequency is equal to the eigenfrequency of the system the amplitude diverges to infinity, i.e., the resonance.

Now, the retarded Green's function mentioned above is closely related to the temperature Green's function defined by

$$\mathscr{G}[\tau, \tau'] = - \langle T_\tau A(\tau) B(\tau') \rangle \tag{3.2.15}$$

where τ and τ' are real quantities varying in the range $0 \leq \tau,\, \tau' \leq \beta$ $(\beta = 1/k_{\mathrm{B}}T)$ and $A(\tau)$ is given by

$$A(\tau) = \exp(\tau\mathscr{H})\, A \exp(-\tau\mathscr{H}) \;.$$

Also, T_τ in (3.2.15) is Wick's operator mentioned in Sect. 1.8.1. In the present problem, we use it in the sense of the Bose type, i.e., $T_\tau A(\tau)B(\tau')$ is equal to $A(\tau)B(\tau')$ if $\tau > \tau'$ and to $B(\tau')A(\tau)$ if $\tau' > \tau$.

Returning our discussion somewhat circuitously, we derive an equation corresponding to (3.2.14) at finite temperatures. For this purpose, we assume that the system obeys the canonical distribution. Then w_n is given by

$$w_n = \frac{\exp(-\beta\varepsilon_n)}{Z}\;, \quad Z = \sum_n \exp(-\beta\varepsilon_n)$$

with the partition function Z of the system. Repeating the same procedure as in the case of absolute zero discussed previously, we obtain

$$G_r[t, t'] = - \frac{i}{\hbar}\, \theta(t - t') \frac{1}{Z} \sum_{nn'} \exp(-\beta\varepsilon) \Big\{ \langle n|A|n'\rangle\langle n'|B|n\rangle$$

$$\cdot \exp\!\Big[\frac{i}{\hbar}(\varepsilon - \varepsilon')(t - t')\Big] - \langle n|B|n'\rangle\langle n'|A|n\rangle \exp\!\Big[\frac{i}{\hbar}(\varepsilon - \varepsilon')(t' - t)\Big]\Big\}$$

where to simplify the notation we put $\varepsilon = \varepsilon_n$ and $\varepsilon' = \varepsilon_{n'}$. As is seen from the above equation, $G_r[t, t']$ is a function of $t - t'$. Taking $t - t'$ as a new variable t and making a change of variables $n \rightleftarrows n'$ in the first term in the brackets on the right-hand side, we find

$$G_r[t] = - \frac{i}{\hbar}\, \theta(t) \frac{1}{Z} \sum_{nn'} \exp(-\beta\varepsilon)\, \langle n|B|n'\rangle\langle n'|A|n\rangle$$

$$\cdot \exp\!\Big[-\frac{i}{\hbar}(\varepsilon - \varepsilon')\, t\Big] \{\exp[\beta(\varepsilon - \varepsilon')] - 1\} \;.$$

Substituting this into (3.2.12) and carrying out the integration over t, after a simple calculation we see that $G_r(\omega)$ is expressed as

$$G_r(\omega) = \frac{1}{Z} \sum_{nn'} \exp(-\beta\varepsilon)\langle n|B|n'\rangle\langle n'|A|n\rangle\, \frac{\exp[\beta(\varepsilon - \varepsilon')] - 1}{\hbar\omega - (\varepsilon - \varepsilon') + i\delta}\;,$$

where we denote $\hbar\delta$ simply by δ. By introducing spectral function

$$A(\omega) = \frac{1}{Z} \sum_{nn'} \exp(-\beta\varepsilon)\langle n|B|n'\rangle\langle n'|A|n\rangle[\exp(\beta\omega) - 1]\, \delta\,[\omega - (\varepsilon - \varepsilon')] \;,$$

it follows that

$$G_r(\omega) = \int_{-\infty}^{\infty} \frac{A(\omega')}{\hbar\omega + i\delta - \omega'} \, d\omega' \ . \tag{3.2.16}$$

On the contrary, the temperature Green's function given by (3.2.15) is shown to be a function of $\tau - \tau'$ by repeating exactly the same procedure as above. So, let $\tau - \tau'$ be τ. The range for τ thus defined is $-\beta \leq \tau \leq \beta$. If we put $\tau' = 0$ in (3.2.15), we have[4] for $\tau > 0$

$$\mathcal{G}[\tau] = -\langle A(\tau)B\rangle = -\frac{1}{Z} \operatorname{tr}\{\exp(-\beta\mathcal{H})\exp(\tau\mathcal{H})\,A\,\exp(-\tau\mathcal{H})\,B\}$$

$$= -\frac{1}{Z}\sum_{nn'}\exp[-\beta\varepsilon' + \tau(\varepsilon' - \varepsilon)]\langle n'|A|n\rangle\langle n|B|n'\rangle \tag{3.2.17a}$$

and for $\tau < 0$

$$\mathcal{G}[\tau] = -\langle BA(\tau)\rangle = -\frac{1}{Z} \operatorname{tr}\{\exp(-\beta\mathcal{H})\,B\,\exp(\tau\mathcal{H})\,A\,\exp(-\tau\mathcal{H})\}$$

$$= -\frac{1}{Z}\sum_{nn'}\exp[-\beta\varepsilon + \tau(\varepsilon' - \varepsilon)]\,\langle n|B|n'\rangle\langle n,|A|n\rangle \ . \tag{3.2.17b}$$

From these equations, an important property of $\mathcal{G}[\tau]$

$$\mathcal{G}[\tau + \beta] = \mathcal{G}[\tau] \tag{3.2.18}$$

is derived. To prove this, we assume that $-\beta < \tau < 0$. Then, we have $0 < \beta + \tau < \beta$ so that $\mathcal{G}[\tau + \beta]$ is given by replacing τ with $\tau + \beta$ in (3.2.17a). The result is just equal to (3.2.17b). In this way, it is seen that (3.2.18) is valid.

Now expand $\mathcal{G}[\tau]$ in terms of Fourier series

$$\mathcal{G}[\tau] = \frac{1}{\beta}\sum_{l} \mathcal{G}(i\nu_l)\exp(-i\nu_l\tau) \tag{3.2.19}$$

as in Sect. 1.8. Because of the property (3.2.18), ν_l take the values

$$\nu_l = \frac{2l\pi}{\beta} \quad (l = 0, \pm 1, \pm 2, \dots) \ . \tag{3.2.20}$$

Multiplying both sides in the above Fourier series by $\exp(i\nu_{l'}\tau)$ and integrating over τ from 0 to β, we find

$$\mathcal{G}(i\nu_l) = \int_{0}^{\beta} \mathcal{G}[\tau]\exp(i\nu_l\tau)\,d\tau \ . \tag{3.2.21}$$

---

[4] $\sum_n w_n \langle n|X|n\rangle = \frac{1}{Z}\sum_n \exp(-\beta\varepsilon_n)\langle n|X|n\rangle = \frac{1}{Z}\operatorname{tr}\{\exp(-\beta\mathcal{H})\,X\}$.

Substituting (3.2.17a) into this equation, performing the integration over τ, and using the definition of spectral function mentioned above, we have

$$\mathscr{G}(i\nu_l) = \int_{-\infty}^{\infty} \frac{A(\omega')}{i\nu_l - \omega'} \, d\omega' \ . \tag{3.2.22}$$

This equation corresponds to (1.8.4) so that the concept of analytic continuation is also applicable to the present case. That is, if we extend $i\nu_l$ to a general complex number z and consider a function $\mathscr{G}(z)$, it follows that

$$G_r(\omega) = \mathscr{G}(\hbar\omega + i\delta) \ . \tag{3.2.23}$$

This relation is a very useful property from the practical point of view. For, as we saw in Sect. 1.8, when we apply, for example, perturbation theory the actual calculation is much easier for the temperature Green's function than for the retarded Green's function. Anyway, if we derive the temperature Green's function by some approximation method, we can calculate $G_r(\omega)$ by analytic continuation so that we are able to get information about elementary excitations in the system. We will discuss its applications in Sect. 3.4.

3.2.4 The Case of the Grand Canonical Distribution

We have been discussing the problem so far on the basis of the canonical distribution. But the same situation is also true even in the case of the grand canonical distribution. Since the argument is exactly the same as in the canonical distribution, we will only summarize the results. The retarded Green's function is defined by the same equation as (3.2.11) also in this distribution. However, the statistical-mechanical average $\langle \ldots \rangle$ is defined by

$$\langle \ldots \rangle = \frac{\mathrm{tr}\,\{\exp[-\beta(\mathscr{H} - \mu N)] \ldots\}}{\mathrm{tr}\,\{\exp[-\beta(\mathscr{H} - \mu N)]\}}$$

and $A(t)$ by

$$A(t) = \exp\left[\frac{i}{\hbar}(\mathscr{H} - \mu N)\,t\right] A \exp\left[-\frac{i}{\hbar}(\mathscr{H} - \mu N)t\right]$$

where μ is the chemical potential per particle. In the same way, the temperature Green's function is given by (3.2.15), the only difference being that the statistical-mechanical average is defined as above and that $A(\tau)$ is defined by

$$A(\tau) = \exp[\tau(\mathscr{H} - \mu N)] A \exp[-\tau(\mathscr{H} - \mu N)] \ .$$

In short, to obtain results for the grand canonical distribution it is sufficient to replace $\mathscr{H}$ in the canonical distribution by $\mathscr{H} - \mu N$. For this reason, the results obtained in Sect. 3.2.3 remain valid if we expand the quantities in terms of eigenfunctions of $\mathscr{H} - \mu N$ instead of $\mathscr{H}$. Therefore, it is concluded that the property (3.2.23) holds in the grand canonical distribution.

3.3 The Electron Gas

In a previous section we discussed the relationship between stimulus and response in a rather abstract form. To see this relationship physically in a more concrete form, let us consider the problem of the electron gas again. Since the electron has a charge, it is easy to impose an external field on the electron system. For example, we may apply an electric or magnetic field. However, for the convenience of later discussion, we will deal with the external field with a slightly different form.

3.3.1 Test Charge as the External Field

Let us suppose that we add a charge from the outside to the electron gas under consideration. Furthermore, we assume that the charge varies in time and space. Of course, it is impossible to embed the charge in an actual metal at our will. However, it may be possible to do so conceptually (Gedanken-experiment). This charge has an interaction with electrons, so that it behaves as the external field. As it were, the charge added from the outside has a role of a "probe" and this charge will be called the "test charge".

Let us now assume that the charge density of the test charge has a form

$$- er_q \exp(i\boldsymbol{q}\cdot\boldsymbol{r} - i\omega t + \delta t) \qquad (3.3.1)$$

as a function of spatial position $\boldsymbol{r}$ and time t. Here $- e$ is the electronic charge and r_q a constant. Let the position vector of the ith electron in the electron gas be $\boldsymbol{r}_i$ $(i = 1,2, \ldots, N)$. Then the Coulomb interaction between the test charge and the electron gas is expressed as

$$\mathscr{H}'(t) = \sum_i \int \frac{e^2 r_q}{|\boldsymbol{r} - \boldsymbol{r}_i|} \exp(i\boldsymbol{q}\cdot\boldsymbol{r} - i\omega t + \delta t)\, d\boldsymbol{r} \ .$$

This $\mathscr{H}'(t)$ is the Hamiltonian of the external field acting on the electron gas. It has the same form as (3.2.9) and B is given by

$$B = \sum_i \int \frac{e^2 r_q \exp(i\boldsymbol{q}\cdot\boldsymbol{r})}{|\boldsymbol{r} - \boldsymbol{r}_i|}\, d\boldsymbol{r} \ . \qquad (3.3.2)$$

By the use of the defining equation (3.1.5) for the Fourier transform and (3.1.7) for the Coulomb interaction, it follows that

$$\frac{e^2}{|r - r_i|} = \frac{1}{V} \sum_{q'} \frac{4\pi e^2}{q'^2} \exp[iq' \cdot (r - r_i)] \ .$$

Substituting the above equation into (3.3.2) and carrying out the integration over r, we have

$$B = \frac{4\pi e^2 r_q}{q^2} \sum_i \exp(iq \cdot r_i) \ .$$

Since the number density $\rho(r)$ of electrons at the position r is given by $\rho(r) = \sum_i \delta(r - r_i)$, by means of the Fourier transform

$$\rho(r) = \frac{1}{V} \sum_q \rho_q \exp(iq \cdot r), \quad \rho_q = \int \rho(r) \exp(-iq \cdot r) \, dr \ ,$$

ρ_q is expressed as

$$\rho_q = \sum_i \exp(-iq \cdot r_i) \ .$$

Therefore, it turns out that

$$B = \frac{4\pi e^2 r_q}{q^2} \rho_{-q} = \nu(q) r_q \rho_{-q} \tag{3.3.3}$$

where $\nu(q)$ is the Fourier transform of the Coulomb interaction given by (3.1.7).

Because of the test charge, physical quantities associated with the electron gas will change in time. According to (3.2.13) in the previous section, within the limitation of linear responses, the average value of a physical quantity A at time t is expressed as

$$\langle A \rangle_t = A(\omega) \exp(-i\omega t + \delta t) \ .$$

In (3.2.13) we have written $G_r(\omega)$ instead of $A(\omega)$, but in the present case it is convenient to define the retarded Green's function except for the factor $\nu(q) r_q$. For this reason, we used the notation $A(\omega)$ in the above equation. From (3.2.11, 12) the expression for $A(\omega)$ is given by

$$A(\omega) = -\frac{i}{\hbar} \int_0^\infty \langle [A(t), B] \rangle \exp(i\omega t - \delta t) \, dt$$

where we put $t' = 0$ in (3.2.11).

Although we may take a number of quantities as the physical quantity A, it is most convenient to consider the Fourier transform ρ_q of the number density. The reason for this is as follows. If there is no test charge, electrons distribute uniformly so that the average value of ρ_q becomes zero if $q \neq 0$. In other words,

ρ_q ($q \neq 0$) describes the fluctuation of the number density. When the test charge is added to the electron gas, it is expected that the distribution of electrons is no longer uniform and fluctuates spatially. Taking ρ_q as A corresponds to observing this spatial variation, i.e., the number density fluctuation. The average value of ρ_q at time t is expressed as

$$\langle \rho_q \rangle_t = \rho_q(\omega) \exp(-i\omega t + \delta t) \; ,$$

so that by means of (3.3.3) it follows that

$$\rho_q(\omega) = v(q) \, r_q G_r(q, \omega) \tag{3.3.4}$$

where $G_r(q, \omega)$ is the Fourier transform of the retarded Green's function

$$G_r[q, t] = - \frac{i}{\hbar} \, \theta(t) \langle [\rho_q(t), \rho_{-q}] \rangle \; . \tag{3.3.5}$$

3.3.2 Dielectric Constants

If we put the test charge in the electron gas, as mentioned above, the number density of electrons deviates from its value at thermal equilibrium. Physically speaking, the induced charge is produced by the test charge. In order to describe this situation mathematically, let us apply the Maxwell equation. If we denote the electric displacement vector by D, we have

div $D = 4\pi \times$ (test charge density).

Since we have been assuming that the system is electrically neutral, the test charge is just the true charge; the above equation represents this situation. On the other hand, for the electric field vector E the following relation holds

div $E = 4\pi \times$ (free charge density)

$= 4\pi \times$ (test charge density + induced charge density).

Now suppose the Fourier transform:

$$D(r, t) = \frac{1}{V} \sum_{q'} D(q', \omega) \exp(iq' \cdot r - i\omega t) \; .$$

Since in the present problem we are assuming the time and space dependence in the form $\exp(iq \cdot r - i\omega t)$, the above equation takes the form

$$D(r, t) = \frac{1}{V} D(q, \omega) \exp(iq \cdot r - i\omega t) \; .$$

Substituting this into the Maxwell equation and putting $\delta = 0$ in (3.3.1) for simplicity, we have

$$iqD(q, \omega) = - 4\pi Ver_q$$

where D implies the component of D along q direction (longitudinal component).

On the other hand, by means of the Fourier transform mentioned earlier, the induced charge density is calculated from[5]

$$\frac{1}{V} \langle \rho_q \rangle, \exp(iq \cdot r) = \frac{1}{V} \rho_q(\omega) \exp(iq \cdot r - i\omega t)$$

where we put $\delta = 0$. By the use of this equation, the Maxwell equation for E is written as

$$iqE(q, \omega) = - 4\pi Ver_q - 4\pi e \rho_q(\omega)$$

if we repeat the same discussion as for $D(q, \omega)$. The ratio of D to E is the dielectric constant ϵ, i.e.,

$$\epsilon(q, \omega) = \frac{D(q, \omega)}{E(q, \omega)} .$$

From the above two equations together with (3.3.4), we obtain

$$\frac{1}{\epsilon(q, \omega)} = 1 + \frac{\nu(q)}{V} G_r(q, \omega) . \tag{3.3.6}$$

In this way, it turns out that the dielectric constant (or dielectric function) dependent on the wave number q and angular frequency ω is expressed in terms of the retarded Green's function given by (3.3.5). It should be noted that (3.3.6) is an exact relation without any approximations.

If the relation $\epsilon(q, \omega) = 0$ is satisfied for suitable q and ω, as can be seen from the definition, it is possible to have $E \neq 0$ even if $D = 0$. This means that the oscillating electric field appears in the system, even if there is no test charge. Thus, this oscillation is characteristic of the system and represents an elementary excitation existing in the system. To see what kind of elementary excitations occur, it is necessary to know the explicit form of $G_r(q, \omega)$. We will discuss this point in Sect. 3.4. For the time being, it is sufficient to understand that the information about the elementary excitations which can appear in the system is obtained from the zeros of the dielectric constant.

In the previous section we derived the general expression for the retarded Green's function. The result is applicable also to the present problem. For simplicity considering the case of absolute zero, we put $A \rightarrow \rho_q$ and $B \rightarrow \rho_{-q}$ in (3.2.14). Then, the Fourier transform of (3.3.5) is expressed as

---

[5] The following equation represents the number density of induced charge. One should multiply this by $-e$ to obtain the induced charge density.

$$G_r(q, \omega) = \frac{1}{\hbar} \sum_n \left(\frac{\langle 0|\rho_q|n\rangle\langle n|\rho_{-q}|0\rangle}{\omega - \omega_{n0} + i\delta} - \frac{\langle 0|\rho_{-q}|n\rangle\langle n|\rho_q|0\rangle}{\omega + \omega_{n0} + i\delta} \right) .$$

Since $\rho_q^* = \rho_{-q}$, we have $\langle 0|\rho_{-q}|n\rangle = \langle n|\rho_q|0\rangle^*$. Also, since our system is isotropic spatially, the directions $-q$ and q are completely equivalent. Therefore, it follows that $|\langle n|\rho_{-q}|0\rangle|^2 = |\langle n|\rho_q|0\rangle|^2$. Noting these relations and substituting the above equation into (3.3.6), we obtain

$$\frac{1}{\epsilon(q, \omega)} = 1 - \frac{\nu(q)}{V\hbar} \sum_n |\langle n|\rho_q|0\rangle|^2 \left(\frac{1}{\omega + \omega_{n0} + i\delta} - \frac{1}{\omega - \omega_{n0} + i\delta} \right) .$$

$$(3.3.7)$$

This equation is essentially the same as the result obtained in Sect. 2.2.

3.3.3 The Correlation Energy

As stated in Sect. 3.3.2, in the case of the electron gas the relationship between the stimulus and the response is conveniently represented by a physical quantity such as the dielectric constant. It may be understood that the dielectric constant $\epsilon(q, \omega)$ describes the dynamic behavior of the electron gas for the disturbance with wave number q and angular frequency ω. Moreover, this $\epsilon(q, \omega)$ yields important information about the static property of the system. As an example, let us consider the correlation energy of the electron gas.

We first observe the Coulomb interaction $\mathscr{H}'$ between electrons. By the use of the Fourier transform, $\mathscr{H}'$ is written as

$$\mathscr{H}' = \frac{1}{2} \sum_{i \neq j} \frac{e^2}{|r_i - r_j|} = \frac{1}{2V} \sum_{q, i \neq j} \nu(q) \exp[iq \cdot (r_i - r_j)] .$$

The restriction $i \neq j$ in the above equation comes from the fact that the electron does not interact with itself. However, writing $\sum_{i \neq j} = \sum_{i,j} - \sum_{i=j}$ and using the defining equation for ρ_q, we obtain

$$\mathscr{H}' = \frac{1}{2V} \sum_q \nu(q) (\rho_{-q}\rho_q - N) . \tag{3.3.8}$$

If we denote the ground state of system by $|0\rangle$, the average value of $\mathscr{H}'$ in this state is expressed as

$$\langle 0|\mathscr{H}'|0\rangle = \frac{1}{2V} \sum_q \nu(q) (\langle 0|\rho_{-q}\rho_q|0\rangle - N) \tag{3.3.9}$$

where we have assumed that $|0\rangle$ is normalized. It will be shown that the quantity $\langle 0|\rho_{-q}\rho_q|0\rangle$ in the above equation is closely related to the dielectric constant.

To see this relation, let us return to (3.3.7) and note the following formula

$$\lim_{\delta \to +0} \frac{1}{x \pm i\delta} = P\left(\frac{1}{x}\right) \mp i\pi\delta(x) \; .$$

Here the symbol P implies taking the principal value of the integral. If we apply the above formula to (3.3.7) and take the imaginary parts of both sides, we see that the terms involving P do not contribute. As a result, we have

$$\text{Im} \left\{\frac{1}{\epsilon(q, \omega)}\right\} = \frac{\pi\nu(q)}{V\hbar} \sum_n |\langle n|\rho_q|0\rangle|^2 [\delta(\omega + \omega_{n0}) - \delta(\omega - \omega_{n0})] \; . \quad (3.3.10)$$

Furthermore, if we integrate the above equation over ω from 0 to ∞, since by definition $\omega_{n0} = (\varepsilon_n - \varepsilon_0)/\hbar$ and $\omega_{n0} > 0$, the δ function in the first term on the right-hand side yields no contribution. Therefore, the relation

$$\int_0^\infty \text{Im} \left\{\frac{1}{\epsilon(q, \omega)}\right\} d\omega = - \frac{\pi\nu(q)}{V\hbar} \sum_n |\langle n|\rho_q|0\rangle|^2$$

is obtained. Or, by using

$$\sum_n |\langle n|\rho_q|0\rangle|^2 = \sum_n \langle 0|\rho_{-q}|n\rangle \langle n|\rho_q|0\rangle = \langle 0|\rho_{-q}\rho_q|0\rangle$$

we have

$$\nu(q)\langle 0|\rho_{-q}\rho_q|0\rangle = - \frac{V\hbar}{\pi} \int_0^\infty \text{Im} \left\{\frac{1}{\epsilon(q, \omega)}\right\} d\omega \; . \quad (3.3.11)$$

Substituting the above equation into (3.3.9), we obtain

$$\langle 0|\mathscr{H}'|0\rangle = - \sum_q \left\{\frac{\hbar}{2\pi} \int_0^\infty \text{Im} \left[\frac{1}{\epsilon(q, \omega)}\right] d\omega + \frac{\nu(q)}{2} n\right\} \quad (3.3.12)$$

where n stands for the average number density N/V of the electron gas.

In this way, the average value of the Coulomb interaction is shown to be closely related to the dielectric constant. However, it is necessary to know the average value of kinetic energy, in order to compute the total energy of the system. This average value is unfortunately not directly connected with the dielectric constant. However, in an indirect way the relation between total energy and dielectric constant can be derived. Since the Feynman theorem is convenient to show it, let us discuss this theorem in the following. Suppose that a given Hamiltonian depends on a parameter λ. Let it be $\mathscr{H}(\lambda)$. The wave function and the energy eigenvalue should depend on λ, so that we write them $\psi(\lambda)$ and $E(\lambda)$, respectively. Thus, the Schrödinger equation is written as

$$\mathscr{H}(\lambda) \psi(\lambda) = E(\lambda) \psi(\lambda) \; .$$

If we differentiate this equation by λ, we have

$$\frac{\partial \mathcal{H}}{\partial \lambda} \psi + \mathcal{H} \frac{\partial \psi}{\partial \lambda} = \frac{\partial E}{\partial \lambda} \psi + E \frac{\partial \psi}{\partial \lambda}$$

whence it follows that[6]

$$\int \psi^* \frac{\partial \mathcal{H}}{\partial \lambda} \psi d\tau + \int \psi^* \mathcal{H} \frac{\partial \psi}{\partial \lambda} d\tau = \frac{\partial E}{\partial \lambda} + E \int \psi^* \frac{\partial \psi}{\partial \lambda} d\tau \ .$$

Here we have assumed that ψ is normalized. By the use of the following relation which comes from the fact that $\mathcal{H}$ is the Hermitian operator:

$$\int \psi^* \mathcal{H} \frac{\partial \psi}{\partial \lambda} d\tau = \left(\int \frac{\partial \psi^*}{\partial \lambda} \mathcal{H} \psi d\tau \right)^* = E \int \psi^* \frac{\partial \psi}{\partial \lambda} d\tau \ ,$$

the Feynman theorem

$$\frac{\partial E}{\partial \lambda} = \int \psi^* \frac{\partial \mathcal{H}}{\partial \lambda} \psi d\tau \tag{3.3.13}$$

is derived.

In applying the theorem to the problem of the electron gas, it is convenient to write the Hamiltonian as $\mathcal{H} = \mathcal{H}_0 + e^2 \mathcal{H}''$, where $\mathcal{H}_0$ represents the kinetic energy, $e^2 \mathcal{H}''$ the Coulomb interaction. If we choose e^2 as the parameter λ, the right-hand side of (3.3.13) is just equal to (3.3.12) devided by e^2. If $\lambda = 0$, i.e., when there is no Coulomb interaction, we have from (3.1.11)

$$E(0) = \frac{3}{5} N \frac{\hbar^2 k_F^2}{2m} \ .$$

Therefore, if the right-hand side of (3.3.12) is known as a function of e^2, by integrating (3.3.13) with respect to e^2 under the initial condition given above, the total energy can in principle be calculated.

We have discussed so far the relationship between total energy and dielectric constant. However, the calculation of correlation energy is not always carried out by the above procedure. Historically speaking, the correlation energy was calculated by taking a partial sum of the higher-order terms in the perturbation expansion. We have already mentioned the first-order perturbation term in Sect. 3.1. If we improve the approximation and proceed to the second-order term, there appears a part which diverges to infinity. Similarly, there appears divergence in the higher-order terms and the degree of divergence depends on intermediate states. It turns out that the sum of the most-divergent terms leads to a finite

---

[6] The integration by τ implies both the integrals over the spatial coordinates of all particles and the sums over their spin coordinates.

result. This means that the sum of something like $\infty - \infty + \infty - \infty + \ldots$ yields a finite result. A similar situation exists also in the classical electron gas. In this case, the summation over all the ring clusters is employed to overcome the divergence difficulty. What is mentioned above is a quantum-mechanical version of this classical method. In any case, if we express the result in the same form as (3.1.16), it reads

$$\varepsilon = \frac{2.21}{r_s^2} - \frac{0.916}{r_s} + 0.0622 \ln r_s - 0.096 + \ldots \ . \tag{3.3.14}$$

The third term and what follows are the correlation energy. It should be noted that the term $\ln r_s$ having the logarithmic singularity appears in the above equation. This implies a divergence of perturbation expansion. For, if each term in the perturbation series were finite, the correlation energy would be a power series in r_s. Now, (3.3.14) leads to a good result at the high density limit ($r_s \ll 1$). In an actual metal (for example alkali metals) r_s takes the value $2 \sim 6$, so that it is impossible to apply the above result in its original form. To derive the equation for the correlation energy applicable to a real system is a very difficult problem. However, we would like to add that several attempts are being made at present to deal with this problem.

3.3.4 Dynamic Structure Factors

We have already remarked in Sect. 1.6 that the dynamic structure factor is important in discussing the inelastic scattering of a neutron. It is also closely related to the dielectric constant. To see the relation between them, as in Sect. 1.5 let us use an integral representation for the δ function:

$$\delta(x) = \frac{1}{2\pi} \int_{-\infty}^{\infty} \exp(ixt)\, dt = \frac{1}{2\pi} \int_{-\infty}^{\infty} \exp(-ixt)\, dt \ .$$

Since $\omega_{n0} = (\varepsilon_n - \varepsilon_0)/\hbar$ by definition, we have

$$\sum_n |\langle n|\rho_q|0\rangle|^2 \delta(\omega + \omega_{n0}) = \frac{1}{2\pi} \sum_n \int_{-\infty}^{\infty} \langle 0|\rho_{-q}|n\rangle\langle n|\rho_q|0\rangle \exp\Big[-i\omega t$$

$$-\frac{i}{\hbar}(\varepsilon_n - \varepsilon_0)\, t\Big] = \frac{1}{2\pi} \int_{-\infty}^{\infty} \exp(-i\omega t)\, \langle 0|\rho_{-q}(t)\rho_q|0\rangle\, dt \ .$$

In the same way

$$\sum_n |\langle n|\rho_q|0\rangle|^2 \delta(\omega - \omega_{n0}) = \frac{1}{2\pi} \int_{-\infty}^{\infty} \langle 0|\rho_{-q}(t)\rho_q|0\rangle \exp(i\omega t)\, dt \ .$$

Therefore, from (3.3.10) we obtain

$$\mathrm{Im}\left\{\frac{1}{\epsilon(\boldsymbol{q},\omega)}\right\} = \frac{v(\boldsymbol{q})}{2V\hbar}\int_{-\infty}^{\infty}\langle 0|\rho_{-\boldsymbol{q}}(t)\rho_{\boldsymbol{q}}|0\rangle\,[\exp(-i\omega t) - \exp(i\omega t)]\,dt \ . \quad (3.3.15)$$

As we have mentioned in Sect. 1.6, if we put $S(\boldsymbol{q}; t) = \langle 0|\rho_{-\boldsymbol{q}}(t)\rho_{\boldsymbol{q}}|0\rangle$ [see (1.6.3)], its Fourier transform $S(\boldsymbol{q}, \omega)$ defined by

$$S(\boldsymbol{q}, \omega) = \int_{-\infty}^{\infty}\frac{dt}{2\pi\hbar}\,S(\boldsymbol{q}; t)\exp(i\omega t)$$

is the dynamic structure factor. By means of the above equation, (3.3.15) is expressed as

$$\mathrm{Im}\left\{\frac{1}{\epsilon(\boldsymbol{q}, \omega)}\right\} = \frac{\pi v(\boldsymbol{q})}{V}\,[S(\boldsymbol{q}, -\omega) - S(\boldsymbol{q}, \omega)] \ . \quad (3.3.16)$$

From this equation it is seen that the quantity $\mathrm{Im}\{1/\epsilon(\boldsymbol{q}, \omega)\}$ which is necessary to calculate $\langle 0|\mathscr{H}'|0\rangle$ is also derived from the dynamic structure factor. That is, the stimulus such as neutron scattering describes not only the dynamic properties of the system but also the static properties. We have already mentioned in Sect. 3.2 that we are able to obtain important information about elementary excitations through the relationship between stimulus and response. In addition to this, the same method yields information about the ground state.

3.4 Individual Excitation and Collective Excitation

Elementary excitations which can be excited in the Fermi liquid are roughly classified into two classes: individual excitation and collective excitation. The former reflects the individuality of each particle and the latter is connected with the assembly of all particles. In the system of free particles, only the former excitation takes place since each particle moves independently. In other words, the existence of interactions between particles is inevitable in having the collective excitation. An explicit form of collective excitation depends on the property of interaction, especially on whether it is of long range or of short range. In this section, we will discuss mainly the collective excitation. However, to develop a unified treatment for a general Fermi liquid, it is convenient to generalize the argument for the electon gas mentioned in a previous section. Let us start with this point.

3.4.1 Density Fluctuation Due to the External Field

In the case of the electron gas, it is intuitively understandable to apply an external field since the constituent particle has the charge. However, if the particle has no charge as in the case of liquid ^{3}He, it is impossible to express the external field in such a way that we apply the electric field from the outside. Nevertheless, the method discussed in Sect. 3.3.1 can be extended to a general system in the following way. Suppose we add conceptually the same particles from the outside as the constituent particles of the Fermi liquid under consideration. We assume as in (3.3.1) that the number density of added particles has the form

$$R_q \exp(i\mathbf{q} \cdot \mathbf{r} - i\omega t + \delta t) \; .$$

These particles interact with particles in the Fermi liquid, so that they will behave as the external field. That is, the Hamiltonian $\mathcal{H}'(t)$ of the external field is expressed as

$$\mathcal{H}'(t) = \sum_i \int v(\mathbf{r} - \mathbf{r}_i) \, R_q \exp(i\mathbf{q} \cdot \mathbf{r} - i\omega t + \delta t) \, d\mathbf{r} \; ,$$

where v represents the interaction potential between particles. By the use of the Fourier transform, we have $B = v(\mathbf{q})R_q \rho_{-q}$ corresponding to (3.3.3). On the other hand, the fluctuation of number density induced by the external field is written by

$$\frac{1}{V} v(\mathbf{q}) \, R_q G_r(\mathbf{q}, \omega) \, \exp(i\mathbf{q} \cdot \mathbf{r} - i\omega t)$$

as in Sect. 3.3. A sum of the above equation with the number density of added particles yields the net number density in the presence of the external field. Therefore, if we define $1/\epsilon(\mathbf{q}, \omega)$ analogously to the electron gas as the ratio of net number density to that of added particles, we obtain the same equation as (3.3.6)

$$\frac{1}{\epsilon(\mathbf{q}, \omega)} = 1 + \frac{v(\mathbf{q})}{V} \, G_r(\mathbf{q}, \omega) \; . \tag{3.4.1}$$

As previously, if $\epsilon(\mathbf{q}, \omega) = 0$, the density fluctuation can occur in the system even if we do not add the particles from the outside. Thus, it is supposed to correspond to the elementary excitation characteristic of the system.

3.4.2 The Zeroth-Order Approximation for the Retarded Green's Function

In order to find the zeros of $\epsilon(\mathbf{q}, \omega)$ by means of (3.4.1), it is of course necessary to derive an explicit form for $G_r(\mathbf{q}, \omega)$. Although the retarded Green's function of interest is given by (3.3.5), i.e.

$$G_r[\mathbf{q}, t] = -\frac{i}{\hbar} \, \theta(t) \langle [\rho_q(t), \rho_{-q}] \rangle \; , \tag{3.4.2}$$

it is convenient, as was mentioned in Sect. 3.2, to introduce the temperature Green's function corresponding to (3.4.2):

$$\mathscr{G}[q, \tau] = - \langle T_\tau \rho_q(\tau) \, \rho_{-q} \rangle \; . \tag{3.4.3}$$

To express the above equation in a form of second quantization, we note that the number density of particle is written as

$$\rho(r) = \sum_s \psi^\dagger(x) \, \psi(x)$$

in terms of field operators. Remember that x stands for both r and s. Substitution of (3.1.2) leads to

$$\rho(r) = \frac{1}{V} \sum_{k_1 k_2 \sigma} a^\dagger_{k_1 \sigma} a_{k_2 \sigma} \exp[i(k_2 - k_1) \cdot r] \; ,$$

so that by means of the Fourier transform, we obtain

$$\rho_q = \int \rho(r) \exp(-iq \cdot r) \, dr = \sum_{k\sigma} a^\dagger_{k-q, \sigma} a_{k\sigma} \; . \tag{3.4.4}$$

Substituting the above equation into (3.4.3), we find

$$\mathscr{G}[q, \tau] = - \sum_{kk'\sigma\sigma'} \langle T_\tau a^\dagger_{k-q, \sigma}(\tau) a_{k\sigma}(\tau) a^\dagger_{k'+q, \sigma'} a_{k'\sigma'} \rangle \; . \tag{3.4.5}$$

So far we have been rewriting the equation and no approximation has been involved. However, we have to rely on some approximation procedure in order to calculate (3.4.5) explicitly. The most systematic way to do so is perturbation expansion, but we will content ourselves with the lowest-order approximation here. Even under this simplest approximation, some characteristic features of elementary excitation in the Fermi liquid will be revealed.

First of all, we approximate the statistical-mechanical average $\langle \; \rangle$ in (3.4.5) by the average $\langle \; \rangle_0$ for free particles. If we consider the grand canonical distribution, the Hamiltonian minus μN under this approximation is written as

$$\mathscr{H}_0 = \sum_s \varepsilon_s a^\dagger_s a_s, \quad \varepsilon_s = \frac{\hbar^2 k^2}{2m} - \mu \tag{3.4.6}$$

where s means both wave vector k and spin state σ. Next, we approximate the Hamiltonian appearing in the defining equation for $a_r(\tau)$ by that of free particles. This amounts to

$$a_r(\tau) = \exp(\tau \mathscr{H}_0) \, a_r \exp(-\tau \mathscr{H}_0) \; .$$

Differentiating the above equation by τ and using the anticommutation relations for Fermi operators, we have

$$\frac{da_r(\tau)}{d\tau} = \exp(\tau \mathscr{H}_0) \, (\mathscr{H}_0 a_r - a_r \mathscr{H}_0) \exp(-\tau \mathscr{H}_0)$$

$$\mathscr{H}_0 a_r - a_r \mathscr{H}_0 = \sum_s \varepsilon_s (a^\dagger_s a_s a_r - a_r a^\dagger_s a_s) = - \varepsilon_r a_r$$

whence it follows that

$$\frac{da_r(\tau)}{d\tau} = - \varepsilon_r a_r(\tau) \ .$$

Solving the above differential equation under the initial condition that $a_r(\tau) = a_r$ at $\tau = 0$, we find

$$a_r(\tau) = \exp(-\varepsilon_r \tau) \, a_r \ .$$

In the same way, the relation

$$a_r^\dagger(\tau) = \exp(\varepsilon_r \tau) \, a_r^\dagger$$

is derived.

The above approximation corresponds to the zeroth order in the perturbation expansion, so that to clarify it we will add the suffix (0) in the following. Then, from (3.4.5) for $\tau > 0$ we have

$$\mathscr{G}^{(0)}[\boldsymbol{q}, \tau] = - \sum_{kk'\sigma\sigma'} \exp[(\varepsilon_{k-q} - \varepsilon_k)\tau] \, \langle a_{k-q,\sigma}^\dagger a_{k\sigma} a_{k'+q,\sigma'}^\dagger a_{k'\sigma'} \rangle_0 \ . \qquad (3.4.7)$$

To calculate $\langle \ \rangle_0$ in this equation we may apply the Bloch–de Dominicis theorem. Thus, choose pairs among creation and annihilation operators, make a product of their average values and sum up over all possible ways of pairings. However, we must associate a $-$ sign if the number of permutations of operators is odd in making pairs and a $+$ sign if it is even. In the present case, if $\boldsymbol{q} \neq 0$ the $\langle a_{k-q,\sigma}^\dagger a_{k\sigma} \rangle_0$ vanishes, so that only remaining pairs are: the pair of $a_{k-q,\sigma}^\dagger$ with $a_{k'\sigma'}$ and the pair of $a_{k\sigma}$ with $a_{k'+q,\sigma'}^\dagger$. In order that the average values of these pairs do not vanish, the conditions $k - q = k'$ and $\sigma = \sigma'$ are required. Thus, if we omit the spin state σ, it follows that

$$\langle a_{k'}^\dagger a_{k'+q} a_{k'+q}^\dagger a_{k'} \rangle_0 = \langle a_{k'}^\dagger a_{k'} \rangle_0 \langle a_{k'+q} a_{k'+q}^\dagger \rangle_0$$
$$= f_{k'}(1 - f_{k'+q})$$

where f_k is the Fermi distribution function given by $f_k = [\exp(\beta\varepsilon_k) + 1]^{-1}$. In this way, if we use the variable $\boldsymbol{k}$ instead of $\boldsymbol{k}'$, (3.4.7) is expressed as·

$$\mathscr{G}^{(0)}[\boldsymbol{q}, \tau] = - \sum_{k\sigma} \exp[(\varepsilon_k - \varepsilon_{k+q}) \, \tau] f_k (1 - f_{k+q}) \ . \qquad (3.4.8)$$

Now, according to (3.2.21), the Fourier coefficient $\mathscr{G}(\boldsymbol{q}, \mathrm{i}\nu_l)$ of $\mathscr{G}[\boldsymbol{q}, \tau]$ is given by

$$\mathscr{G}(\boldsymbol{q}, \mathrm{i}\nu_l) = \int_0^\beta \mathscr{G}[\boldsymbol{q}, \tau] \exp(\mathrm{i}\nu_l \tau) \, d\tau \ .$$

Substituting (3.4.8) into the right-hand side in this equation and integrating by τ, we obtain

$$\mathcal{G}^{(0)}(q, i\nu_l) = - \sum_{k\sigma} \frac{f_k(1 - f_{k+q})\{\exp[\beta(\varepsilon_k - \varepsilon_{k+q})] - 1\}}{\varepsilon_k - \varepsilon_{k+q} + i\nu_l} \; .$$

If we substitute the equations for Fermi distribution functions, it turns out that the numerator in this equation is equal to

$$(1 - f_k) f_{k+q} - f_k(1 - f_{k+q}) = f_{k+q} - f_k \; .$$

Here, let us apply the result of analytic continuation and extend $i\nu_l$ to a general complex number z. Since according to (3.2.23) $G_r(q, \omega) = \mathcal{G}(q, \hbar\omega + i\delta)$ holds if we put $z = \hbar\omega + i\delta$, the zeroth-order approximation for $G_r(q, \omega)$ is expressed as follows

$$G_r^{(0)}(q, \omega) = \sum_{k\sigma} \frac{f_{k+q} - f_k}{\varepsilon_{k+q} - \varepsilon_k - \hbar\omega - i\delta} \; . \tag{3.4.9}$$

3.4.3 Individual Excitation and Collective Excitation

If we replace $G_r(q, \omega)$ by its zeroth order term in (3.4.1) and assume that the correction term is small as compared to 1, we have

$$\epsilon(q, \omega) = \left[1 + \frac{\nu(q)}{V} G_r^{(0)}(q, \omega)\right]^{-1} \approx 1 - \frac{\nu(q)}{V} G_r^{(0)}(q, \omega) \; .$$

To be more precise, the situation is as follows. If we extend the calculation mentioned in Sect. 3.4.2 to higher-order terms in the perturbation expansion, it is shown that they include the terms $X^2, X^3, \ldots$, where $X = \nu(q)G_r^{(0)}(q, \omega)/V$. Summing up these terms, we have $1/\epsilon(q, \omega) = 1 + X + X^2 + X^3 + \ldots = 1/(1 - X)$ so that we are led to the above result. In the case of the electron gas or liquid ^{3}He, the particle has the spin 1/2. Therefore, taking account of the factor 2 coming from the freedom of spin and putting $\epsilon(q, \omega) = 0$, we obtain

$$\frac{2\nu(q)}{V} \sum_k \frac{f_{k+q} - f_k}{\varepsilon_{k+q} - \varepsilon_k - \hbar\omega} = 1 \tag{3.4.10}$$

as the equation to determine the angular frequency of elementary excitation (we put $\delta = 0$).

If we solve (3.4.10), ω is determined as a function of q so that the dispersion relation $\omega = \omega(q)$ will be derived. Before solving this equation quantitatively, let us consider a qualitative feature of the dispersion relations. Since we are dealing with the excited states, it follows that $\omega > 0$. If we consider the point where the denominator in (3.4.10) vanishes, it leads to $\hbar\omega = \varepsilon_{k+q} - \varepsilon_k$ whence we have $\varepsilon_{k+q} > \varepsilon_k$. Now consider the term $f_{k+q} - f_k$ in (3.4.10). In the case of absolute zero (or $k_B T \ll \mu$), this term vanishes if the magnitudes of both $k + q$ and k are larger or smaller than the Fermi wave number k_F. Therefore, the conditions $|k + q| > k_F$ and $|k| < k_F$ are obtained and we have $f_{k+q} - f_k = -1$ in this case. Noting this point we assume that the size of system is finite Then, ε_k takes discrete values so that if we plot the left-hand side in (3.4.10) as a function of $\hbar\omega$, the result will be as shown in Fig. 3.8. Note that this quantity is proportional to $1/\omega$ in the limit $\omega \to \infty$. Taking the value 1 on the ordinate, drawing the horizontal line across it and looking for points of intersection with lines as depicted in Fig. 3.8, we can obtain the values of ω.

There are two kinds of solutions in this case. If we let the volume of system tend to infinity, the interval on the $\hbar\omega$ axis in Fig. 3.8 becomes infinitesimally small, so that $\hbar\omega = \varepsilon_{k+q} - \varepsilon_k$ is actually the solution of (3.4.10). This leads to the individual excitation of the system. That is, the above excitation energy is associated with a process in which the particle with k inside the Fermi surface is brought to the state $k + q$ outside the Fermi surface, as is shown in Fig. 3.9. On the other hand, the point of intersection with a curve decreasing as $1/\omega$ describes an excitation different from individual excitation. This is the collective excitation. It is called plasma oscillation in the case of an electron gas and zero sound in the case of liquid ^{3}He; the details will be discussed in Sects. 3.4.4, 5.

If we substitute (3.4.6) into the equation for individual excitation, the chemical potentials cancel out and we are left with

$$\omega = \frac{\hbar}{m} k \cdot q + \frac{\hbar}{2m} q^2 .$$

For an infinite volume of the system, ω takes a continuous distribution for fixed q, since k varies continuously. However, it is shown that

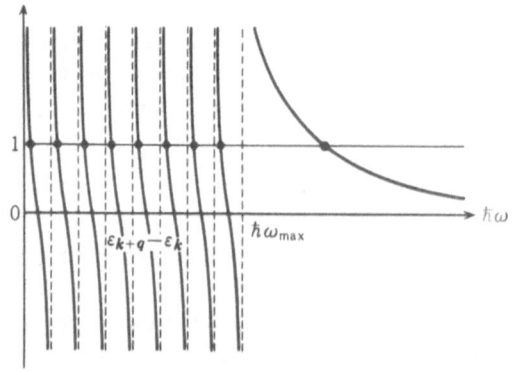

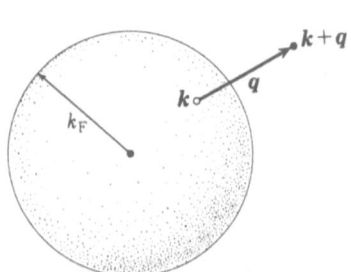

Fig. 3.8. The ordinate represents the left-hand side in (3.4.10)

Fig. 3.9. Individual excitations

$0 \leq \omega \leq \omega_{\max}(q)$ in this case. In other words, there exists an upper bound to ω. In order to derive it let us take the following procedure. To make the right-hand side in the above equation as large as possible, we let k and q be parallel. Furthermore, noting an inequality $k \leq k_F$, we obtain

$$\omega_{\max}(q) = \frac{\hbar}{m} k_F q + \frac{\hbar q^2}{2m} = v_F q + \frac{\hbar q^2}{2m} \tag{3.4.11}$$

with v_F the Fermi velocity defined by $v_F = \hbar k_F/m$. From the foregoing discussion, it is seen that the dispersion relation for individual excitation is represented by the continuous spectrum (shaded part in Fig. 3.10). Here, we add the dispersion relation of plasma oscillation or zero sound for the sake of reference. As discussed in Chap. 1, in the case of a phonon, only a finite number of elementary excitations are possible for a given wave number. On the other hand, it is a characteristic feature of Fermi liquid that a continuously infinite number of elementary excitations are allowed for a given wave number. As is seen from Fig. 3.8, if $v(q) < 0$, i.e., for the attractive potential, there exists no solution corresponding to the collective excitation. In this case, there appear the Cooper pairs in the Fermi liquid and the system attains a superconducting state.

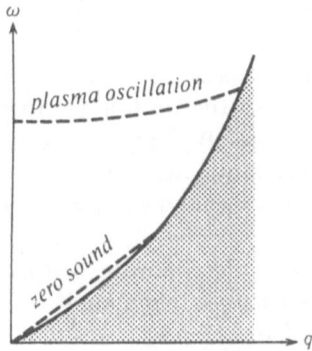

Fig. 3.10. Individual excitations and collective excitation

3.4.4 Plasma Oscillation

In the electron gas, ω of collective excitation determined from (3.4.10) approaches a constant independent of temperature in the limit $q \to 0$. The collective oscillation in this case is called plasma oscillation as was mentioned in Sect. 2.2. If we put $k + q \to -k'$ in the term including f_{k+q} in (3.4.10), make a change of variable from k to k' and after that put k' a new k, we find after a simple calculation

$$\frac{2v(q)}{V} \sum_k \frac{2(\varepsilon_{k+q} - \varepsilon_k)}{(\hbar\omega)^2 - (\varepsilon_{k+q} - \varepsilon_k)^2} f_k = 1. \tag{3.4.12}$$

In this equation, the relation

$$\varepsilon_{k+q} - \varepsilon_k = \frac{\hbar^2}{2m}(2k \cdot q + q^2) \tag{3.4.13}$$

holds. However, as we will see later on, ω is finite in the limit $q \to 0$, so that $(\varepsilon_{k+q} - \varepsilon_k)^2$ in the denominator in (3.4.12) can be neglected as compared to $(\hbar\omega)^2$ in this limit. Thus, the relation

$$\frac{2v(q)}{V} \sum_k \frac{(\hbar^2/m)(2k \cdot q + q^2)}{(\hbar\omega)^2} f_k = 1 \tag{3.4.14}$$

is valid for $q \to 0$. In this equation the term including $k \cdot q$ vanishes when we take a sum over k since the multiplying factor is spherically symmetric with respect to k. Substituting $v(q) = 4\pi e^2/q^2$ for the Coulomb interaction and using

$$2\sum_k f_k = N \;,$$

we obtain from (3.4.14)

$$\omega^2 = \frac{4\pi n e^2}{m} = \omega_p^2 \;. \tag{3.4.15}$$

The ω_p is the plasma frequency. For a usual alkali metal $n \approx 10^{23} \text{cm}^{-3}$, $m \approx 10^{-27}$g and $e^2 \approx 10^{-19}$ in cgs electrostatic units, so that it follows that $\omega_p \approx 10^{16}\text{s}^{-1}$. If we convert this to energy, by using $\hbar \approx 10^{-27}$ erg·s we find $\hbar\omega_p \approx 10^{-11}$ erg ≈ 10 eV. Thus, the elementary excitation associated with plasma oscillation, which is often called the plasmon, has an energy of the order of eV.

In order to observe plasma oscillation experimentally, the following method is usually employed. We prepare a thin film of a given sample, irradiate it with an electron beam with a known energy of the order of 10 keV and measure the energy of transmitted electron beam. The incident electron beam excites a plasmon in the sample and therefore suffers an energy loss. Conversely, by measuring the energy loss we can get the energy of the plasmon or the plasma frequency. By means of this method, the plasma frequencies of several metals, semiconductors, alloys, alkali halides, etc., have been experimentally measured.

Although we considered the limit $q \to 0$ in deriving (3.4.15), it is possible to obtain the results exact up to the order of q^2. For this purpose, assuming that $(\varepsilon_{k+q} - \varepsilon_k)^2$ is sufficiently small as compared to $(\hbar\omega)^2$ in (3.4.12), we make an expansion in the following way

$$\frac{2v(q)}{V} \sum_k \left[\frac{2(\varepsilon_{k+q} - \varepsilon_k)}{(\hbar\omega)^2} + \frac{2(\varepsilon_{k+q} - \varepsilon_k)^3}{(\hbar\omega)^4} + \ldots\right] f_k = 1.$$

Substituting (3.4.13) into the above equation and using the symmetry with respect to k, we obtain

$$\omega^2 = \frac{8\pi e^2}{mV} \sum_k \left[1 + \frac{3\hbar^2(k\cdot q)^2}{m^2\omega^2} + O(q^4)\right]f_k .$$

Furthermore, by means of the relations:

$$\sum_k (k\cdot q)^2 f_k = \sum_k (k_x^2 q_x^2 + k_y^2 q_y^2 + k_z^2 q_z^2)f_k = \frac{q^2}{3}\sum_k k^2 f_k ,$$

$$2\sum_k k^2 f_k = \frac{3k_F^2}{5} N ,$$

ω^2 is expressed as

$$\omega^2 = \omega_p^2 + \frac{3\hbar^2 k_F^2}{5m^2} q^2 .$$

The term corresponding to q^2 in this equation is also observed experimentally.

Let us now turn our attention to the physical meaning of (3.4.15) for the plasma frequency. Since this equation does not contain Planck's constant, the same results should be derived from the viewpoint of classical mechanics. Thus, we will discuss the problem on the basis of Newton's equation of motion. First, we differentiate the relation $\rho_q = \sum \exp(-iq\cdot r_j)$ by time to obtain

$$\dot{\rho}_q = -i\sum_j q\cdot v_j \exp(-iq\cdot r_j)$$

where v_j is the velocity of the jth electron. The Coulomb potential between electrons is

$$U = \frac{1}{2}\sum_{j\neq k} \frac{e^2}{|r_j - r_k|} = \frac{1}{2V}\sum_{j\neq k,q'} \frac{4\pi e^2}{q'^2} \exp[iq'\cdot(r_j - r_k)] .$$

Therefore, differentiating the equation for $\dot{\rho}_q$ by t again and using Newton's equation $m\dot{v}_j = -\partial U/\partial r_j$, we obtain

$$\ddot{\rho}_q = -\sum_j (q\cdot v_j)^2 \exp(-iq\cdot r_j) - \sum_{jkq'} \frac{4\pi e^2}{mq'^2 V} q\cdot q' \exp[iq'\cdot(r_j - r_k) - iq\cdot r_j] .$$

Let us observe a term $\exp[i(q' - q)\cdot r_j]$ which appears on the second term in the above equation. If $q' \neq q$ when we sum up over j, it is expected that these exponential functions cancel out with each other in a suitable way leading to the result that the sum vanishes. To assume that this is really true is called random phase approximation or briefly RPA. Under this approximation, taking the limit $q \to 0$ we have

$$\ddot{\rho}_q = -\omega_{\mathrm{p}}^2 \rho_q \ .$$

That is, ρ_q oscillates harmonically with the angular frequency ω_{p}. As is seen from the above procedure, the plasma oscillation exists even in the classical electron gas. Historically, the plasma oscillation in an ionized gas had been discovered long before that of electrons in a solid was observed.

3.4.5 Zero Sound

In a system with a short-range force as in liquid ^{3}He, the dispersion relation of collective excitation has a property different from the plasma oscillation. As we mentioned in Sect. 3.1.5, the $\nu(q)$ for a short-range force is considered to be a constant independent of q. That is, according to (3.1.17) we can assume that $\nu(q) = \nu$. As a result, it follows from (3.4.14) that $\omega \propto q$. Thus, the collective oscillation exhibits a dispersion relation characteristic of an acoustic phonon. This is called "zero sound". In the following, we will study its dispersion relation in some detail. We first note that $\hbar\omega$ and $\varepsilon_{k+q} - \varepsilon_k$ in the denominator in (3.4. 12) are of the same order of magnitude in the present case. Because of this situation, we cannot expand the denominator as in the case of plasma oscillation. Therefore, we will return to (3.4.10) and let it be a starting-point of the following discussion.

If we make the Taylor expansion of f_{k+q} with respect to q in (3.4.10), we have

$$f_{k+q} = f_k + \frac{\partial f_k}{\partial k} \cdot q + O(q^2) \ .$$

For simplicity, if we consider the case of absolute zero, it follows that $\partial f_k/\partial \varepsilon_k = -\delta(\varepsilon_k)$. Therefore

$$\frac{\partial f_k}{\partial k} = \frac{\partial f_k}{\partial \varepsilon_k} \frac{\partial \varepsilon_k}{\partial k} = -\frac{\hbar^2 k}{m} \delta\left(\frac{\hbar^2 k^2}{2m} - \mu\right) \ .$$

Since we can put $\mu = \hbar^2 k_{\mathrm{F}}^2/2m$ at absolute zero, by the use of a property $\delta(ax) = \delta(x)/a$ for the δ function, we find

$$f_{k+q} - f_k = -2k \cdot q\delta(k^2 - k_{\mathrm{F}}^2)$$

neglecting the term of the order of q^2. Substituting the above equation into (3.4.10) and neglecting the term of the order of q^2 in (3.4.13), we obtain

$$\frac{2\nu}{V} \sum_k \frac{2k \cdot q\delta(k^2 - k_{\mathrm{F}}^2)}{\hbar\omega - (\hbar^2 k \cdot q/m)} = 1 \tag{3.4.16}$$

as the equation to determine ω. In calculating this equation, we transform the sum over k into the integral and introduce the polar coordinates taking the q

direction as the z axis. If we let the angle between $\boldsymbol{q}$ and $\boldsymbol{k}$ be θ, we have $\boldsymbol{k} \cdot \boldsymbol{q} = kq \cos\theta$. Also, the relation $\delta(k^2 - k_F^2) = \delta(k - k_F)/2k_F$ holds. If we take account of these relations and use an integral variable t defined by $t = \cos\theta$, (3.4.16) is expressed as

$$\frac{2vk_F^2}{(2\pi)^2} \int_{-1}^{1} \frac{qt}{\hbar\omega - (\hbar^2 k_F qt/m)} \, dt = 1 \ . \tag{3.4.17}$$

If we define γ by

$$\gamma = \frac{\omega}{qv_F} \tag{3.4.18}$$

(v_F = Fermi velocity = $\hbar k_F/m$) and introduce the state density $N(0)$ at the Fermi surface per unit volume given by

$$N(0) = \frac{mk_F}{\pi^2\hbar^2} \ ,$$

(3.4.17) is written as

$$\frac{vN(0)}{2} \int_{-1}^{1} \frac{t}{\gamma - t} \, dt = 1 \ .$$

This integral is calculated by elementary integration. In this way, as the equation to determine γ, we are led to

$$\frac{\gamma}{2} \ln\frac{\gamma + 1}{\gamma - 1} - 1 = \frac{1}{vN(0)} \ . \tag{3.4.19}$$

If we write (3.4.18) as $\omega = \gamma v_F q$, we see that γv_F represents the sound velocity of zero sound. As a function of γ, the left-hand side in (3.4.19) is graphically shown by a curve in Fig. 3.11. Taking the value $1/vN(0)$ on the ordinate and

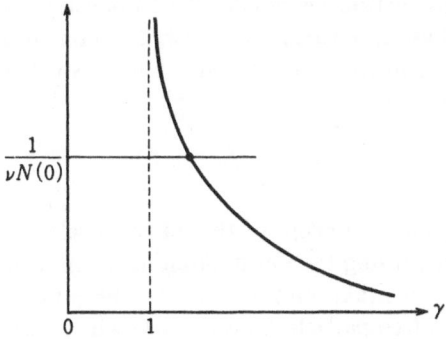

Fig. 3.11. Graphical solution of (3.4.19) (the ordinate represents the left-hand side of this equation)

finding the point of intersection with the curve as shown in the figure, we obtain the solution of (3.4.19). As can be seen from this method, γ is larger than 1 as long as ν is positive and nonzero. Therefore, as shown in Fig. 3.10, when q is small the zero sound always appears in the upper part of the continuous spectra of individual excitations. Also, as in the electron gas, there is no collective excitation if $\nu < 0$.

It should be noted that (3.4.19) has a form quite similar to the one obtained by Landau. The concept of zero sound was first proposed by Landau on the basis of a rather phenomenological Fermi-liquid theory. What we have discussed above is supposed to be an explanation of its existence on the basis of first principle, although the approximations involved are quite rough.

3.5 General Property of Fermi Liquid

To summarize what we have been discussing so far in this chapter, it may be stated roughly as follows. 1) By applying a suitable external field to the Fermi liquid and considering its response, useful information is obtained about elementary excitations existing in the system. 2) In examining the response, it is convenient to use the Green's functions as a mathematical tool.

In actual calculation of the Green's functions, however, we have used a simple perturbation calculation as discussed in a previous section. As we have remarked several times so far, it is not applicable to the real physical system such as electrons in metals or liquid ^{3}He. So how to deal with the actual Fermi liquid becomes an important problem. One of the ways is to rely on Landau's Fermi-liquid theory. If we consider that parameters appearing in this theory (Landau parameters) are quantities to be determined from experiment, it may be posisble to treat experimental data in such a way that the various observed quantities are interrelated with each other. Another way may be to improve the approximations of perturbation expansion step by step. Then, the following questions will arise. Are there any general properties common to all the perturbation terms even if they are not calculable in explicit forms? In what physical properties are they reflected? Is there really some ideal Fermi liquid to which a simple perturbation theory is applicable? In this section we would like to develop an argument concentrating on answering these questions. However, to enter into great details of problems seems to be beyond the scope of this volume, so that we will briefly sketch the outlines of the results.

3.5.1 Energy of the Quasiparticle

In Landau's Fermi-liquid theory, the physical concept of the quasiparticle has an important meaning. Each particle constituting the Fermi liquid has a motion interacting with all other particles, but if we take into account all these interactions the particle behaves as if it were a free particle. This particle which puts on, so to speak, the dress of interactions is the quasiparticle and plays an important role in describing elementary excitations of system. Needless to say, the energy of free particle with mass m and wave number k is expressed as $\hbar^2 k^2/2m$. Then, how should we determine the energy of the quasiparticle?

To study the above problem, let us consider the distribution of free particles in k space at absolute zero. If we denote the average number of particles with wave number k by $\langle n_k \rangle$, the following equations hold in this case:

$$\langle n_k \rangle = \begin{cases} 1 & (\varepsilon_k < 0) \\ 0 & (\varepsilon_k > 0) \end{cases} \tag{3.5.1}$$

where $\varepsilon_k = (\hbar^2 k^2/2m) - \mu$. We note that the above equation is expressed as

$$\langle n_k \rangle = \frac{1}{2\pi i} \int_{-i\infty}^{i\infty} \frac{\exp(z0^+)}{z - \varepsilon_k} dz \tag{3.5.2}$$

in the form of integral on the complex plane. Here 0^+ implies a positive infinitesimal which we let go to zero after the calculation is completed. Because of the factor $\exp(z0^+)$, the integral does not change even if, as shown in Fig. 3.12, we add the path of integration which is a semicircle with sufficiently large radius R. If $\varepsilon_k > 0$, since there are no singular points of integrand in the domain enclosed by the path, the integral vanishes according to the Cauchy theorem. On the other hand, if $\varepsilon_k < 0$, by means of the residue theorem we see that the right-hand side in (3.5.2) is equal to 1. Thus, the equivalence of (3.5.2) with (3.5.1) is proved.

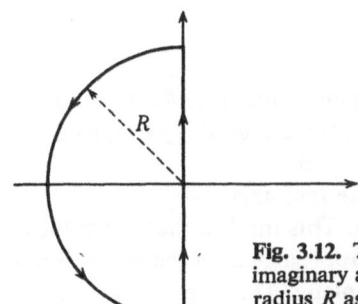

Fig. 3.12. The path of integration in (3.5.2) is the one along the imaginary axis on the complex plane. If we add the semicircle with radius R as shown in the figure, take the path indicated by arrow, and let $R \to \infty$, then the integral (3.5.2) on the complex plane will be calculated

It is proved that for a system in which particles have interactions the following relation holds corresponding to (3.5.2):

$$\langle n_k \rangle = \frac{1}{2\pi i} \int_{-i\infty}^{i\infty} \frac{\exp(z0^+)}{z - \varepsilon_k - \sum(k, z)} dz \; . \tag{3.5.3}$$

Here the $\sum(k, z)$ is called the self-energy part and is represented by appropriate Feynman diagrams in the scheme of the perturbation expansion. Not only in the present problem but also in general, there remains a mathematical question of whether or not the perturbation series converges when we consider the perturbation expansion to infinite order. Formally, for a given expansion parameter λ in perturbation, if the physical quantity is a regular function of λ, the series in powers of λ yields a correct answer within its radius of convergence. However, if the point $\lambda = 0$ is a singular point of the function, the power series in λ of the perturbation expansion will be meaningless. In the case of a Fermi liquid, when the system becomes superconducting, the singularity of the type $\exp(-1/\lambda)$ appears so that the perturbation expansion is not applicable. But it would be expected that the perturbation expansion converges if the system does not become superconducting, i.e., in the case of a normal Fermi liquid. In other words, a convergent perturbation series is a mathematical criterion for describing the normal Fermi liquid. In the following, we will proceed assuming that this is really the case.

We now return to (3.5.2) and observe the term $1/(z - \varepsilon_k)$ in the integrand. If we find the pole of this function, we have $z = \varepsilon_k$ so that the energy of the particle less the chemical potential μ is derived. Corresponding to this, the pole of the integrand in (3.5.3) which is calculated from

$$z - \varepsilon_k - \sum(k, z) = 0 \qquad (3.5.4)$$

yields the energy of the quasiparticle with wave number k less the chemical potential. As will be stated later on, the specific heat of the Fermi liquid at low temperatures is determined by the above-mentioned energy of the quasiparticle.

3.5.2 Lifetime of the Quasiparticle

In order to discuss the behavior of energy determined from (3.5.4), it is necessary to know the mathematical properties of $\sum(k, z)$. If we regard $\sum(k, z)$ as a function of z, the following properties have been derived.
1) $\sum(k, z)$ is an analytic function except on the real axis.
2) The real axis is a branch cut of the function. This implies that the values of the function differ according to whether z is approached on the real axis from the upper half plane or from the lower half plane. This is expressed as

$$\sum(k, x \pm i\delta) = K(k, x) \mp iJ(k, x), \quad (x: \text{real}, \ \delta \to 0)$$
$$J(k, x) \geqq 0 \ .$$

3) The imaginary part $J(k, x)$ in the above equation vanishes when $x \to 0$. That is

$$J(k, x) = C(k)x^2, \quad C(k) > 0 \quad (x \to 0) \ .$$

Among these properties, 1) and 2) are general consequences but 3) is the property which has been proved for the term of each order in the perturbation expansion. Since $\sum(k, z)$ has in general an imaginary part as seen from 2), the energy of the quasiparticle which is the solution of (3.5.4) contains also in general an imaginary part. It means that the quasiparticle has a finite lifetime. This phenomenon is called "damping" of the quasiparticle. We will discuss the damping in some detail in Chap. 6.

If the quasiparticle has the finite lifetime τ, due to the uncertainty principle in quantum mechanics, the energy has a width of the order of $\Delta E \approx \hbar/\tau$. The above-mentioned 3) implies that the lifetime of the quasiparticle is very long near the Fermi surface so that the width of energy is small there. This is one reason why the description by elementary excitation such as the quasiparticle is so successful. When stated more precisely, the situation is as follows. The quasiparticle will have a clear meaning as an elementary excitation provided that its energy width is smaller than the width $k_B T$ arising from thermal motion. In liquid ^{3}He, from the experimental data for viscosity coefficient, diffusion constant and thermal conductivity, etc., it is possible to estimate the lifetime of the quasiparticle. The order of magnitude of τ thus determined is given by

$$\tau \approx 10^{-12}/T^2 \text{ [s]}$$

when we express T in terms of absolute temperature. Thus, the above condition is written as $10^{12}\hbar T^2 < k_B T$ whence it follows that $T < 0.1$ K. In other words, in the case of liquid ^{3}He, the concept of the quasiparticle becomes effective at low temperatures below 0.1 K. On the contrary, at temperatures above it, the width of the quasiparticle is the same as or larger than that due to thermal motion so that the physical characteristics of a particle will be lost. This fact leads to one limitation for the concept of elementary excitation. On the other hand, it is expected that at very low temperatures (of the order of 10^{-3}K) liquid ^{3}He becomes a superfluid, and quite recently this has been verified experimentally. Thus, the superfluid transition temperature yields another limitation for the Fermi-liquid theory on the low temperature side.

3.5.3 Existence of the Fermi Surface. Specific Heat and Magnetic Susceptibility at Low Temperatures

Although the Fermi surface of a free particle is spherical, that of electrons in a solid has a shape different from it because of the periodic potential of lattice points composing the crystal. However, the original concept of the Fermi surface was introduced under the one-body approximation. Then, one may raise the following question. Does the Fermi surface exist when we take into account interactions between particles? What physical meaning does it have, if any? Among the answers to these questions, the following points are known at present.

First, let us consider the case in which the Hamiltonian is isotropic and invariant under the spatial translation as in an electron gas and liquid ^{3}He. In this case, we have $\sum(\boldsymbol{k}, z) = \sum(k, z)$ where $k = |\boldsymbol{k}|$. On the basis of properties 1), 2), and 3) mentioned in Sect. 3.5.2, the following facts are proved.

1) The Fermi surface is determined from the equation $\varepsilon_k + \sum(k, 0) = 0$ or $\varepsilon_k + K(k, 0) = 0$. The $\langle n_k \rangle$ given by (3.5.3) shows a discontinuity as a function of k and it occurs at $k = k_F$ where k_F is the Fermi wave vector of a free particle, i.e., $k_F = (3\pi^2 n)^{1/3}$ with n number density.

2) At low temperatures ($k_B T \ll \mu$), the specific heat of a Fermi liquid is determined from the energy of the quasiparticle near the Fermi surface. For example, if the energy E_k of the quasiparticle is written as $E_k = (\text{const}) + \hbar^2 k^2 / 2m_t$ for $k \approx k_F$, the specific heat C at low temperatures is expressed as

$$C = \frac{k_B^2 m_t k_F V}{3\hbar^2} T$$

where V is the volume of the system. The above equation coincides with the expression for free particles where the mass of the particle is replaced by the "thermal mass" m_t.

3) When an external magnetic field H is imposed, the energy of the quasiparticle has the form

$$E_{k\sigma} = E_k - \mu_B \gamma(k) H \sigma$$

with μ_B the magnetic moment. The spin paramagnetic susceptibility at low temperatures is given by the expression for free particles where the mass m is replaced by the effective mass m_s. Also, the ratio of m_s to m_t is determined by the value of $\gamma(k)$ at the Fermi surface and it follows that

$$\frac{m_s}{m_t} = \gamma(k_F) \ .$$

For example, in liquid ^{3}He the above ratio takes a value around 4.

The above 1), 2), and 3) are applicable to the isotropic and translationally invariant system. However, the situation differs slightly for the electrons in a solid. In this case, it is known that under the approximation where interactions between electrons are neglected, i.e., the one-body approximation, the electrons exhibit a band structure. By using procedures leading to 1), 2), and 3), the following is proved.

4) For electrons in a solid, there exists the Fermi surface even if the interactions are taken into account. The Fermi surface is defined as the surface where the distribution of electrons is discontinuous. This Fermi surface where all the interactions are considered is called the true Fermi surface. The true Fermi surface has a shape different from the one obtained under the one-body approximation. However, even if the shape changes, its enclosed volume does not

change. The Fermi surface observed by various experiments (e.g. de Haas–van Alphen effect) is the true Fermi surface.

3.5.4 Dilute Solution of ³He in Liquid ⁴He

Changing our subject slightly, we shall discuss an example of a system which is supposed to be an ideal Fermi liquid. If the temperature of pure liquid ⁴He is lowered, it undergoes the so-called λ transition at around $T_\lambda = 2.2$ K and below it the system becomes a superfluid. This temperature is called the λ point. When ³He is put in pure liquid ⁴He, the temperature of the λ point decreases as the concentration of ³He increases. Denoting the numbers of ⁴He atoms and ³He atoms in this ⁴He–³He mixture by N_4 and N_3, respectively, we introduce x defined by $x = N_3/(N_3 + N_4)$; x is the molar fraction of ³He. If we plot the λ point as a function of x, the result is represented by a curve AB (λ line) in Fig. 3.13. As will be stated below, phase separation takes place at the point B. The absolute temperature T_B and the molar fraction x_B of ³He at this point have been measured as

$$T_B = 0.87 \text{ K}, \qquad x_B = 0.66.$$

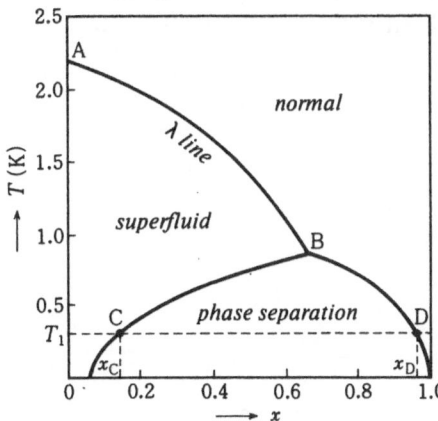

Fig. 3.13. Phase separation line and λ line in ⁴He-³He mixture

At the temperatures below T_B, the phenomenon called phase separation is observed. Consider the temperature T_1 below T_B as indicated by a dotted line in Fig. 3.13 and assume that we prepare a ⁴He-³He mixture whose molar fraction of ³He is between x_C and x_D. In this case, according to experiment, ³He and ⁴He are not mixed uniformly, but separated into two parts: one with the molar fraction x_C containing more ⁴He and the other with x_D containing more ³He. Since the part with more ³He is lighter than the part with more ⁴He, the former floats on the latter and a distinct boundary is observed between them. This phenomenon is the phase separation. However, if the molar fraction of ^{3}He is smaller than x_c, ^{3}He is mixed up with ^{4}He uniformly and the dilute solution of ^{3}He is realized.

According to recent investigations, when T_1 approaches absolute zero, x_c approaches the value around 0.06. Therefore, it is experimentally possible to obtain a dilute solution of ^{3}He in which a small amount of ^{3}He is dissolved in ^{4}He near absolute zero. This dilute solution is expected to behave as an ideal Fermi liquid. For, if we focus our attention on only ^{3}He atoms, their number density is much smaller than that in pure liquid ^{3}He so that ^{3}He is at a low-density state leading to an expectation that we can apply the perturbational method discussed in Sect. 3.1.5. However, as a matter of fact, since there exist ^{4}He atoms besides ^{3}He atoms, the situation is not so simple. One of the ways dealing with this problem is to assume that ^{3}He atom is a quasiparticle in the sense of Fermi-liquid theory and to introduce a suitable interaction between quasiparticles. As for this interaction, an appropriate form is employed and parameters in it are determined from experimental data.

An interesting feature of the dilute solution of ^{3}He is that we can change to some extent the Fermi temperature by varying the molar fraction of ^{3}He. This is not the case in pure liquid ^{3}He. On the other hand, the dilute solution of ^{3}He has an additional adjustable parameter. As a result, much more information is expected to be obtained about the properties of a Fermi liquid. For this reason, the dilute solution of ^{3}He has recently been studied extensively both experimentally and theoretically.

4. Phase Transitions and Elementary Excitations

Depending on temperature, pressure, and other parameters, the macroscopic system of particles exhibits a variety of phases. The transition from one phase to another is probably one of the most dramatic phenomena shown by the macroscopic system. The purpose of this chapter is to study the phase transition as a quantum mechanism of elementary excitations. The nature of elementary excitations changes in general on the phase transition, and conversely, the mechanism of the phase transition may often be described in terms of elementary excitations. We will begin with the qualitative aspect of the problem, i.e., the relation between the symmetry of the macroscopic system and its elementary excitations.

4.1 Phase Transition and Broken Symmetry

The phase transition of the macroscopic system takes place in some cases between two phases which are distinguishable from each other only quantitatively, such as liquid and gas phases in the case of boiling. In many cases, however, the phase transition is accompanied by a change of symmetries; the two phases are distinguishable also by the difference in symmetry, such as an anisotropic crystal and an isotropic gas in the case of sublimation. Among various symmetry changes, the easiest to understand are perhaps those arising from the changes of atomic configuration; the structual change of a crystal from one phase to another, for example. The change of atomic configuration may be either discontinuous (first-order transition) or continuous (second-order transition) at the transition point.

   In general, however, the symmetry change is not necessarily related to the change of atomic configuration. In the case of a ferromagnetic insulator, for instance, we are concerned with the orientation of atomic magnetic moments. When the temperature is higher than the so-called Curie point, the crystal is in the paramagnetic state and its macroscopic moment, i.e., the expectation value of the sum of atomic moments, vanishes unless an external magnetic field is applied. Below the Curie point, atomic moments align themselves in a certain direction with the aid of their mutual interaction and the crystal, even in the absence of an external field, possesses a finite macroscopic moment, which we call the *spontaneous magnetization*. Being a vector, the nonvanishing macroscopic magnetization of the crystal singles out the privileged direction in our otherwise isotropic space. In fact, we are concerned here also with the symmetry against time reversal, i.e., the formal replacement of time t by $-t$ in the laws of dynamics; the magnetic moment changes its sign upon time reversal. We will be concerned with even more abstract symmetry when we deal with the so-called λ transition of liquid ^{4}He into the *superfluid* state and the transition of the electron fluid in a metal into the *superconducting* state. What matters in these cases is the symmetry against the gauge transformation, i.e., the transformation of the phase of the state function in quantum mechanics.

Now, particles in a macroscopic system, however large their total number, do obey the same laws of mechanics as particles in a small system. The Hamiltonian possesses a number of fundamental symmetries irrespective of the size of the system. For example, reflecting the reversibility of dynamical laws, the Hamiltonian is invariant against time reversal. (We ignore the possible breakdown of fundamental symmetries, which is to be discussed in high energy physics). Similarly, reflecting the homogeneity of space, the Hamiltonian is invariant against any translation (the parallel displacement of coordinate axes); the interaction potential between atoms, in particular, depends only upon their relative positions. When a macroscopic system of such atoms, say one mole of argon, is in the gas (or liquid) phase, the expectation value of the atomic density is uniform in space (except in the vicinity of the surface), so that macroscopic properties directly reflect the translational symmetry of the Hamiltonian. When atoms form a crystal, however, the expectation value of the atomic density is a periodic function of the spatial position and therefore invariant only against a discrete set of translations which agree with the periods of the expectation value of the density. Mathematically speaking, the translations against which the expectation value of the atomic desnity in a crystal is invariant form a subgroup of the whole group of translations against which the Hamiltonian is invariant. We characterize such a situation by saying that we have a *broken symmetry* of the macroscopic state.

Needless to say, the motion of microscopic particles remains to be described by one and the same Hamiltonian, when the symmetry of the macroscopic system changes by the phase transition, say, from one crystal structure to another. Therefore, even if the phase of higher symmetry reflects the full symmetry of the Hamiltonian, the phase of lower symmetry reflects the partial symmetry of the Hamiltonian. In other words, we have a broken symmetry whenever the symmetry changes by the phase transition.

We will show some examples and discuss the relation with elementary excitations.

4.2 Order Parameters

The change of symmetries caused by the phase transition is usually connected with the manifestation of a certain order parameter. Suppose that the expectation value $\langle f \rangle$ of some physical quantity f of the macroscopic system in thermal equilibrium vanishes in one of the two phases in question and is nonvanishing in the other phase. We may then take $\langle f \rangle$ as the order parameter to characterize the latter phase. The spontaneous magnetization of the ferromagnetic crystal is such an example.

As another example, let us think of the order parameter to characterize the periodic arrangement of crystal atoms. Suppose that the crystal consists of N atoms, whose positions are denoted by $r_1, r_2, \ldots, r_N$. The atomic density at the point r in space is represented by the operator

$$\rho(r) = \sum_{j=1}^{N} \delta(r - r_j) . \tag{4.2.1}$$

Its Fourier transform is given by

$$\rho_q = \sum_{j=1}^{N} \exp(-i q \cdot r_j) . \tag{4.2.2}$$

In the case of gas and liquid phases, $\langle \rho(r) \rangle$ does not depend on r except in the vicinity of the surface, so that $\langle \rho_q \rangle$ vanishes unless $q = 0$. In the case of the crystal, on the other hand, we have at least one nonvanishing wave vector q, for which $\langle \rho_q \rangle \neq 0$. We denote such a wave vector by Q. The density–density correlation function discussed in Sect. 1.6 can then be factorized for $q = Q$ as $\langle \rho_q(t)\rho_{-q}(0) \rangle \cong \langle \rho_q \rangle \langle \rho_{-q} \rangle$ and therefore gives rise to Bragg reflection peaks in x-ray and neutron elastic scatterings. Conversely, the existence of Bragg reflections experimentally implies the order parameter $\langle \rho_q \rangle$ characteristic of the crystal.

It should be emphasized that the general definition of the crystal given above refers nowhere to the primitive image of the crystal, in which each atom is supposed to be localized in the vicinity of each lattice point. This kind of image does not apply to solid helium (nor to the classical solid near its melting point), as we have mentioned in Sect. 1.9. Consequently the vector period Q may not necessarily be expressed in the form (1.3.5) in terms of the single set of fundamental periods b_j. An example is the *charge density wave* (CDW) state shown by some layer metallic compounds (Nb or Ta chalcogenides) and quasi-one-dimensional organic conductors (TTF-TCNQ, etc.). The density of conduction electrons, $\langle \rho_q^{(e)} \rangle$, takes a nonvanishing value for the wave vector Q different from the reciprocal lattice vectors K of the host crystal and induces the variation with the wave vector Q in the atomic arrangement through the electron–phonon interaction. When the ratio of the new period Q to the original K is a simple rational number, we have a commensurate CDW. Otherwise, we have an incommensurate CDW.

The third example is the magnetic order in the system of localized spins. Suppose that at each lattice point R_j we have the spin angular momentum $\hbar S_j$ of the magnitude $\hbar S$ and the associated magnetic moment βS_j. We assume that the spin ordering is described by the so-called Heisenberg model, whose Hamiltonian takes the form

$$\mathscr{H}_{\text{ex}} = -\frac{1}{2} \sum_{j \neq l} J_{jl} \mathbf{S}_j \cdot \mathbf{S}_l \ . \tag{4.2.3}$$

When the exchange integrals J_{jl} are all positive, the ground state is the ferromagnetic state, in which all spins are aligned in one and the same direction. In a more general case, the magnetic order is defined in the following way.

We assume that there are N unit cells, each of which contains one atom. Corresponding to (4.2.2), we define

$$\mathbf{S}(\mathbf{k}) = N^{-1/2} \sum_{j=1}^{N} \mathbf{S}_j \exp(i\mathbf{k} \cdot \mathbf{R}_j) \tag{4.2.4}$$

where $\mathbf{k}$ is the reduced wave vector. The inverse transformation takes the form

$$\mathbf{S}_j = N^{-1/2} \sum_{\mathbf{k}} \mathbf{S}(\mathbf{k}) \exp(-i\mathbf{k} \cdot \mathbf{R}_j) \ . \tag{4.2.5}$$

The sum is taken over N values of the reduced wave vector. The state of spin ordering is characterized by the nonvanishing expectation values $\langle \mathbf{S}(\mathbf{k}) \rangle$ for $\mathbf{k} = \mathbf{Q}$ and $\mathbf{k} = -\mathbf{Q}$ and $\langle \mathbf{S}(\mathbf{Q}) \rangle = \langle \mathbf{S}(-\mathbf{Q}) \rangle^*$ may be taken as the order parameter. The ferromagnetic state is the special case of $\mathbf{Q} = 0$.

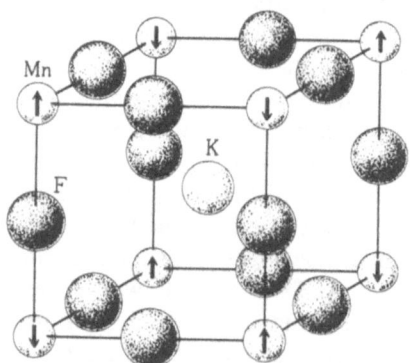

Fig. 4.1. The antiferromagnetic spin ordering in $KMnF_3$

Figure 4.1 shows the antiferromagnetic spin ordering in $KMnF_3$, in which $\langle \mathbf{S}_j \rangle$ of neighboring Mn atoms have the same magnitude and opposite directions. The lattice vectors of Mn atoms are $\mathbf{R}_j = (n_1 a, n_2 a, n_3 a)$ and form a simple cubic lattice. In this case, $\mathbf{Q} = \mathbf{K}/2$, where $\mathbf{K} = (2\pi/a, 2\pi/a, 2\pi/a)$ is a reciprocal lattice vector. Note that $\exp(i\mathbf{Q} \cdot \mathbf{R}_j) = +1$ or -1 according to whether $n_1 + n_2 + n_3$ is even or odd. Assuming that $\langle \mathbf{S}(\mathbf{Q}) \rangle$ is along the z axis, we write the expectation value of the spin as

$$\langle S_{jz} \rangle = A \cos(\mathbf{Q} \cdot \mathbf{R}_j) = \pm A \ . \tag{4.2.6}$$

If Q is slightly shorter than $K/2$, so that $Q = K/2 - q$, $q = (2\pi/pa, 0, 0)$, $p \gg 1$, we have

$$\langle S_{jz} \rangle = \pm A \cos(2\pi n_1/p) \ . \tag{4.2.7}$$

Thus the amplitude of (4.2.6) is now modulated with the wavelength pa. This kind of spin ordering is called *the spin density wave* (SDW). The famous example is metallic chromium, but spins are carried here by conduction electrons and we cannot apply the localized spin model.

Figure 4.2 shows the so-called *screw spin structure*, which is characterized by the nonvanishing expectation values of $S_{j\pm} = S_{jx} \pm iS_{jy}$, or their Fourier transforms $S_{\pm}(k) = S_x(k) \pm iS_y(k)$. If $\langle S_+(Q) \rangle = \langle S_-(-Q) \rangle^* \neq 0$, $\langle S_-(Q) \rangle = \langle S_+(-Q) \rangle^* = 0$, for instance, we have

$$\langle S_{jx} \rangle = A \cos(Q \cdot R_j), \quad \langle S_{jy} \rangle = A \sin(Q \cdot R_j) \ . \tag{4.2.8}$$

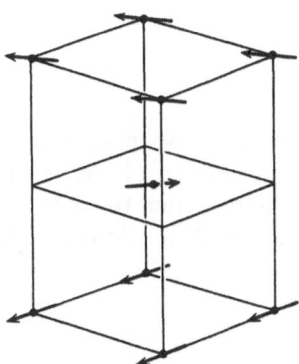

Fig. 4.2. The screw spin ordering in a rutile type crystal

It is possible to prove that the ground state of the Hamiltonian (4.2.3) will be a screw structure, if we regard S_j as classical vectors and assume that J_{jl} depends only on the relative position $R_{jl} = R_j - R_l$. The wave vector Q of the screw structure is such that the Fourier transform

$$J(q) = \sum_j J_{jl} \exp(iq \cdot R_{jl}) \tag{4.2.9}$$

is the maximum for $q = Q$. When $J(q)$ is the maximum either at $q = 0$ or $q = K/2$, however, we have the ferromagnetic or antiferromagnetic structures as the ground state. Since Q is determined by the way in which the exchange integral varies as the function of R_{jl}, it is not necessarily commensurate with reciprocal lattice vectors of the host crystal. Anyway, when magnetic ordering other than the ferromagnetic one appears, the translational symmetry of the crystal is lowered.

Since we are familiar with crystals and magnets and can intuitively understand the orders in them without much difficulty, it appears to us that no problem arises concerning their definitions. If we apply the well-known principles of statistical mechanics without caution, however, we are led to the conclusion, for instance in the case of a magnet, that the expectation value $\langle S_j \rangle$ in thermal equilibrium should always vanish in the absence of an external magnetic field. That is to say that the magnetic Bragg peaks should not be observed in the neutron scattering experiment.

In fact, under time reversal, the spin angular momentum changes its sign, whereas the Hamiltonian remains invariant. Therefore, if the quantum mechanical expectation value of S_j is equal to M in a certain eigenstate of the Hamiltonian, the time-reversed state has the same energy eigenvalue and gives $- M$ as the expectation value of S_j. Statistical mechanics tells us that the two states possessing equal energy should be equally probable in thermal equilibrium. Hence we conclude $\langle S_j \rangle = 0$, which contradicts the existence of the permanent magnet.

4.3 Magnons

Before solving this paradox, we need to study the Heisenberg model defined by (4.2.3) in more detail. We assume that the Fourier transform (4.2.9) is the maximum at $q = 0$ (the ferromagnetic case) and also suppose that the uniform external magnetic field H is applied along the z axis. We therefore add to (4.2.3) the Zeeman term

$$\mathcal{H}_z = - \sum_j \beta H S_{jz} \tag{4.3.1}$$

and write the total Hamiltonian as $\mathcal{H} = \mathcal{H}_{ex} + \mathcal{H}_z$.

All spins are parallel to the z axis in the ground state, which we denote by $|0\rangle$ in Dirac's notation. Since the operators $S_{j\pm1}$ change the eigenvalue of S_{jz} by ± 1 and $|0\rangle$ is the eigenstate belonging to the maximum eigenvalue S of S_{jz}, we have

$$S_{jz}|0\rangle = S|0\rangle , \quad S_{j+}|0\rangle = 0 . \tag{4.3.2}$$

The order parameter in the present case is the expectation value of the total spin $S^{tot} = \sum_j S_j$ and therefore the vector of the magnitude NS and the direction parallel to the z axis. In fact, the expectation values of S_{jx}, S_{jy} should vanish since $\langle 0|S_{j+}|0\rangle = \langle 0|S_{j-}|0\rangle^* = 0$ from the second equation of (4.3.2). From (4.3.2) and

$$S_j \cdot S_l = S_{jz}S_{lz} + \frac{1}{2}(S_{j+}S_{l-} + S_{j-}S_{l+}) \tag{4.3.3}$$

we obtain the lowest eigenvalue of $\mathcal{H}$ as

$$E_0 = -\frac{1}{2}NS^2J(0) - NS\beta H \ . \tag{4.3.4}$$

4.3.1 Magnons

Let us now turn to the low-lying excited states of our spin system. The state, which is obtained from $|0\rangle$ by reducing the eigenvalue of S_{jz} by unity, may be written as $|j\rangle = [2S]^{-1/2}S_{j-}|0\rangle$ including the normalization factor. With use of well-known commutation relations of angular momenta, we see

$$
\begin{aligned}
|j\rangle &= [2S]^{-1/2}\{[\mathcal{H}, S_{j-}] + S_{j-}\mathcal{H}\}|0\rangle \\
&= [E_0 + SJ(0) + \beta H]|j\rangle - S\sum_{l\neq j}J_{jl}|l\rangle \ .
\end{aligned}
\tag{4.3.5}
$$

But for the last term on the right, $|j\rangle$ would have been the eigenstate of $\mathcal{H}$ with the excitation energy $\langle j|\mathcal{H}|j\rangle - E_0 = SJ(0) + \beta H$. This may be regarded as the excitation energy of the localized "spin defect" analogous to the point defect in a crystal. Alternatively it may be regarded as the excitation energy of the spin in the "molecular field". The exchange interaction in (4.2.3) is equivalent to the effective magnetic field $\beta^{-1}\Sigma J_{jl}S_l$ acting upon S_j. In the *molecular field approximation* (*mean field approximation*) we replace the effective field by its expectation value. Adding the external field, we have the effective field $\beta H + \beta^{-1}SJ(0)$ along the z axis acting on S_j. Hence we need the energy $\beta H + SJ(0)$ in order to decrease S_{jz} from S to $S - 1$.

If the nondiagonal terms $\langle l|\mathcal{H}|j\rangle = -SJ_{jl}$ in (4.3.5) are taken into account, the spin defect will propagate as the magnon through the crystal, just like the defecton described in Sect. 1.9. In fact, with use of the Fourier transform of (4.2.4), we define $|k\rangle = S_{-}(k)|0\rangle$ and easily check that it is the eigenstate of $\mathcal{H}$ with the energy $E_0 + \varepsilon(k)$, where

$$\varepsilon(k) = \beta H + S[J(0) - J(k)] \ . \tag{4.3.6}$$

Thus $|k\rangle$ is the state in which one magnon of the reduced wave vector k and excitation energy $\varepsilon(k)$ is excited (Fig. 4.4).

The $k = 0$ mode is the precession of all the spins around the external field in one and the same phase. The exchange energy (4.2.3) depends only on the relative orientation of spins and therefore makes no contribution to the uniform precession, so that $\varepsilon(0)$ agrees with the excitation energy of the single spin put in the external field. The exchange energy makes a contribution when a phase difference exists between precessions of neighboring spins ($k \neq 0$). Because of the point symmetry of the lattice, the second term on the right of (4.3.6) is proportional to k^2 for small k. The range r_0 of the exchange interaction is defined as the distance $|R_J - R_I|$, beyond which J_{JI} decreases appreciably. The excitation energy $\varepsilon(k)$ will saturate for $k \gtrsim r_0^{-1}$.

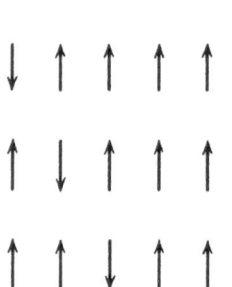

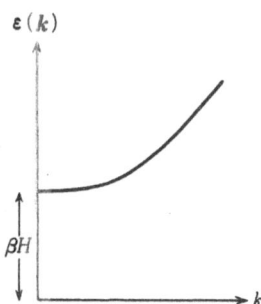

Fig. 4.3. The propagation of the spin defect　　Fig. 4.4. The energy spectrum of the magnon

4.3.2 Spin Wave Approximation

Even if many magnons are excited, the spin system may be regarded as a gas of magnons, of which the excitation energy is approximately equal to the sum of (4.3.6), provided that the density of magnons is sufficiently small. In particular, the energy of exciting n magnons at $k = 0$ is rigorously given by $n\beta H$ and the excited state so obtained is represented by

$$|n\rangle = [{}_{2NS}C_n(n!)^2]^{1/2}(S_-^{tot})^n|0\rangle \tag{4.3.7}$$

where ${}_mC_n$ is the binomial coefficient.

In order to deal with more general excited states, we need an approximation method. In the *Holstein-Primakoff method,* for instance, we express spin operators in terms of creation and destruction operators of Bose particles. Thus, assuming that B_J^+, B_J satisfy the usual commutation rules of Bose type, we write

$$S_{Jz} = S - B_J^+B_J, \quad S_{J+} = [2S - B_J^+B_J]^{1/2}B_J$$
$$S_{J-} = B_J^+[2S - B_J^+B_J]^{1/2} . \tag{4.3.8}$$

We restrict ourselves to linear combinations of those eigenfunctions of $B_J^+B_J$ which belong to eigenvalues not larger than $2S$. Under this restriction, operators (4.3.8) possess the same algebraic properties as spin operators. In practice, however, we expand the square root as $S_{J+} = (2S)^{1/2}[B_J - (4S)^{-1}B_J^+B_JB_J + \ldots]$ and retain only lower-order terms, provided that the spin defect $\langle B_J^+B_J \rangle$ is smaller than $2S$.

Substituting (4.3.8) in the Hamiltonian $\mathscr{H}_{ex} + \mathscr{H}_z$ and ignoring terms of fourth-order and higher in B operators, we obtain the Hamiltonian $\mathscr{H}_0$ which corresponds to the harmonic approximation of phonons. It is diagonalized by the Fourier transformation

$$B_j = N^{-1/2} \sum_k b_k \exp(-i\boldsymbol{k} \cdot \boldsymbol{R}_j) \tag{4.3.9}$$

where b_k^+, b_k also satisfy the commutation relations of the Bose type. They are creation and destruction operators of the magnon of the reduced wave vector $\boldsymbol{k}$, respectively, and the eigenvalue of $b_k^+ b_k$ is the number of the magnons. In fact, with use of (4.3.6), we have

$$\mathscr{H}_0 = E_0 + \sum_k \varepsilon(\boldsymbol{k}) \, b_k^+ b_k \; . \tag{4.3.10}$$

The expectation value of the magnon number in thermal equilibrium is then given by the Planck distribution

$$\langle b_k^+ b_k \rangle = \frac{1}{\exp[\varepsilon(\boldsymbol{k})/k_B T] - 1} \; . \tag{4.3.11}$$

The spin defect at the jth lattice point is given by

$$\langle B_j^+ B_j \rangle = N^{-1} \sum_k \langle b_k^+ b_k \rangle$$

$$= \left(\frac{V}{N} \right) \frac{1}{(2\pi)^3} \int \frac{d\boldsymbol{k}}{\exp[\varepsilon(\boldsymbol{k})/k_B T] - 1} \; . \tag{4.3.12}$$

The integral is proportional to $\exp[-\beta H/k_B T]$ when the external field H is finite and to $T^{3/2}$ when $H = 0$. The harmonic approximation is legitimate at low temperatures where the spin defect (4.3.12) is much smaller than $2S$.

4.3.3 Antiferromagnets

The spin wave approximation may be applied also to the case of magnetic ordering characterized by a nonvanishing wave vector $\boldsymbol{Q}$. Let us take the antiferromagnetic ordering as an example. Assume that we have a negative exchange integral J between nearest-neighbor spins on the simple cubic lattice such as shown in Fig. 4.1. Then $J(\boldsymbol{q}) = 2J \, (\cos aq_x + \cos aq_y + \cos aq_z)$ takes the maximum value $6|J|$ when $\boldsymbol{q}$ is equal to one-half of the reciprocal lattice vector $\boldsymbol{K} = (2\pi/a, 2\pi/a, 2\pi/a)$. When $\boldsymbol{S}_j$ are regarded as classical vectors of the magnitude S, therefore, the state of lowest energy is the antiferromagnetic spin ordering shown in Fig. 4.1. Lattice points are classified into A and B sublattices according to whether $\boldsymbol{K} \cdot \boldsymbol{R}_j/2\pi$ is even or odd; we denote the A sublattice point by the index j and the B sublattice point by the index l. Let $|0\rangle$ be the quantum-mechanical state which corresponds to the classical antiferromagnetic ordering shown in

Fig. 4.1. Then

$$S_{jz}|0\rangle = S|0\rangle, \qquad S_{lz}|0\rangle = -S|0\rangle; \\ S_{j+}|0\rangle = 0, \qquad S_{l-}|0\rangle = 0 \Bigg\} . \tag{4.3.13}$$

In contrast to the ferromagnetic case, the state $|0\rangle$ defined by (4.3.13) is *not* the eigenstate of $\mathscr{H}_{ex}$. Operating on $|0\rangle$, $\mathscr{H}_{ex}$ gives rise to states belonging to eigenvalues of $S-1$ and $-S+1$ of S_{jz}, S_{lz}, respectively. The true ground state $|\rangle$ will be a linear combination of $|0\rangle$ and those states which we obtain therefrom by repeatedly operating $\mathscr{H}_{ex}$. Intuitively, we know that spins perform the zero-point precession around the classical antiferromagnetic state. In the present case, the order parameter is $S(\boldsymbol{K}/2)$, which is proportional to the difference between A sublattice spin $\boldsymbol{S}_A = \sum_j \boldsymbol{S}_j$ and B sublattice spin $\boldsymbol{S}_B = \sum_l \boldsymbol{S}_l$. It is *not* commutable with $\mathscr{H}_{ex}$. The operator $S_{Az} - S_{Bz}$ has the maximum value $2NS$ in the state $|0\rangle$, but its expectation value in the true ground state $|\rangle$ is somewhat smaller because of the zero-point precession.

One method of obtaining $|\rangle$ is to deal with the zero-point precession by the spin wave approximation starting from $|0\rangle$. Here we introduce the anisotropy energy

$$\mathscr{H}_A = -\frac{1}{2}D(\sum_j S_{jz}^2 + \sum_l S_{lz}^2) \tag{4.3.14}$$

which favors the spin ordering along the z axis ($D > 0$, $S > 1/2$). In terms of spin defect destruction operators A_j, B_l, we write fluctuations from (4.3.13) as

$$S_{jz} = S - A_j^\dagger A_j, \qquad S_{lz} = -S + B_l^\dagger B_l \\ S_{j+} \cong (2S)^{1/2}A_j, \qquad S_{l-} \cong (2S)^{1/2}B_l \Bigg\} \tag{4.3.15}$$

where we have retained lowest-order terms in the expansion of square roots in (4.3.8). Fourier transforms cooresponding to (4.3.9) are given by

$$A_j = (2/N)^{1/2} \sum_{\boldsymbol{k}} a_{\boldsymbol{k}} \exp(i\boldsymbol{k}\cdot\boldsymbol{R}_j) \\ B_l = (2/N)^{1/2} \sum_{\boldsymbol{k}} b_{\boldsymbol{k}} \exp(i\boldsymbol{k}\cdot\boldsymbol{R}_l) \Bigg\} . \tag{4.3.16}$$

Since j, l are restricted to their respective sublattices, $\boldsymbol{k}$ runs over half of the original simple-cubic Brillouin zone ($-\pi/2a < k_x < \pi/2a$, etc., for instance). The harmonic Hamiltonian now takes the form

$$\mathscr{H} = E_0 + \sum_{\boldsymbol{k}} [(\xi + \beta H)\, a_{\boldsymbol{k}}^\dagger a_{\boldsymbol{k}} + (\xi - \beta H)\, b_{\boldsymbol{k}}^\dagger b_{\boldsymbol{k}} \\ + \eta_{\boldsymbol{k}}(a_{\boldsymbol{k}}^\dagger b_{\boldsymbol{k}}^\dagger + b_{\boldsymbol{k}} a_{\boldsymbol{k}})] \tag{4.3.17}$$

where E_0 is the energy of the classical antiferromagnetic state and

$$\xi = D - J(0)S, \quad \eta_{\boldsymbol{k}} = -J(\boldsymbol{k})S . \tag{4.3.18}$$

The harmonic Hamiltonian can be diagonalized by the so-called Bogoliubov transformation

$$a_k = u_k \alpha_k - v_k \beta_k^\dagger, \quad b_k = u_k \beta_k - v_k \alpha_k^\dagger \tag{4.3.19}$$

where $u_k = \cosh(\theta_k/2)$, $v_k = \sinh(\theta_k/2)$, $\theta_k = \tanh^{-1}(\eta_k/\xi)$. Thus

$$\mathcal{H} = E_0 + \sum_k (\varepsilon_k - \xi) + \sum_k [(\varepsilon_k + \beta H) \alpha_k^\dagger \alpha_k + (\varepsilon_k - \beta H) \beta_k^\dagger \beta_k] \tag{4.3.20}$$

$$\varepsilon_k = (\xi^2 - \eta_k^2)^{1/2} . \tag{4.3.21}$$

Hence $\alpha_k^\dagger \alpha_k$, $\beta_k^\dagger \beta_k$ represent magnon numbers and $\varepsilon_k \pm \beta H$ are excitation energies. The ground state is defined by demanding that no magnon should exist

$$\alpha_k| \, \rangle > 0, \; \beta_k| \, \rangle = 0 \tag{4.3.22}$$

provided that the excitation energy of the magnon is positive. At $k = 0$, ε_k takes the minimum value $\beta H_f = [D(D + 12S|J|)]^{1/2}$, so that the external magnetic field H should be weaker than H_f. When H exceeds H_f, the excitation energy of the magnon described by β_k becomes negative, and therefore the state with a macroscopic number of magnons excited can have lower energy. This implies that the system undergoes the phase transition to a new state (of antiferromagnetic ordering in a direction perpendicular to the z axis).

The second term on the right of (4.3.20) means that the energy decreases somewhat from the classical value because of the zero-point precession. The amplitude of the zero-point precession is given by

$$\langle B_l^\dagger B_l \rangle = (2/N) \sum_k \langle b_k^\dagger b_k \rangle = (2/N) \sum_k v_k^2 . \tag{4.3.23}$$

For $D \neq 0$, ε_k, v_k^2 remain finite as $k \to 0$. This means that the anisotropy suppresses the zero-point precession and stabilizes the antiferromagnetic order.

When $D = 0$, $\varepsilon_k \cong 6|J| \, Sak$ and therefore u_k^2, v_k^2 are divergent like $(2ak)^{-1}$ as $k \to 0$. But (4.3.23), when the sum over k is replaced by the integral, is convergent for three-dimensional (and also two-dimensional) systems. For the present model, it is about 1 percent.

4.4 Hilbert Space of the Macroscopic Systems and Coherent States

In order to solve the paradox raised in Sect. 4.2, we should notice that the Hilbert space to describe the macroscopic system by quantum mechanics shows some peculiar structure, which is alien to the system of a small number of particles. We will explain it by taking the Heisenberg ferromagnet in the preceding section as an example.

4.4.1 Hilbert Space

In the limit of the vanishing external magnetic field, the Hamiltonian of the Heisenberg model reduces to the isotropic $\mathcal{H}_{ex}$ and the z axis is no longer privileged. The state, in which spins are aligned in any direction other than the z axis, has the same energy as the state $|0\rangle$. The lowest energy is thus degenerate, but this degeneracy is somewhat different from the one we find in the system of a small number of particles.

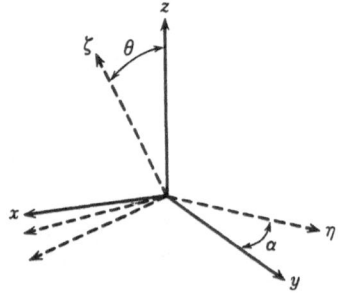

Fig. 4.5. The rotation of coordinate axes

Now suppose that the η axis is obtained by rotating the y axis around the z axis by the angle α, and then the ζ axis is obtained by rotating the z axis around the η axis by the angle θ (Fig. 4.5). As is well known in quantum mechanics, spins are all aligned along the ζ axis in the state $|\theta\alpha\rangle$, which is obtained by applying the operator $U_\eta(\theta)U_z(\alpha)$ to the state $|0\rangle$. Here $U_\mu(\theta) = \exp(-i\theta \sum_J S_{J\mu})$ is the operator, which rotates all the spins round the μ axis by the angle θ. Since $\mathcal{H}_{ex}$ is isotropic, it commutes with U operators and the state $|\theta\alpha\rangle$ has the identical energy as $|0\rangle$. Despite this, it is as much unrealistic to take a linear combination of these two states as to think of a superposition of live and dead states of Schrödinger's cat. States $|0\rangle$ and $|\theta\alpha\rangle$ are *macroscopically distinct* states, which we can distinguish by their directions of spontaneous magnetization.

Alternatively we may take the following point of view. Thus, it would make no sense to talk of the direction of an isolated permanent magnet. To find its orientation, we would need another magnet (the terrestrial magnetism, for instance) to interact. Through the interaction the relative orientation of the two magnets will be fixed. To find the orientation of the system of two magnets as a whole, however, we need the third magnet. Our world of macroscopic bodies is in a state of broken symmetry, so that all the magnets have definite relative orientations to each other. An "isolated" magnet should be taken as the idealized limit of making the magnetic field produced by other magnets vanishingly weak.

This physical argument was formulated by Bogoliubov as the mathematical concept of *quasi-average*. For instance, the magnetic order parameter in Sect.

4.2 is defined in the following way. We first suppose that the spatially varying external magnetic field $H(r) = H_0 \cos Q \cdot r$ is applied to the spin system. Hence the usual statistical average in thermal equilibrium, $\langle S(Q) \rangle$, is nonvanishing in general. We then take the thermodynamic limit, which in the present example means taking the limit of an infinite number of lattice points, $N \to \infty$, with the lattice constant being kept constant. Finally we take the limit $H_0 \to 0$. If the limit

$$\lim_{\substack{H_0 \to 0 \\ N \to \infty}} \lim \langle S(Q) \rangle \tag{4.4.1}$$

is nonvanishing, we call it the order parameter to characterize the magnetic *long-range order*.

In order to see the significance of taking the limit $N \to \infty$, let us explicitly write down the relation between ferromagnetic states $|\theta\alpha\rangle$ and $|0\rangle$ in the case of $S = 1/2$. We may ignore the operator $U_z(\alpha)$ since it gives only a constant phase factor to $|0\rangle$. Expanding the exponential in $U_\eta(\theta)$, we obtain

$$|\theta\alpha\rangle = \prod_{j=1}^{N} \left[\cos\frac{\theta}{2} + \exp(i\alpha) \sin\frac{\theta}{2} S_{j-} \right] |0\rangle . \tag{4.4.2}$$

Taking the inner product $\langle 0|\theta\alpha\rangle = \langle 0|1|\theta\alpha\rangle$ and remembering that $\langle 0|S_{j-}|0\rangle = 0$, we have

$$\langle 0|\theta\alpha\rangle = \left(\cos\frac{\theta}{2} \right)^N = \exp\left(N \ln \cos\frac{\theta}{2} \right) . \tag{4.4.3}$$

Since $0 < \theta < \pi$, $\ln \cos(\theta/2) < 0$, and N is a macroscopic number of the order of 10^{22}, (4.4.3) is an extremely minute number, comparable to the probability that a kettle on ice comes to the boil. As we usually put the latter equal to zero, we should also take the limit $N \to \infty$ in (4.4.3). Then, however small θ may be, $|0\rangle$ and $|\theta\alpha\rangle$ are orthogonal to each other, so that the spin rotation operator $U(\theta)$ cannot be an infinitesimal transformation. We cannot regard $|0\rangle$ and $|\theta\alpha\rangle$ as two elements which belong to the same Hilbert space.

Let us denote by $\Omega(0)$ the whole set of linear combinations of those states which we obtain by exiting magnons from $|0\rangle$. This is the Hilbert space to describe thermally excited states as well as the ground state of the ferromagnet whose spontaneous magnetization is along the z axis. Similarly, we denote by $\Omega(\theta)$ the whole set of linear combinations of those states which we obtain by exciting magnons from $|\theta\alpha\rangle$. This is the Hilbert space to describe the ferromagnet whose spontaneous magnetization is in the θ direction. Magnons in this case represent the spin deviations from the θ direction, so that their creation and destruction operators should also be redefined with reference to the new ground state $|\theta\alpha\rangle$.

4.4.2 Condensation of Magnons

We may also take N, though macroscopic, as a finite number. Then the state $|\theta\alpha\rangle$ also belongs to $\Omega(0)$. In fact, expanding (4.4.2) into powers of $\exp(i\alpha)$ and remembering $S_{j-}^2 = 0$ in the case of $S = 1/2$, we can write

$$|\theta\alpha\rangle = \sum_{n=0}^{\infty} \exp(in\alpha)\, w_n^{1/2} |n\rangle \ . \tag{4.4.4}$$

Here $|n\rangle$ is the state (4.3.7), in which n magnons of $k = 0$ are excited, and w_n is the binomial probability

$$w_n = {}_N C_n \cos^{2(N-n)}\frac{\theta}{2} \sin^{2n}\frac{\theta}{2} \tag{4.4.5}$$

Since N is macroscopic, $N \gg N^{1/2} \gg 1$, so that (4.4.5) shows an extremely sharp peak at $n = N_m = N \sin^2(\theta/2)$, around which w_n may be approximated by the Gaussian distribution of the width $n = N^{1/2} \sin(\theta/2) \cos(\theta/2)$. Thus, in the state $|\theta\alpha\rangle$, the number of magnons of $k = 0$ has the macroscopic expectation value N_m and the fluctuation from it is only of the order of $N_m^{1/2}$. We say then that we have the *condensation* of magnons. It is the condensation in k space (or momentum space) and sometimes called the quantum condensation. Note that the number of thermally excited magnons given by (4.3.11) for fixed k is of the order of unity. We obtain a number of the order of N only after taking the sum over k.

Now, from (4.3.7),

$$\langle n + 1 | S_-^{tot} | n \rangle = [(n + 1)(N - n)]^{1/2}$$
$$\cong (1/2)N \sin\theta \tag{4.4.6}$$

where the second expression on the right holds when n is in the vicinity of N_m. From (4.4.4, 6),

$$\langle \theta\alpha | S_-^{tot} | \theta\alpha \rangle = \sum_n \exp(-i\alpha)(w_{n+1}w_n)^{1/2} \langle n + 1 | S_-^{tot} | n \rangle$$
$$\cong (1/2)N \sin\theta \exp(-i\alpha) \sum_n w_n$$
$$= (1/2N \sin\theta \exp(-i\alpha) \ . \tag{4.4.7}$$

This is the value when the classical vector of the magnitude $N/2$ is directed along the ζ axis. We have thus reached practically the same conclusion as in Sect. 4.4.1, but this is not at all surprising since (4.4.7) is the asymptotic expression, in which we have ignored N^{-1} in comparison with unity.

4.4.3 Coherent States

The state (4.4.4) is an example of the so-called *coherent representation*. Needless to say, every quantum-mechanical system shows the particle–wave duality; the particle number n when it is regarded as a system of particles and the phase α when it is regarded as a wave are canonically conjugate variables, whose fluctuations should statisfy the uncertainty relation $\Delta n \, \Delta \alpha \geq 1$. When the expectation value N of the particle number is macroscopic in the sense that $N \gg N^{1/2} \gg 1$, however, we can form a wave packet with $\Delta n \approx N^{1/2}$, $\Delta \alpha \approx N^{-1/2}$ by superposing states of particle numbers n distributed around N. This is the state, in which the phase as well as the particle number are almost definite on the macroscopic scale, and is called a *coherent state*.

The simplest example is the system of photons, all of which have the same momentum and polarization. The coherent state in this case is nothing else but the state to be identified with the coherent plane electromagnetic wave in classical electromagnetism. Let $b^\dagger$, b be creation and destruction operators of these photons, respectively, and $|n\rangle$ be the normalized eigenfunction belonging to the eigenvalue n of the photon number operator $b^\dagger b$

$$|n\rangle = (n!)^{-1/2} (b^\dagger)^n |0\rangle \ . \tag{4.4.8}$$

The coherent state is the wave packet obtained by superposing the $|n\rangle$ with the same phase factor as

$$|\alpha\rangle = \sum_{n=0}^{\infty} \exp(in\alpha) \, w_n^{1/2} |n\rangle \ . \tag{4.4.9}$$

Here w_n is a probability distribution, which has a sharp peak at a macroscopic particle number N with a width of the order of $N^{1/2}$ (see Fig. 4.6). Neglecting N^{-1} in comparison with unity, as in (4.4.7), we obtain the asymptotic expression

$$\langle \alpha | b | \alpha \rangle \cong N^{1/2} \exp(i\alpha) \ . \tag{4.4.10}$$

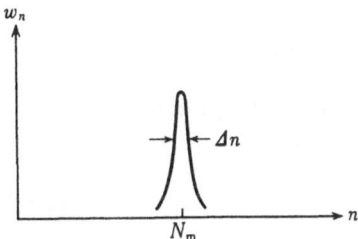

Fig. 4.6. The distribution of the magnon number

The coherent state of the form (4.4.9) is used in quantum optics to represent the state of laser light. We can also apply it to other Bose systems. For instance, in the case of the displacive ferroelectric crystal such as $BaTiO_3$, positive and negative ions in the unit cell undergo relative displacements without external electric field below a certain transition temperature and thus produce a spontaneous electric polarization. This transition may be regarded as the condensation of optic phonons at $k = 0$, which results in the nonvanishing expectation value of (1.4.15) in thermal equilibrium. This idea can be applied whenever the crystal undergoes a spontaneous deformation.

4.5 Coherence of de Broglie Wave and Superfluidity

The most abstract of broken symmetries we encounter on phase transitions is probably the one concerned with the gauge transformation of the first kind, i.e., the shifting of the phase of state functions. In this case, as we will see in Sect. 4.5.2, the phase below the transition temperature, i.e., the state of broken gauge symmetry is characterized by its superfluidity—superconductivity in the case of charged particles. That is to say that the flow of matter can take place in thermal equilibrium without pressure and temperature gradients.

Examples known so far are superfluid states of liquid ⁴He and liquid ³He, and superconducting states of the electron fluid in many metals. We also expect that the dilute (less than 6 percent) mixture of ³He in ⁴He will show superfluidity at ultralow temperatures and that the neutron fluid inside the neutron star may be in the superfluid state.

4.5.1 Coherent States of the de Broglie Wave

The ⁴He atom is a Bose particle with spin zero, so that creation and destruction operators $b_k^\dagger$, b_k of ⁴He atoms with momentum $\hbar k$ satisfy the same commutation rules as phonon operators. In the energy range we are concerned with in condensed-matter physics, however, atoms are not really created nor annihilated. The Hamiltonian is therefore commutable with the operator

$$\mathcal{N} = \sum_k b_k^\dagger b_k \tag{4.5.1}$$

which represents the total number of atoms. In other words, the Hamiltonian is invariant against the gauge transformation of first kind

$$b_k \to b_k \exp(i\gamma), \quad b_k^\dagger \to b_k^\dagger \exp(-i\gamma) \tag{4.5.2}$$

where γ is a real constant.

The assembly of ⁴He atoms may also be regarded as the de Broglie wave. The wave is quantized and described by operators $\psi(r)$ defined at each point r in space. The relation between particle and wave pictures is given by the Fourier transformation

$$\psi(r) = V^{-1/2} \sum_k b_k \exp(i k \cdot r) \tag{4.5.3}$$

where V is the total volume, and we impose the periodic boundary condition upon ψ.

As in the cases of photons and phonons, there exists the possibility of coherent states, in which $\langle b_k \rangle$ is of the order of $N^{1/2}$ for a particular k. Then $\langle b_k^\dagger b_k \rangle$ is of the order of N and we say that ^{4}He atoms are in the state of *Bose condensation*. Generalizing this concept, we can think of the state, in which $\langle \psi(r) \rangle$ is of the order of unity, and say that de Broglie's wave is in a coherent state. The order parameter to characterize it may be assumed to be proportional to $\langle \psi \rangle$,

$$\Psi(r) \propto \langle \psi(r) \rangle \; . \tag{4.5.4}$$

The order parameter Ψ is *not* an operator, but a macroscopic, thermodynamic variable. It is complex in general and its phase represents the phase of coherence of de Broglie's wave. As the rotational symmetry is broken by the spontaneous magnetization in the case of the ferromagnet, the gauge symmetry is broken by the nonvanishing order parameter Ψ.

Admittedly the coherent state of de Broglie's matter wave is more difficult to understand than crystal or ferromagnetic states, or coherent states of photons and phonons. This is partly because the gauge symmetry is more abstract, but the main reason seems to be our intuition that even a macroscopic system must have a definite number of particles as far as it is isolated. In fact, since ψ is the destruction operator of particles, its expectation value (diagonal matrix element) in the state of a definite particle number should vanish. In order to make the expectation value in (4.5.4) nonvanishing, we need to allow for the particle number to fluctuate around its expectation value N up to the order of $N^{1/2}$ and need also to think the wave packet to be of the form (4.4.4). Thus, corresponding to the grand ensemble in statistical mechanics, we suppose that our system is in contact and exchanges particles with a "large" source of particles. Alternatively we may suppose that our system is a subsystem of a "large" system.

In contrast to the control of the photon distribution by laser, it is difficult to control the distribution of atoms in gaseous or liquid helium. We can only change the temperature and pressure. Fortunately, at $T_\lambda = 2.17$ K (under the saturated vapor pressure), liquid ^{4}He undergoes a second-order phase transition called the λ *transition* and shows superfluidity below T_λ. The liquid must undergo the Bose condensation at T_λ, below which it is in a coherent state. (Einstein pointed out long time ago that the perfect Bose gas without interparticle interaction undergoes the Bose condensation at low temperature.)

Now, in the case of the Fermi fluid such as liquid ^{3}He, the electron fluid in a metal, and the dilute mixture of ^{3}He in ^{4}He (Chap. 3), any one-particle state cannot be occupied by two or more particles (Pauli principle) and we have no correspondence to the Bose condensation. However, the condensation of pairs characterized by the nonvanishing order parameter of the form

$$\Psi(x_1, x_2) \propto \langle \psi(x_1) \psi(x_2) \rangle \tag{4.5.5}$$

is possible. Here $\psi(x) = \psi(r, \sigma)$ is the destruction operator, which annihilates the particle with spin σ at the point r in space. In what follows, we consider the particle of spin $\hbar/2$, so that we have two possible spin orientations $\sigma = \uparrow, \downarrow$. We call the condensed pair the *Cooper pair*.

The state, which gives a nonvanishing value to (4.5.5), was first discovered by Bardeen, Cooper, and Schrieffer and is therefore called the *BCS state*. Let $a_{k\sigma}$ be the Fourier transform of $\psi(x)$, i.e., the destruction operator of the electron with momentum $\hbar k$ and spin σ. Then $B_k^\dagger = a_{-k\downarrow}^\dagger a_{k\uparrow}^\dagger$ is the operator which creates a pair of Fermi particles with opposite momenta and spins. As we can easily check by the use of anticommutation rules (3.1.3), $B^\dagger$ operators with different k's commute with each other, but $(B_k^\dagger)^2 = 0$. Hence these operators are somewhat similar to the spin components S_{j-} in the case of $S = 1/2$. In analogy with the magnon condensed state (4.4.2), we therefore try the state

$$|\alpha\rangle = \prod_k [u_k + \exp(i\alpha) v_k B_k^\dagger]|\text{vac}\rangle \tag{4.5.6}$$

where α is real, $u_k = \cos(\theta_k/2)$, $v_k = \sin(\theta_k/2)$, and $|\text{vac}\rangle$ is the normalized state function of the electron vacuum. Thus $a_{k\sigma}|\text{vac}\rangle = 0$ for all k, σ. Hence $\langle B_k\rangle = \langle\alpha|B_k|\alpha\rangle = u_k v_k \exp(i\alpha)$, so that

$$\langle\psi(r_1, \uparrow)\,\psi(r_2, \downarrow)\rangle = V^{-1} \sum_k u_k v_k \exp[ik\cdot(r_1 - r_2) - i\alpha] . \tag{4.5.7}$$

The ground state of the Fermi gas without interparticle interaction may also be expressed in the form of (4.5.6), but $u_k v_k \equiv 0$ in (4.5.7) since θ_k is equal either to π or to 0 according to whether k is less or greater than the Fermi wave number k_F. Only by the presence of an appropriate interaction between particles, is the variation of θ_k at $k = k_F$ smoothed out and we obtain the BCS coherent state (Fig. 4.7). Note that the kinetic energy of particles takes the minimum value for the ground state of the Fermi gas. Hence the gain in the interaction energy should surpass the loss in the kinetic energy when the system takes the BCS coherent state.

In the case of metallic electrons, emission and absorption of the phonon give rise to the interaction between two electrons and this is attractive for electrons

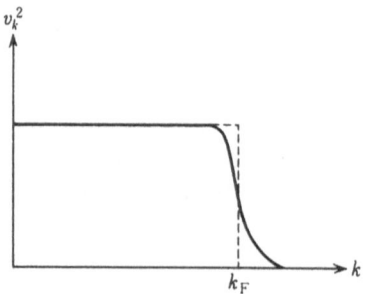

Fig. 4.7. The momentum distribution in the BSC state

near the Fermi surface. The magnitude is of the same order as the Coulomb repulsion between electrons. The electron fluid undergoes the transition to the BCS coherent state at a low temperature when the phonon mediated attraction surpasses the Coulomb repulsion. It is to be identified with the superconducting state, in which the stationary electric current can flow without the voltage drop.

Since (4.5.7) depends only on the distance $|r_1 - r_2|$, the relative orbital angular momentum of the two particles vanishes. The pair is obviously in the spin singlet state and therefore in the state denoted by ^1S in spectroscopy. The Cooper pair in this case has no degeneracy with respect to its internal motion. In general, however, the condensed pair described by (4.5.5) may be either in the spin singlet state ($S = 0$) with even orbital angular momentum ($L = 0, 2, 4, \ldots$) or in the spin triplet state ($S = 1$) with odd orbital angular momentum ($L = 1,3,5, \ldots$). In fact it has recently been discovered that liquid ^{3}He undergoes the transition to the superfluid state at around $2\,\text{mK} = 2 \times 10^{-3}$K. The Cooper pair in this case is in the state $^3P(S = 1, L = 1)$. Since the interaction between He atoms is strongly repulsive at short distances (the hard core, see Fig. 1.1), it is energetically favorable to form the pair in a state of the non-vanishing orbital angular momentum, in which the two particles can avoid the strong repulsion, classically speaking, by the presence of centrifugal force. But it is not quite clear why the ^3P state is most favorable (see Sect. 4.10). Anyway, in contrast to superfluid ^{4}He and superconductors which possess the degeneracy only against the Abelian group of gauge transformations, superfluid ^{3}He has the additional degeneracy against the non-Abelian group of spin and orbital rotations.

4.5.2 Ginzburg–Landau Theory of Superconductivity

The relation between coherence of de Broglie's wave and superfluidity or superconductivity is most clearly shown by the Ginzburg–Landau (GL) theory of superconductivity. It is a phenomenological theory independent of specific microscopic models since we assume only the existence of the order parameter of the type (4.5.5) and apply thermodynamics in the vicinity of the second-order transition point, where the magnitude of the order parameter is still small. As we will see below, it may be regarded as the simplest case of so-called "gauge theory". It has also played the role of a very useful model in the recent development of statistical mechanics on critical phenomena near the second-order phase-transition point.

Now, the order parameter Ψ in GL theory of superconductivity is the state function of the center of mass motion of the Cooper pair and obtained by taking $r_1 = r_2$ in (4.5.7). The center of mass motion has the vanishing momentum in the state (4.5.7), but Ψ in general depends on the position r of the center of mass. Note that GL theory is applicable not only to the case of Cooper pairs, but also to the condensation of charged Bose particles, for instance (and therefore to superfluid ^{4}He in the limit of the vanishing charge).

The important point is that, in the presence of stationary magnetic field, the gauge transformation of the vector potential

$$A \to A + (\partial \Lambda / \partial r) \tag{4.5.8}$$

should be accompanied by the "local" gauge transformation of the order parameter

$$\Psi \to \Psi \exp [(ie^*/\hbar c)\Lambda]. \tag{4.5.9}$$

For the superconducting system of electrons, we should take $e^* = 2e$, where e is the electronic charge.

Now let F be the free energy of our system calculated by some means under the subsidiary condition that the expectation value of the operator $\psi(r\uparrow) \, \psi(r\downarrow)$ shall have a given functional form $\Psi(r)$. Then F is a function of Ψ and its spatial derivatives (which means the nonlocal dependence of F on Ψ at different points of space). Since the transition to the superconducting state is of second order, Ψ has a small absolute value and varies slowly in space near the transition point. Remembering further that F should be real and gauge-invariant, we assume F in GL theory in the form

$$F = \int dr[a|\Psi|^2 + \frac{1}{2}b|\Psi|^4 + \frac{\hbar^2}{4m}\left|\left(\frac{\partial}{\partial r} - \frac{2ie}{\hbar c}A\right)\Psi\right|^2 + \frac{1}{8\pi}(\text{rot } A)^2] . \tag{4.5.10}$$

We have taken the coefficient of the third term on the right arbitrarily as $(\hbar^2/4m)$, where m is the electronic mass, since the proportionality constant in (4.5.5) may be chosen at our disposal. The coefficient of A in the same term is of course to make F gauge-invariant. The constant $\phi_0 = (\pi\hbar c/|e|)$ is called the flux quantum, since the magnetic flux enclosed by the (bulk) superconducting circuit is restricted to an integral multiple of ϕ_0 (*flux quantization*). By adding the last term on the right of (4.5.10), we can derive the Maxwell equation for the stationary magnetic field from the thermodynamic principle of minimum energy. Parameters a, b are to be calculated by microscopic theory. Here we assume only that near the transition point T_c the parameter b is positive and independent of the temperature T and that the parameter a is positive for $T > T_c$ and negative for $T < T_c$.

The probability for the order parameter and the vector potential to have functional forms $\Psi(r)$, $A(r)$, respectively, is proportional to $\exp(-F/k_B T)$ at temperature T. Its "integral" over all possible functional forms is the so-called sum over states. We need to fix the gauge of the vector potential; otherwise the sum over states will diverge. In practice it is convenient to take the Coulomb gauge, in which

$$\text{div } A = 0 \tag{4.5.11}$$

and A itself vanishes in the absence of the magnetic field.

In the thermodynamic limit, F is an extensive quantity proportional to the normalization volume V and Ψ, A will almost certainly take the functional forms for which F is a minimum (the maximum probability). When $A \equiv 0$, in particular, F takes the mimum value at $\Psi = 0$ (the normal state) for $T > T_c$ and at $|\Psi| = \Psi_0 \equiv (-a/b)^{1/2}$ (the superconducting state) for $T < T_c$. All states with $\Psi = \Psi_0 \exp(i\alpha)$ are degenerate, where α is a real constant. This phase constant α is essentially the same parameter as we have introduced in (4.5.6). For the present, we will take $\alpha = 0$.

We now turn to the case when A is nonvanishing, but small. Under an appropriate boundary condition, A is expressed uniquely as the sum of transverse component A_t and longitudinal component A_1, so that $\text{div} A_t = 0$ and rot $A_1 = 0$. The longitudinal component may be written as $A_1 = \partial \Lambda / \partial r$ with use of a suitable scalar Λ. We also decompose the order parameter $\Psi - \Psi_0$ into real and imaginary parts as $\Psi_1 + i\Psi_2$. The increase of F due to the presence of A is given, up to second order, by

$$\Delta F = \int dr \left[\frac{\hbar^2}{4m} (\nabla \Psi_1)^2 - 2a\Psi_1^2 + \frac{\hbar^2}{4m} \left(\nabla \Psi_2 - \frac{2e}{\hbar c} \Psi_0 A_t \right)^2 \right.$$
$$\left. + \frac{e^2}{mc^2} \Psi_0^2 A_t^2 + \frac{1}{8\pi} (\text{rot} A_t)^2 \right]. \tag{4.512}$$

The Euler equation obtained by varying A_t is nothing else but the Maxwell equation and takes the form $[\lambda^{-2} - \nabla^2] A_t = 0$ with the finite range $\lambda = [mc^2/8\pi \Psi_0^2 e^2]^{1/2}$. The magnetic flux can only penetrate the depth λ from the surface of the superconductor (*Meissner effect*). In the terminology of quantum theory of fields, we might say that the condensate Ψ_0 of the matter field endowed the magnetic field with a finite "mass". The real part Ψ_1 of the order parameter also satisfies the Euler equation of a similar form with the range $\xi = (\hbar^2 / 4m|a|)^{1/2}$, which is called the coherence length. Inside the bulk superconductor whose linear dimension is much bigger than λ, ξ, both A_t and Ψ_1 vanish. Finally, as regards Ψ_2, F takes the minimum value for $\Psi_2 = [2e/\hbar c] \Psi_0 \Lambda$. This is the first-order term in the expansion of the gauge transformation (4.5.9) into powers of Λ. In short, in the presence of a weak magnetic field, only the phase of the order parameter is shifted by the longitudinal component of the vector potential and $|\Psi|$ remains invariant. This property is sometimes referred to as the "rigidity of the macroscopic wave function".

Remember that the longitudinal component of the vector potential has "physical" reality in the following case. Suppose that a certain amount of magnetic flux is enclosed by a bulk superconducting ring and take a closed loop which lies deeper than the penetration depth λ inside the ring and encircles the flux. As we have seen above, A_t vanishes on the loop, but A_1 is finite since the line integral of A_1 along the loop should be equal to the magnetic flux enclosed by the loop. From this together with the demand that the order parameter shall be single-valued, we can readily conclude the quantization of the magnetic flux in units of $\phi_0 = (\pi\hbar c/|e|)$. The quantization of the magnetic flux shows the wave nature of matter and the reality of the vector potential unambiguously on the macroscopic scale.

The superconducting ring is characterized topologically by its double connectivity. The phase of Ψ can be infinitely many valued round the "hole" of the ring. In the case of the type II superconductor ($\sqrt{2}\lambda > \xi$ in GL theory), even if the conductor itself is simply connected, there appear *vortex lines* when the magnetic field exceeds a certain critical value and Ψ spontaneously takes a multiply connected structure. The vortex line is a one-dimensional singularity of the Ψ field; Ψ vanishes on the line, round which its phase is infinitely many-valued. When the line is isolated, one flux quantum ϕ_0 is produced by the superconducting current flowing within the distance λ from the line. The tension of the vortex line arises mainly from magnetic field energy as well as kinetic energy of superflow in the limit $\lambda \gg \xi$.

4.5.3 Josephson Effect

The phase α of Ψ may take any value in so far as the superconductor is isolated. When two superconductors are in contact and can exchange electrons with each other, however, their relative phase is fixed, just like the relative orientation of two magnets being fixed by the dipolar coupling. In GL theory, the contact means the existence of the boundary energy proportional to the product of order parameters on both sides, $\Psi_1^*\Psi_2 + \Psi_2^*\Psi_1$, so that the coherent state of the combined system is established. Here we may concern ourselves only with the adjustment of the relative phase $\alpha = \alpha_2 - \alpha_1$, ignoring the change in absolute values of order parameters. Let us assume the boundary energy in the simplest form

$$\Delta E = - E_0 \cos \alpha \tag{4.5.13}$$

since it must be a periodic and even function of α.

The quantum-mechanical description will take the following form. Take a metal–insulator–metal tunnel junction, in which two metals are superconductors of the same kind, and regard the electron tunneling through the insulating film as perturbation. The unperturbed energy is the sum of bulk energies of two superconductors, $E(N_1 + n) + E(N_2 - n)$. Here N_1, N_2 are average electron numbers in respective metals and n is the fluctuation caused by tunneling. Demanding that the bulk energy shall be a minimum with respect to n, we obtain the well-known equilibrium condition (at $T = 0$) of thermodynamics that the chemical potential $\mu = \partial E/\partial N$ takes the same value in both metals. Ordinarily, under this equilibrium condition, the macroscopic flow of particles vanishes between two systems in contact. This common sense does not apply to super-fluids nor to superconductors. Here a coherent state with the coherence energy (4.5.13) is established by superposing unperturbed states of different n values into the wave packet of the form (4.4.9) with a definite relative phase α. Since α is the phase of Cooper pairs, $\hbar\alpha$ and $n/2$ are canonically conjugate with each other.

One of canonical equations of motion, thus takes the form

$$\frac{dn}{dt} = \frac{2}{\hbar} \frac{\partial \Delta E}{\partial \alpha} = \frac{2}{\hbar} E_0 \sin \alpha \ . \tag{4.5.14}$$

This means that the macroscopic particle flow can exist between two systems of equal chemical potential if the combined system is in the coherent state with definite α (dc *Josephson effect*). Essentially the same mechanism must prevail inside bulk superfluids as well as bulk superconductors. Here the particle flow is proportional to the infinitesimal phase difference between neighboring volume elements. In fact, in GL theory, the electric current in the Maxwell equation obtained from (4.5.10) by varying A takes the form

$$j = \frac{e}{m} |\Psi|^2 \left(\hbar \frac{\partial \alpha}{\partial r} - \frac{2e}{c} A \right) \tag{4.5.15}$$

where we have written $\Psi = |\Psi| \exp(i\alpha)$. The expression (4.5.15) is the gauge-invariant statement that the current is essentially proportional to the space gradient of the phase.

The other canonical equation of motion, to be combined with (4.5.14), takes the form

$$\frac{1}{2} \hbar \frac{d\alpha}{dt} = \mu_1 - \mu_2 \tag{4.5.16}$$

which is referred to as the ac Josephson effect. Usually the difference in chemical potentials of two superconductors is equal to eV, where V is the bias voltage. Then (4.5.16) does not depend on any material constants, but only contains the universal constant $\hbar/e$. When V is a dc voltage, α is a linear function of time and the current (4.5.14) oscillates with the frequency $(eV/\pi\hbar)$.

4.6 Broken Symmetry and Elementary Excitation

When some symmetry is broken in the ground state, there appears a mode of elementary excitations, which may be interpreted as the collective motion of the macroscopic system trying to restore the broken symmetry. Before going into abstract general arguments, let us show some examples.

4.6.1 The Heisenberg Ferromagnet

The simplest example is the Heisenberg ferromagnet described by (4.2.3) without external field. Since the model is isotropic in spin space, the state with all spins aligned in any direction has the same energy as the state $|0\rangle$, in which all spins are along the z axis. In Sect. 4.4, we have described the situation by saying that the excitation energy of the $k = 0$ magnon vanishes. Logically, however, this is tantamount to the statement of the degeneracy of the ground state due to the isotropy of the Hamiltonian. The statement in terms of magnons will become a nontrivial one only when the Fourier transform of the exchange interaction, (4.2.9), is assumed to be continuous at $q = 0$. This is equivalent to saying that the exchange interaction is of a short-range type. Then we can, without checking the explicit expression (4.3.6), conclude that the excitation energy $\varepsilon(k)$ of the magnon with the wave vector k vanishes in the limit $k \to 0$.

We notice in this case that the expectation value of the number of magnons in equilibrium, (4.3.11), is divergent as $k \to 0$. This divergence may be taken as the violent thermal fluctuations through which spins are trying to restore the rotational symmetry broken in the ground state $|0\rangle$. Fortunately, in the three-dimensional system, the spin defect at each lattice point, given by (4.3.12), is convergent since the volume element in k space is proportional to $k^2 dk$. In one- and two-dimensional systems, on the other hand, "volume" elements are proportional to dk and $k dk$, respectively, and therefore (4.3.12) is divergent. This implies that the long-range order in the ground state is destroyed by thermal fluctuations. Strictly speaking, the divergent integral only means the inconsistency of the spin wave approximation based on the assumption that the spin defect is small. It is not a mathematical proof of the absence of the ferromagnetic long-range order in one- and two-dimensional Heisenberg models at finite temperature. Here we mention only that such a proof is in fact possible; with use of the so-called Bogoliubov inequality (a kind of Schwartz inequality) it has been shown that the quasiaverage (4.4.1) vanishes for one- and two-dimensional systems at finite temperature.

Let us now generalize (4.2.3) to the anisotropic exchange interaction

$$\mathscr{H}_{\text{ex}} = -\frac{1}{2} \sum_{j \neq l} \sum_{\mu = x, y, z} J_{jl}^{(\mu)} S_{j\mu} S_{l\mu} . \tag{4.6.1}$$

It reduces to the *Ising model* when $J^{(x)} = J^{(y)} = 0$. Since we are concerned only with z components of spins, we may regard S_{jz} as the c number which takes values $\pm 1/2$. While the Heisenberg model has the continuous symmetry represented by the rotation group, the Ising model has the discrete symmetry against time reversal and, in the case of $J^{(z)} > 0$, possesses only two ground states, one with all spins up and the other with all spins down. Take Ising spins on one-dimensional lattice points with the nearest neighbor exchange interaction $J^{(z)} = J > 0$. We start with the ground state with all spins up and then reverse all spins to the right of an arbitrarily chosen spin. The system is thus divided into two *domains*, one with all spins up and the other with all spins down. The creation of the domain boundary costs the energy $(1/2)J$, which arises from the antiparallel pair of spins at the boundary. The creation energy does not depend on the total number N of spins; the domain boundary is a "point" defect in the one-dimensional system. Including the entropy due to the random distribution of domain boundaries, we obtain the free energy per boundary as $(1/2)Jp + k_B T[p \ln p + (1 - p) \ln (1 - p)]$, where Np is the total number of boundaries. The free energy is a minimum when $p = [\exp (J/2k_B T) + 1]^{-1}$ and its minimum value is equal to $- k_B T \ln [1 + \exp (- J/2k_B T)]$, which agrees with the well-known result of statistical mechanics. In the one-dimensional system, the long-range order is impossible because we have a finite concentration of domain boundaries, however low the temperature is. In the two-dimensional system, on the other hand, the boundary energy is proportional to the length of the boundary curve, so that the thermal fluctuation cannot create domains to destroy the long-range order. In fact, the two-dimensional Ising model shows the phase transition at a finite temperature, as has been proved rigorously by Onsager.

4.6.2 The Spin Model of Liquid ^{4}He

Let us turn to the so-called *XY model*, in which the spin has the magnitude $1/2$ and $J^{(x)} = J^{(y)} (= J > 0)$, $J^{(z)} = 0$. It may also be regarded as a lattice model of superfluid ^{4}He. Thus ^{4}He atoms are not allowed to assume any points in space, but restricted to occupy discrete lattice points. In view of the hard core of the atom, we assume that two or more atoms cannot occupy the same lattice point. The occupancy of each lattice point is represented by the orientation of the spin ($S_{jz} = \pm 1/2$), so that $S_{j\pm} = S_{lx} \pm S_{jy}$ are creation and destruction operators and the rotation of spins round the z axis means the gauge transformation (4.5.2). The XY exchange interaction between spins proportional to $S_{j+}S_{l-} + S_{j-}S_{l+}$ represents the tunneling of the atom from one site to another. The invariance of the interaction against the spin rotation round the z axis, $[\mathcal{H}_{ex}, \sum_j S_{jz}] = 0$, means the conservation of the number of atoms.

In the ground state, spins are parallel to a certain direction on the xy plane. Let α be the angle between the spin direction and the x axis. In contrast to the Heisenberg model, the ground state is no longer an eigenstate of the α component of the total spin, $S_\alpha = \sum_j S_{j\alpha}$, since the Hamiltonian does not commute with x and y components of the total spin. Thus, spins are doing the "zero-point precession". In spite of it, the expectation values $\langle S_\alpha \rangle$ and therefore $\langle S_{j\pm} \rangle$ do not vanish. This means that liquid ^{4}He is in the coherent state. The angle α represents the phase of the matter wave and ground states are degenerate with respect to α. We expect to obtain a branch of elementary excitations, whose excitation energy $\varepsilon(k)$ vanishes for the vanishing wave vector, since we assume that exchange interaction of a short range.

Let us assume that spins are all parallel to the x axis ($\alpha = 0$) in the ground state, and also that the spin fluctuation round the x axis is small, so that $\langle S_{jx}\rangle \cong$ 1/2. We linearize the equation of motion $dS_{jy}/dt = \sum_l J_{jl}S_{jz}S_{lx}$ and the one for S_{jz} by replacing S_{jx} with its expectation value (RPA). Similarly we replace the commutation relation approximately as

$$[S_{jy}, S_{lz}] = i\delta_{jl}S_{jx} \cong \frac{i}{2}\delta_{jl} . \tag{4.6.2}$$

Then Fourier transforms (4.2.4) are expressed as

$$S_y(k) = \left[\frac{\varepsilon(k)}{[J(0) - J(k)]}\right]^{1/2} (b_k + b_{-k}^\dagger)$$

$$S_z(k) = \left[\frac{\varepsilon(k)}{J(0)}\right]^{1/2} i(b_{-k}^\dagger - b_k) \tag{4.6.3}$$

$$\varepsilon(k) = \left\{\frac{1}{2} J(0) [J(0) - J(k)]\right\}^{1/2} \tag{4.6.4}$$

where b_k, $b_k^\dagger$ are destruction and creation operators of Bose type, respectively. They satisfy equations of motion

$$\hbar \frac{d}{dt} b_k = - i\varepsilon(k) b_k \tag{4.6.5}$$

Hence $b_k^\dagger$ creates the elementary excitation, whose excitation energy is given by (4.6.4).

For small k, $\varepsilon(k)$ is proportional to k, so that the elementary excitation in this limit may be identified with the phonon observed in superfluid ^{4}He. The coefficient of b operators in (4.6.3) gives the zero-point amplitude of spins. The amplitude of the y mode is inversely proportional to $k^{1/2}$ and therefore divergent as $k \to 0$. This mode thus represents the fluctuation of the phase α (*the phase mode* or *phason*). The spin defect $\langle S_{jy}^2\rangle = N^{-1} \sum_k \langle S_y(k)S_y(- k)\rangle$ is convergent in the case of the three dimensional system, thanks to the integration in k space, but divergent for the two-dimensional system at finite temperature. In this case, too, it is possible to prove that the long-range order in the usual sense, i.e., the nonvanishing quasiaverage $\langle S_{j\pm}\rangle$, does not exist. In contrast to the Heisenberg model, however, a certain phase transition seems to occur in the two-dimensional XY model and Bose fluid (in accordance with the experimental observation that very thin ^{4}He films exhibit superfluidity at low temperature).

It is believed that the phase transition in this case is connected with vortex lines. In the three-dimensional system, the energy of the vortex line is proportional to its length and therefore, like the dislocation line in a crystal, it cannot be regarded as the elementary excitation which is created by thermal excitation. In the two-dimensional case, on the other hand, the "line" has a finite length independent of the total size of the system, even if we take into account the finite thickness of the physical system. However, at least in a neutral superfluid such as ^{4}He, the velocity of the vortex, being inversely proportional to the distance from the center, decreases so gradually that the hydrodynamic energy is proportional to the logarithm of the linear dimension of the system. Hence the single vortex line cannot be regarded as the ordinary elementary excitation even in the two-dimensional system. We should notice, however, that the entropy arising from the random distribution of vortex centers is also proportional to the logarithm of the system size; free vortices must be thermally excited when the temperature exceeds a certain critical value. Below this critical temperature, a *pair* of two vortices with opposite signs will form a bound state. The vortex pair has an excitation energy independent of the system size and may therefore be regarded as an elementary excitation. The phase transition to occur in the two-dimensional XY model or Bose fluid is believed to be the cooperative dissociation of vortex pairs.

4.6.3 Classical Crystals

Let us consider a crystal, to which we can apply the naive picture of the crystal having one atom in the vicinity of each lattice point. Equilibrium positions of atoms can be specified by giving the center of mass position X of the whole crystal together with three vectors of fundamental periods. They are order parameters of this classical crystal.

Like transverse spin components in the XY model, X is not commutable with the Hamiltonian, so that it is not a constant of motion. The constant of motion is the center of mass momentum P, which is conjugate with X. Needless to say, the state of definite P is represented by a plane wave of X and the expectation value of (4.2.2) in this state vanishes unless $q = 0$. In actuality, the crystal is not in this kind of plane-wave state, but in a wave-packet state, for which X is almost definite. Since the total mass of the crystal is macroscopic, the group velocity of the wave packet and the associated kinetic energy are negligible even if the uncertainty of X is of the same order of magnitude as the atomic size itself.

Now, the long-wavelength acoustic phonon may be regarded as a collective motion of atoms trying to restore the translational symmetry broken in the ground state. The normal mode of the frequency ω_k makes contribution to $\langle \xi^2 \rangle$ by the amount $\hbar/2NM\omega_k$, where ξ is the atomic displacement from the lattice point in the zero-point vibration (Sect. 1.4). In the case of the acoustic phonon, it is proportional to k^{-1} and therefore divergent as $k \to 0$. This divergence implies that the zero-point vibration is trying to make the distribution of the center of mass uniform in space. When integrated over k, $\langle \xi^2 \rangle$ is convergent in three- and two-dimensional systems, but divergent in the one-dimensional system. This implies that the one-dimensional crystal is not stable even at $T = 0$. In the two-dimensional systems, too, $\langle \xi^2 \rangle$ is divergent when $T > 0$. Computer simulations and experiments on the inert gas adsorbed on the cleavage surface of graphite, however, show that the phase transition corresponding to the melting of the three-dimensional crystal occurs also in the two-dimensional system. The situation may be similar to the one we have described concerning the XY model.

4.7 Goldstone's Theorem

In the preceding section, we have seen some examples, in which a certain branch of elementary excitations have vanishing excitation energies in the long-wavelength limit in close relation with the broken symmetry. The relation is often referred to as *Goldstone's theorem*. It is the nonrelativistic version of a theorem bearing the same name in the relativistic quantum theory of fields. In our nonrelativistic case, however, theorem is the synonym of the degeneracy of ground states once we assume the continuity in k space, as we have emphasized already in Sect. 4.6.1. It is difficult, on the other hand, to prove the continuity in general terms without specifying interaction, order parameter, etc. Anyway, let us first state the theorem in abstract form and then discuss the case of superconductivity as an example, to which the theorem (i.e., the continuity) does not apply.

4.7.1 Conditions for the Theorem to Apply

We assume that the Hamiltonian $\mathcal{H}$ of our macroscopic system is invariant against the similarity transformation $\mathcal{H} \to G \mathcal{H} G^{-1}$, so that $\mathcal{H}$ commutes with G. Here G represents one of symmetry operations, which form a *continuous* group. Let us write the infinitesimal transformation as $1 + i\varepsilon D$, where ε is the infinitesimal parameter and D is the generator of the symmetry group. Thus D is the total momentum in the case of the translation group and the total angular momentum in the case of the rotation group. It commutes with $\mathcal{H}$ and is a constant of motion.

Now, let a set of dynamical variables ϕ_1, ϕ_2, ... ϕ_n be bases of an irreducible representation of the symmetry group. We write their infinitesimal transformation as $[D, \phi_\alpha] = \sum_\beta C_{\alpha\beta}\phi_\beta$, so that the determinant formed by $C_{\alpha\beta}$ does not vanish. When $|0\rangle$ is one of lowest energy states of $\mathcal{H}$, we have

$$\sum_n (\langle 0|D|n\rangle\langle n|\phi_\alpha|0\rangle - \langle 0|\phi_\alpha|n\rangle \langle n|D|0\rangle)$$
$$= \sum_\beta C_{\alpha\beta} \langle 0|\phi_\beta|0\rangle \ . \tag{4.7.1}$$

Intermediate states $|n\rangle$ on the left are orthonormalized eigenstates belonging to eigenvalues E_n of $\mathcal{H}$, but only states with $E_n = E_0$ make a contribution to the sum over n, since $[D, \mathcal{H}] = 0$. If the ground state is nondegenerate, therefore, the single term with $n = 0$ remains on the left hand side, which thus vanishes. Hence all the $\langle 0|\phi_\beta|0\rangle$ vanish. Conversely, if at least one $\langle 0|\phi_\alpha|0\rangle$ is nonvanishing and thus the symmetry is broken, the ground state should necessaryily have degeneracy.

To be more specific, we now assume that D is the sum of operators D_j defined at lattice points or the space integral of $D(r)$ defined at all points inside the system. We write the Fourier transform of D_j [or $D(r)$] as $D(k)$ and define

$$L_\alpha(k, \omega) = \int_{-\infty}^{\infty} dt \, \exp(i\omega t) \, \langle 0|[D(k, t), \phi_\alpha]|0\rangle \tag{4.7.2}$$

where $D(k, t) = \exp(it\mathscr{H}/\hbar) D(k) \exp(-it\mathscr{H}/\hbar)$ is the Heisenberg representation. We write the integrand in the form similar to the left-hand side of (4.7.1) and obtain

$$L_\alpha(k, \omega) = 2\pi\hbar \sum_n [\langle 0|D(k)|n\rangle \langle n|\phi_\alpha|0\rangle \, \delta(\hbar\omega - E_{n0})$$

$$- \langle 0|\phi_\alpha|n\rangle \langle n|D(k)|0\rangle \, \delta(\hbar\omega + E_{n0})] \tag{4.7.3}$$

where $E_{n0} = E_n - E_0$.

At $k = 0$, on the other hand, $D(k) = D$, so that from (4.7.1) we have

$$L_\alpha(0, \omega) = 2\pi\hbar\delta(\hbar\omega) \sum_\beta C_{\alpha\beta} \langle 0|\phi_\beta|0\rangle \ . \tag{4.7.4}$$

If we assume that (4.7.3) is continuous at $k = 0$, therefore, we can conclude that (4.7.3) should contain those excited states $|n\rangle$ which are excited from the ground state $|0\rangle$ by the operator $D(k)$ and have vanishing excitation energies in the limit $k \to 0$. This is the nonrelativistic Goldstone theorem, and its proof is thus reduced to the proof of continuity. The continuity at $k = 0$ physically means that the surface effect can be neglected by making the volume of the system V infinite in the thermodynamic limit.

To see this, let us take a region of volume V_0 ($< V$) inside the system and define $D^{(0)}$ by taking the sum of D_j over lattice points j in V_0. If we first take the limit $V_0 \to V$ and then the limit $V \to \infty$, $D^{(0)}$ is of course equal to D, which is a constant of motion. If we first take the limit $V \to \infty$ and then the limit $V_0 \to \infty$, on the other hand, $D^{(0)}$ is equal to the limit of $D(k)$ as $k \to 0$. In this case, $D^{(0)}$ is a constant of motion *only* when the contribution to $dD^{(0)}/dt$ from particles outside V_0 vanishes in the limit $V_0 \to \infty$.

4.7.2 The Case of Superconductivity

Such an example is given by the system of electrons in a metal. When we deal with charge density fluctuations of wavelengths comparable to the system size, we should include the surface charge in order to satisfy the charge conservation law, because of the long-range nature of the Coulomb interaction between electrons. Consequently Goldstone's theorem does not apply to the superconductor, as we will show below.

When we deal with fluctuations of wavelengths longer than the coherence length, we can assume that the absolute value of the superconducting order parameter is kept constant. We need to concern ourselves only with fluctuations of the phase α of the order parameter. Let us denote the phase at space point r by $\alpha(r) = (2/\hbar)\varphi(r)$. Let us also denote the average electron density by n_0 and its fluctuation at space point r by $\nu(r)$. As we have mentioned in Sect. 4.5, $\nu(r)$ and $\varphi(r)$ are canonically conjugate with each other:

$$[\nu(r), \varphi(r')] = i\hbar\delta(r - r') . \tag{4.7.5}$$

We now apply the theory of Josephson effect in Sect. 4.5 to the phase fluctuation inside the bulk superconductor, so that the dynamics of the phase may be described with use of the effective Hamiltonian

$$\mathcal{H}_{\text{eff}} = \int \left\{ \frac{n_0}{2m} \left[\frac{\partial\varphi(r)}{\partial r} - \frac{e}{c} A(r) \right]^2 + \frac{1}{2} \frac{m}{n_0} v^2\nu^2(r) \right\} dr$$

$$+ \frac{1}{2} e^2 \iint \frac{\nu(r)\,\nu(r')}{|r - r'|} dr'\, dr . \tag{4.7.6}$$

We have taken $T = 0$, where the square of the absolute value of the order parameter is assumed to be equal to $n_0/2$. The second term in the brace on the right of (4.7.6) is the quadratic term of the expansion of the ground-state energy of the electron system into powers of the density. If we take the ground-state energy of the free electron gas model, for instance, the parameter v is the Fermi velocity v_F multiplied by $3^{-1/2}$. We have taken the Coulomb gauge of the electromagnetic potentials, but we have not included the energy of the electromagnetic field. The cross term between $\partial\varphi/\partial r$ and A arising from the brace in (4.7.6) vanishes when integrated. The term proportional to A^2 results in the Meissner current $- c\delta\mathcal{H}/\delta A = - (ne^2/mc^2) A$ as we have mentioned in Sect. 4.5.

We now consider the case of $A = 0$ and introduce Fourier transforms

$$\nu(r) = V^{-1/2} \sum_k \nu_k \exp(ik\cdot r), \quad \varphi(r) = V^{-1/2} \sum_k \varphi_k \exp(-ik\cdot r) . \tag{4.7.7}$$

Then (4.7.5) takes the form

$$\mathcal{H}_{\text{eff}} = \sum_{k\neq 0} \left[\frac{n_0}{2m} k^2\varphi_k\varphi_{-k} + \frac{1}{2} \left(\frac{m}{n_0} v^2 + \frac{4\pi e^2}{k^2} \right) \nu_k\nu_{-k} \right] . \tag{4.7.8}$$

This can be diagonalized as (1.4.13) by introducing Bose operators b_k, $b_k^{\dagger}$ through

$$\nu_k = \left(\frac{\hbar n_0 k^2}{2m\omega(k)} \right)^{1/2} (b_k + b_{-k}^{\dagger})$$

$$\varphi_k = \left(\frac{\hbar m\omega(k)}{2n_0 k^2} \right)^{1/2} i(b_k^{\dagger} - b_{-k}) . \tag{4.7.9}$$

Here $\omega(k)$ is the frequency of our phase mode and given by

$$\omega(k) = [(4\pi n_0 e^2/m) + v^2 k^2]^{1/2} . \tag{4.7.10}$$

In contrast to (4.6.4), it has the finite value $\omega_p = [4\pi n_0 e^2/m]^{1/2}$ in the limit $k \to 0$ and ω_p is nothing else but the plasma frequency characteristic of electronic charge fluctuations (Sect. 3.4).

When the longitudinal component of the vector potential is taken as $A = \partial \Lambda/\partial r$, φ_k in (4.7.8) is replaced by $\varphi_k + (e/c)\Lambda_k$, where Λ_k is the Fourier transform of $\Lambda(r)$. This gauge transformation, when Λ_k is infinitesimal, may be written as $i[D(k), \varphi_k] = (e/c)\,\Lambda_k$, where the generator is given as $D(k) = -(e/\hbar c)\Lambda_k v_k$. Form (4.7.9), the Heisenberg representation is given by

$$D(k, t) \propto b_k \exp[-i\omega(k)t] + b_{-k}^\dagger \exp[i\omega(k)t] . \tag{4.7.11}$$

In the limit $k \to 0$, $\omega(k)$ reduces to ω_p, so that Goldstone's theorem does not apply.

In the limit $e \to 0$, on the contrary, we have $\omega(k) = vk$ and $\partial D(k, t)/\partial t$ vanishes in the limit $k \to 0$, so that Goldstone's theorem does hold. When $e = 0$, (4.7.6) reduces to the Hamiltonian of *quantum hydrodynamics* introduced for the first time by Landau to describe superfluid ⁴He, and the parameter v can be identified with the velocity of sound in the fluid.

4.8 Soft Modes

We now turn to the case of second-order phase transition at a finite transition temperature T_c. Let ψ_j be a dynamical variable defined at each lattice point R_j (or at any point inside the system). Its expectation value in thermal equilibrium, $\Psi_j = \langle \psi_j \rangle$, vanishes for $T > T_c$ and is nonvanishing for $T < T_c$, so that it can be taken as the order parameter. We will assume the second-order transition, so that Ψ_j is continuous at T_c, but it is not difficult to extend our arguments to a first-order transition as far as the discontinuity at T_c is small.

For $T > T_c$, $\langle \psi_j \rangle$ can be made finite by perturbation of the form $\sum_j h_j \psi_j$ which is produced by applying an appropriate external force h_j. When the field is weak, we may concern ourselves with the linear response, which we write $\langle \psi_j \rangle = \sum_l \chi_{jl} h_l$ with use of the generalized susceptibility χ_{jl}. It is also possible to introduce the Fourier transform similar to (4.2.5), so that we write $\langle \psi(k) \rangle = \chi(k)h(k)$. Here $\chi(k)$ is the Fourier transform of χ_{jl}, which is assumed to depend only on $R_j - R_l$. We do not ask how to realize the external field varying in space with the wave vector k in our laboratory. In the cases of superfluidity and supercondutivity, ψ is the destruction operator of the particle (or particle pair), but we do not ask what sort of field would have a linear coupling with it.

Generalizing the magnetic long-range order, we assume that the response function $\chi(k)$ becomes divergent as $T \to T_c$ when k is equal to a certain wave vector Q (and also to $-Q$), so that our system shows a finite response against the infinitesimal external field. In other words, the high-temperature state becomes unstable against the field $h(Q)$ and the "spontaneous deformation" to a new state occurs. The simplest mechanism would be the model possessing a mode of elementary excitations, whose excitation energy becomes vanishing when T approaches T_c from above. This type of mode is called the *soft mode*. The name derives from the structure transformation of crystals, in which the frequency of a certain phonon mode becomes vanishing and therefore the crystal is really softened.

4.8.1 Ferroelectrics with Hydrogen Bonds

As an example, we take the dielectric crystal such as KH_2PO_4, in which hydrogen bonds are responsible for the dielectric anomaly. The proton in the hydrogen bond, when classical mechanics is applied, possesses two equivalent equilibrium positions (potential minima), but in quantum mechanics its tunneling from one equilibrium position to the other should be taken into account because of the small mass of the proton. A simple model is the pseudospin formalism similar to the one used to describe superfluid ^{4}He in Sect. 4.6.2. Thus two possible locations of the jth proton are represented by eigenvalues $\pm 1/2$ of the pseudospin component S_{jz} (which has nothing to do with the physical nuclear spin of the proton). The quantum mechanical tunneling of the proton from one location to the other is described by including a term proportional to S_{jx} in the Hamiltonian. We also include the Ising-type exchange interaction, which results in the ordered arrangement of protons below the transition temperature T_c. The Hamiltonian thus takes the form

$$\mathscr{H} = -\sum_j \Gamma S_{jx} - \frac{1}{2}\sum_{j\neq l}\sum J_{jl}S_{jz}S_{lz} \tag{4.8.1}$$

where we assume $\Gamma > 0$ without the loss of generality. We also assume that the Fourier transform $J(k)$ of J_{jl} takes a positive maximum value at $k = 0$. Hence the second term on the right of (4.8.1) tries to make pseudospins parallel to the z axis, whereas the first term counteracts it.

We apply the molecular field (mean field) approximation to the second term, so that we replace one of S_{jz} by its expectation value $\langle S_z\rangle$. We assume that the latter is independent of the lattice site j. The approximation is equivalent to saying that each pseudospin is acted upon by the effective field $\eta = (\Gamma^2 + J^2\langle S_z\rangle^2)^{1/2}$ along the direction of the angle $\tan^{-1}(J\langle S_z\rangle/\Gamma)$ measured from the x axis. Here J is the value of $J(k)$ at $k = 0$. Expectation values of the pseudospin components at temperature T are given by

$$\langle S_z\rangle = (J\langle S_z\rangle/2H)\tanh(H/2k_BT)$$
$$\langle S_x\rangle = (\Gamma/2H)\tanh(H/2k_BT) \ . \tag{4.8.2}$$

The first equation has a nonvanishing solution when $J > 2\Gamma$ and $T < T_c$, where T_c is given by

$$1 = (J/2\Gamma)\tanh(\Gamma/2k_B T_c) \ . \tag{4.8.3}$$

For $T > T_c$, $\langle S_z \rangle = 0$, but the fluctuation of S_z does not vanish. By the Fourier transformation of equations of motion, we have

$$\hbar \frac{d}{dt} S_z(\boldsymbol{k}) = - \Gamma S_y(\boldsymbol{k})$$

$$\hbar \frac{d}{dt} S_y(\boldsymbol{k}) = \Gamma S_z(\boldsymbol{k}) - N^{-1/2} \sum_{\boldsymbol{k}'} J(\boldsymbol{k} - \boldsymbol{k}') S_x(\boldsymbol{k}') S_z(\boldsymbol{k} - \boldsymbol{k}') \tag{4.8.4}$$

$$\cong \{\Gamma - J\langle S_x\rangle\} S_z(\boldsymbol{k})$$

In the second equation, we have replaced $S_x(\boldsymbol{k})$ approximately by its expectation value $\langle S_x(\boldsymbol{k}) \rangle \doteq \delta_{\boldsymbol{k},0} N^{1/2} \langle S_x \rangle$, so that modes with different wave vectors are not coupled with one another. This is the intent of *RPA* (random phase approximation). The frequency $\omega(\boldsymbol{k})$ of the fluctuation is given by

$$[\hbar\omega(\boldsymbol{k})]^2 = \Gamma[\Gamma - (J/2) \tanh (\Gamma/2k_B T)] \tag{4.8.5}$$

We see that $\omega(0)$ is proportional to $(T - T_c)^{1/2}$ as T approaches T_c determined by (4.8.3).

4.8.2 Soft Mode and Central Peak

In the case of the ferroelectric with hydrogen bonds, the ordering in proton locations induces the deformation of surrounding ions and therefore the spontaneous electric polarization. In order to take account of this situation in our pseudospin model, we introduce a linear coupling between S_{jz} and an optic mode of lattice vibration. Even for $T > T_c$, then, the softening of the proton motion will perturb the optic vibration through the assumed coupling. Instead of the method of equations of motion starting from the Hamiltonian (4.8.1), however, we will apply the so-called *TDGL theory*, which is the generalization of GL theory in Sect. 4.5 to include the time dependence of the order parameter in relaxation phenomena. When the frequency becomes low near T_c, the damping can be quite important, but the Hamiltonian formalism is neither mathematically simple nor physically transparent to deal with the damping.

Now, we can assume that our order parameter Ψ is proportional to $\langle S_z(\boldsymbol{k}) \rangle$ at $\boldsymbol{k} = 0$. The equilibrium value of Ψ vanishes for $T > T_c$, and Ψ in this case means the fluctuation of the order parameter. We write the increase of the free energy due to the fluctuation as $\Delta F = (a/2)\Psi^2 + vq\Psi$ up to the second order. If we apply the molecular field approximation to (4.8.1) with $\Gamma = 0$, we obtain $a = 4k_B(T - T_c)$ near T_c. In what follows, however, we need only to assume that a is a positive parameter proportional to $T - T_c$. The term linear in Ψ arises from the coupling of the order parameter with the optic mode of vibration described by the normal coordinate q.

Since the decay of the fluctuation will be proportional to $\partial \Delta F / \partial \Psi$ to the lowest order of approximation, we assume the equation of relaxation of Ψ in the form

$$\frac{\partial \Psi}{\partial t} = -\gamma(a\Psi + vq) + f \qquad (4.8.6)$$

where γ is a positive parameter independent of temperature. We have included the random force $f(t)$ to cause the fluctuation of Ψ (Langevin equation). The frequency spectrum of $f(t)$ should be "white" with the intensity $2\gamma k_B T$ (the *fluctuation-dissipation theorem*). The equation of motion of q, on the other hand, takes the form

$$\frac{d^2 q}{dt^2} = -\Omega^2 q - \frac{v}{M}\Psi \qquad (4.8.7)$$

where Ω is the frequency of the optic mode for $\Psi = 0$ and M is the reduced mass of ions. When T is far above T_c, we have $\gamma a \gg \Omega$, so that Ψ "adiabatically" follows the ionic vibration. The latter therefore reflects the softening of the motion of Ψ. As T approaches T_c, however, it becomes increasingly difficult for Ψ to follow the vibration, and the frequency spectrum splits into the low frequency peak due to the anomalously slow relaxation of Ψ and the high frequency peak due to the optic vibration which is somewhat broaded by the coupling with Ψ. The low frequency peak is called the *central peak*.

Mathematically, we take Fourier transforms of (4.8.6), (4.8.7) with respect to t and apply the fluctuation-dissipation theorem to determine the power spectrum of q. We thus obtain

$$\langle |q|^2 \rangle = (2\Omega k_B T / v) |(x^2 - 1)(\varepsilon + \beta - ix) + \beta|^{-2} \qquad (4.8.8)$$

where $x = \omega/\Omega$, $\beta = (\gamma v^2 / N\Omega^3)$, $\varepsilon = (\gamma/\Omega)a^*$, $a^* = a - (v^2/M\Omega^2)$. The transition temperature is renormalized by the coupling from T_c to T_c^*, which is determined by $a^* = 0$. The dimensionless parameter ε measures the difference between T and T_c^*.

For $\varepsilon \gg 1$, (4.8.8) shows a maximum at $|x| \cong [\varepsilon/(\varepsilon + \beta)]^{1/2} \propto (T - T_c^*)^{1/2}$, which would result in the vanishing phonon energy at $T = T_c^*$. For $\varepsilon \ll 1$, however, (4.8.8) is actually proportional to $[(x - 1 + \delta)^2 + \delta^2]^{-2}$ near $x = 1$, so that the phonon peak shows a temperature-independent shift and damping $\delta = [\beta^2/2(1 + \beta^2)]$. Near $x = 0$, on the other hand, (4.8.8) is proportional to $(x^2 + \varepsilon^2)^{-2}$, which gives the central peak of width ε and height ε^{-2}.

Soft modes and central peaks are observed in many structural transformations, but it should be remembered that the relaxation process giving rise to the
central peak is not always the same.

4.9 Mean Field Approximation

The idea of mean field approximation was introduced for the first time by the
Weiss theory of ferromagnetism to deal with phase transitions. It has been called
the molecular-field approximation in magnetism and statistical mechanics.
Needless to say, Hartree and Hartree–Fock approximations are best known examples in quantum mechanics. Band theory may be regarded as the application
to the electron in a crystal. The electrons are assumed to move independently of
each other in a periodic field, which includes the effect of Coulomb interaction
between electrons on the average. The Coulomb force exerted on one electron
by the other should fluctuate as the second electron moves, but we ignore the
fluctuation in the mean field approximation. Assuming that the periodic field
depends on the electron spin as well, we can apply band theory to magnetic
metals, such as ferromagnetic Fe, Co, Ni. It is a great advantage of the mean
field approximation that we can adapt it to each physical situation and success-
fully describe qualitative features of a variety of phenomena in a wide range of
temperature. Let us see some examples.

4.9.1 Stoner's Model of Ferromagnetic Metals

The magnetic behavior of transition metals is determined primarily by $3d$ electrons, which are rather localized. But the Coulomb interaction between them is
screened by $4s$ electrons. Actually there are five independent d orbitals, but we
ignore this orbital degeneracy. Thus, in our simplified model, we assume one
orbital state at each lattice point and write the Hamiltonian as

$$\mathcal{H} = \sum_{j,l} t_{jl} A_{j\sigma}^{\dagger} A_{l\sigma} + U \sum_{j} N_{j\uparrow} N_{j\downarrow} \; . \tag{4.9.1}$$

Here $A_{j\sigma}$ is the destruction operator of the electron localized near the jth lattice
point (Wannier state) with spin σ, and $N_{j\sigma} = A_{j\sigma}^{\dagger} A_{j\sigma}$ is the electron number. The
first term on the right of (4.9.1) represents the electron tunneling from one site
to another and the second term is the Coulomb repulsion ($U > 0$) between two
electrons on the same site. We have ignored the remaining part of Coulomb
interaction, assuming that this is well screened.

We introduce the destruction operator $a_{k\sigma}$ of the electron in the nonlocalized
Bloch state by

$$A_{j\sigma} = N^{-1/2} \sum_{k} a_{k\sigma} \exp{(i\boldsymbol{k} \cdot \boldsymbol{R}_j)} \tag{4.9.2}$$

where $\boldsymbol{k}$ is the reduced wave vector. Then (4.9.1) takes the form (*Hubbard Hamiltonian*)

$$\mathscr{H} = \sum_k \varepsilon_k a^\dagger_{k\sigma} a_{k\sigma} + N^{-1} U \sum_{k_1 k_2 k_3 k_4} \delta(k_1 + k_2 - k_3 - k_4)\, a^\dagger_{k_1\uparrow} a^\dagger_{k_2\downarrow} a_{k_3\downarrow} a_{k_4\uparrow}$$

$$(4.9.3)$$

where ε_k is the Fourier transform of t_{jl}, which is assumed to depend only on $R_j - R_l$. The delta function $\delta(Q)$ vanishes unless Q is a reciprocal lattice vector, in which case it is equal to unity.

Now the Hartree–Fock (HF) approximation means replacing the product of four a operators in the second term on the right of (4.9.3) as

$$a^\dagger_1 a^\dagger_2 a_3 a_4 \cong \langle a^\dagger_2 a_3 \rangle a^\dagger_1 a_4 + \langle a^\dagger_1 a_4 \rangle a^\dagger_2 a_3 - \langle a^\dagger_2 a_4 \rangle a^\dagger_1 a_3$$
$$- \langle a^\dagger_1 a_3 \rangle a^\dagger_2 a_4 - \langle a^\dagger_1 a_4 \rangle \langle a^\dagger_2 a_3 \rangle + \langle a^\dagger_1 a_3 \rangle \langle a^\dagger_2 a_4 \rangle \ . \qquad (4.9.4)$$

The last two terms on the right are added in order not to double count the interaction when we calculate the expectation value of the energy. The interaction term in the Hamiltonian (4.9.3) is thus replaced by

$$\frac{U}{2N} \sum_q \left\{ \langle \rho_{-q} \rangle \rho_q - \langle \sigma_{-q} \rangle \cdot \sigma_q - \frac{1}{2} \langle \rho_q \rangle \langle \rho_{-q} \rangle + \frac{1}{2} \langle \sigma_q \rangle \cdot \langle \sigma_{-q} \rangle \right\} \qquad (4.9.5)$$

where

$$\rho_q = \sum_{k\sigma} a^\dagger_{k\sigma} a_{k+q\sigma}, \qquad \sigma_q^{(\lambda)} = \sum_{k\sigma\tau} a^\dagger_{k\sigma} \sigma^{(\lambda)}_{\sigma\tau} a_{k+q\tau} \qquad (4.9.6)$$

represent electron number density and spin density, respectively, and $\sigma^{(x)}$, $\sigma^{(y)}$, $\sigma^{(z)}$ are Pauli spin matrices. We can discuss CDW or SDW (or screw spin structure) states by assuming nonvanishing expectation values of ρ_q or σ_q for a nonvanishing wave vector q. Here we will concern ourselves with the ferromagnetic state ($q = 0$) and choose the z axis along the direction of the order parameter $\langle \sigma_0 \rangle$. The term proportional to $\langle \rho_0 \rangle$ in (4.9.5) means the shift of origin to measure ε_k, so that we may ignore it. Writing $\langle \sigma_0^{(z)} \rangle = N\zeta$ and $\sigma = \pm 1$, we write the Hartree–Fock Hamiltonian as

$$\mathscr{H}_m = \frac{1}{4} N\zeta^2 + \sum_{k\sigma} \left(\varepsilon_k - \frac{1}{2} \sigma U\zeta \right) a^\dagger_{k\sigma} a_{k\sigma} \qquad (4.9.7)$$

We substitute the one-electron energy $\varepsilon_k - (\sigma/2)U\zeta$ in the Fermi distribution function to determine the spin polarization ζ self-consistently. Equivalently we seek for the minimum of the free energy calculated with use of (4.9.7). At $T = 0$, assuming the free-electron spectrum $\varepsilon_k = \hbar^2 k^2/2m$, we obtain the total energy

$$E_0 = N \left\{ -\frac{1}{4} U\zeta^2 + \frac{3}{10} \varepsilon_F [(1 + \zeta)^{3/2} + (1 - \zeta)^{3/2}] \right\}$$
$$= N \left(\frac{3}{5} \varepsilon_F + \frac{1}{2} a\zeta^2 + \frac{1}{4} b\zeta^4 + \ldots \right) \ . \qquad (4.9.8)$$

Here ε_F is the Fermi energy for $\zeta = 0$, $a = \chi_0^{-1} - (U/2)$, $b = (27\rho_F)^{-1}$, $\rho_F = (3/4\varepsilon_F)$ is the density of states at the Fermi level, and $\chi_0 = 2\rho_F$ is the Pauli susceptibility (taking the spin magnetic moment to be unity).

If $1 > \rho_F U$, E_0 takes the minimum value at $\zeta = 0$, which means the paramagnetic state. We obtain a nonvanishing ζ, say, by adding the spin Zeeman energy $- NB\zeta$ to (4.9.8), where B is the uniform, steady external magnetic field. For small B, ζ is proportional to B and the coefficient a^{-1} is the susceptibility χ. Thus

$$\chi = \chi_0(1 - \rho_F U)^{-1} \tag{4.9.9}$$

This is an example of Fermi-liquid effects (Sect. 3.5).

When $\rho_F U$ reaches unity, χ diverges and the electron system will show a finite spin polarization even in an infinitesimal external field. This means the onset of the ferromagnetic spin ordering. In fact, when $\rho_F U > 1$, $a < 0$ in (4.9.8) and E_0 takes the minimum value at $\zeta = [|a|/b]^{1/2}$, provided that $|a|$ is still small. One-electron energy spectra for spin-up and spin-down states shift relative to each other by the amount $\Delta = U\zeta$, but their forms remain unchanged. The excitation energy to create a pair consisting of a hole in the up-spin band and an electron in the down-spin band, $\varepsilon_{k+q\downarrow} - \varepsilon_{k\uparrow}$, is of the order of Δ for small q and forms a continuous band. This is the excitation of the individual motion according to the classification mentioned in Chap. 3 and called the *Stoner excitation* in the present case. In addition to this, there also exists a collective motion with spin reversal and that is the spin wave mode. One method to derive its spectrum is to apply RPA; we set up the Heisenberg equation of motion of the operator $a_{k+q\downarrow}^\dagger a_{k\uparrow}$. The equation is nonlinear in the sense that it contains products of four a operators. We linearize it by replacing operators by expectation values (with respect to the HF ground state) in the same manner as (4.9.4). It is perhaps more physical, however, to make use of the dynamical susceptibility, which we will now discuss.

Suppose that we apply a weak magnetic field, which is perpendicular to the spontaneous magnetization and varying in space and time with wave vector q

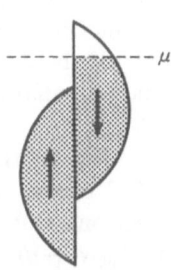

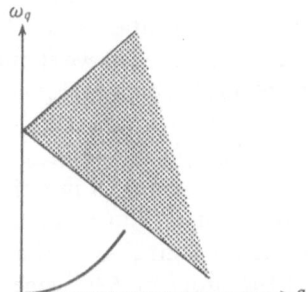

Fig. 4.8. The band structure of a ferromagnetic metal

Fig. 4.9. Individual and collective excitation energy spectra

and frequency ω. Perturbation due to this field is proportional to $\sigma_q^{(\pm)} = (1/2) (\sigma_q^{(x)} \pm i\sigma_q^{(y)})$, but it is too primitive to take Fourier transforms $h^{(\pm)}$ of the external field as proportionality constants. The forced oscillation caused by the external field must also modify the mean field in such a way that the effective field to determine the response is given by $\eta^{(\pm)} = h^{(\pm)} + N^{-1}U \langle\sigma_q^{(\pm)}\rangle$. We add the perturbing term $-\eta^{(\pm)}\sigma_q^{(\pm)}$ to the right of (4.9.7) and calculate the spin response $\langle\sigma_q^{(\pm)}\rangle$ by the method described in Sects. 1.7, 2.2, 3.3. The resulting *transverse* susceptibility $\chi^{(t)}(q, \omega) = \langle\sigma_q^{(+)}\rangle/Nh^{(+)}$ is given by

$$\chi^{(t)}(q, \omega) = \chi_0^{(t)}(q, \omega) [1 - U\chi_0^{(t)}(q, \omega)]^{-1} \tag{4.9.10}$$

$$\chi_0^{(t)}(q, \omega) = \frac{1}{N} \sum_k \frac{n_{k\uparrow} - n_{k+q\downarrow}}{\varepsilon_{k+q\downarrow} - \varepsilon_{k\uparrow} - \hbar\omega - i0^+} . \tag{4.9.11}$$

Here $n_{k\sigma} = \langle a_{k\sigma}^\dagger a_{k\sigma}\rangle$ means the expectation value in the HF approximation. The infinitesimal imaginary part $i0^+$ in the denominator can be ignored when $\hbar\omega$ is outside the Stoner excitation spectrum.

The spin wave is the oscillation which can survive with a finite amplitude in the limit of vanishing external field. Hence its frequency is determined by the condition of divergent transverse susceptibility, $1 = U\chi_0^{(t)}(q, \omega)$. As $q \to 0$, this reduces to $\Delta - \hbar\omega = U\zeta$, i.e., $\omega = 0$, which is another example of Goldstone's theorem. Since ε_k is invariant against $k \to -k$, the spin wave frequency is also invariant against $q \to -q$. For small q, therefore, we can write it as $\omega_q = (\hbar q^2/2m^*)$. When we use the same model as (4.9.11), for instance, $(m^*/m) \cong (27/2\zeta)$, provided that ζ is small.

4.9.2 BCS Model of Superconductors

A simple model of the superconducting electron system is obtained by replacing U in (4.9.3) with a weak attractive coupling $-g$ ($g > 0$). As we have mentioned in Sect. 4.5, we suppose that our system is in contact with a big electron reservoir, so that the electron number of our system may have a fluctuation of the order of $N^{1/2}$ around the expectation value N of the number operator $\mathcal{N}$. As is well known in statistical mechanics, we should then replace the Hamiltonian by $\mathfrak{H} = \mathcal{H} - \mu\mathcal{N}$, where μ is the chemical potential. The expectation value of the electron number is given by $N = -\partial\Omega/\partial\mu$, where the thermodynamic potential Ω is defined by $\exp(-\Omega/k_BT) = \mathrm{tr}\{\exp(-\mathfrak{H}/k_BT)\}$.

From this, we can express μ in terms of N and obtain the usual free energy by $F = \Omega + N\mu$, which reduces to the ground state energy $E_0(N)$ of the N-electron system in the limit $T \to 0$. Since the fluctuation n of the electron number is, at most, of the order of $N^{1/2}$, eigenvalues of $\mathfrak{H}$ are given as $E_0(N + n) - \mu(N + n) \cong E_0(N) - \mu N$ and practically independent of n (note $\mu = \partial E_0/\partial N$ at $T = 0$). Taking advantage of this degeneracy, we will superpose eigenstates of

$\mathfrak{H}$ with different electron numbers to obtain a coherent state of the electron pair field.

To obtain the explicit expression of $\mathfrak{H}$, we need only to replace ε_k in (4.9.3) by $\xi_k = \varepsilon_k - \mu$. We then apply the mean field approximation to the interaction term. In doing so, however, we notice that expectation values of the type we have on the right of (4.9.4) are *not* characteristic of the coherent state. We approximately write $a_1^\dagger a_2^\dagger a_3 a_4 \cong \langle a_1^\dagger a_2^\dagger \rangle a_3 a_4 + \langle a_3 a_4 \rangle a_1^\dagger a_2^\dagger - \langle a_1^\dagger a_2^\dagger \rangle \langle a_3 a_4 \rangle$. Furthermore we assume that the Cooper pair has the vanishing total momentum, so that $\langle a_k a_l \rangle = 0$ unless $k + l = 0$. Thus the mean field Hamiltonian takes the form

$$\mathfrak{H}_m = \sum_{k\sigma} (\xi_k a_{k\sigma}^\dagger a_{k\sigma} + \Delta a_{k\uparrow}^\dagger a_{-k\downarrow}^\dagger + \Delta^* a_{-k\downarrow} a_{k\uparrow}) + g^{-1} N |\Delta|^2 \qquad (4.9.12)$$

where

$$\Delta = - g N^{-1} \sum_k \langle a_{-k\downarrow} a_{k\uparrow} \rangle . \qquad (4.9.13)$$

The absolute value of Δ gives the strength of the mean field, whereas its phase undergoes the gauge transformation $\Delta \to \Delta \exp(2i\gamma)$ against $a_{k\sigma} \to a_{k\sigma} \exp(i\gamma)$.

In order to see the analogy with other problems, we conveniently regard $a_{k\uparrow}$, $a_{-k\uparrow}^\dagger$ as components of a two-dimensional spinor ϕ_k (*Nambu* representation). We denote Pauli matrices operating on this spinor by

$$\hat{t}_1 = \begin{pmatrix} 0 & 1 \\ 1 & 0 \end{pmatrix}, \quad \hat{t}_2 = \begin{pmatrix} 0 & -i \\ i & 0 \end{pmatrix}, \quad \hat{t}_3 = \begin{pmatrix} 1 & 0 \\ 0 & -1 \end{pmatrix} \qquad (4.9.14)$$

which should not be confused with the physical electron spin. For instance, $\phi_k^\dagger \tau_2 \phi_k = a_{k\uparrow}^\dagger a_{k\uparrow} - a_{-k\downarrow} a_{-k\downarrow}^\dagger$.

We can now write (4.9.12) as

$$\mathfrak{H}_m = \sum_k \xi_k + g^{-1} N |\Delta|^2 + \sum_k \phi_k^\dagger \hat{E}_k \phi_k \qquad (4.9.15)$$

where

$$\left. \begin{array}{l} \hat{E}_k = \xi_k \hat{t}_3 + \Delta_1 \hat{t}_1 + \Delta_2 \hat{t}_2 \\ \Delta_\lambda = - (2N)^{-1} g \sum_k \langle \phi_k^\dagger \hat{t}_\lambda \phi_k \rangle \end{array} \right\} . \qquad (4.9.16)$$

The superconducting long-range order is thus represented by the ordering of $\hat{t}$-spins on the XY plane and the correspondence to the spin model of superfluid ^{4}He (Sect. 4.5) is self-evident. Here again, the gauge transformation is represented by the rotation of $\hat{t}$-spins round the Z axis and the Hamiltonian is invariant against it. Hence the order parameter $\Delta = (\Delta_1, \Delta_2)$ can assume any direction on the XY plane. Let us assume that it is along the X axis ($\Delta_1 = |\Delta|$, $\Delta_2 = 0$). Then $\hat{E}_k$ can be diagonalized by rotating spin coordinates by the angle $\theta_k = \tan^{-1}(|\Delta|/\xi_k)$ round the Z axis. The spinor transformation then takes the form $\phi_{k1} = u_k\chi_{k1} - v_k\chi_{k2}$, $\phi_{k2} = v_k\chi_{k1} + u_k\chi_{k2}$, where $u_k = \cos(\theta_g/2)$, $v_k = \sin(\theta_k/2)$. We thus obtain

$$\mathfrak{H}_m = \text{const} + \sum_k \chi_k^\dagger E_k \hat{\tau}_3 \chi_k$$

$$E_k = (\xi_k^2 + |\Delta|^2)^{1/2} .$$

(4.9.17)

Our system can therefore be excited by creating those Fermi particles (called *quasiparticles*) whose creation operators are $\chi_{k1}^\dagger$ and χ_{k2}. The excitation energy of the quasiparticle is given by E_k, which takes the minimum value $|\Delta|$ at the Fermi surface $\xi_k = 0$. Let us assume that this energy gap is much smaller than the Fermi energy $\mu = \hbar^2 k_F^2/2m$ (the weak coupling limit). Note that the vacuum state in which no quasiparticles exist is nothing else but the BCS state (4.5.6) with $\alpha = 0$.

Expectation values of quasiparticle numbers, $\langle \chi_{k1}^\dagger \chi_{k1} \rangle$, $\langle \chi_{k2} \chi_{k2}^\dagger \rangle$, are given by the Fermi distribution $f(E_k) = [\exp(E_k/k_B T) + 1]^{-1}$ at finite temperature T. Substituting this into the definition of Δ, we obtain the BCS gap equation to determine $|\Delta|$ self-consistently;

$$1 = N^{-1}g \sum_k (2E_k)^{-1}[1 - 2f(E_k)] .$$

(4.9.18)

To avoid the nonphysical divergence, we restrict k on the right within a thin shell $|\xi_k| < \hbar\omega (\ll \mu)$ at the Fermi surface, supposing that the attraction between electrons is effective only there. The gap $|\Delta|$ at $T = 0$ is then equal to $\Delta_0 = 2\hbar\omega \exp[-(g\rho_F)^{-1}]$. It decreases with increasing T and vanishes at $T_c = 0.57 (\Delta_0/k_B)$. In the case of superconductors so far known, g is the difference between phonon-mediated attraction and Coulomb repulsion. Hence $g\rho_F$ is of the order of 0.1 and $\hbar\omega$ is an average phonon energy, so that $\Delta_0 \ll \hbar\omega \ll \mu$. We should also mention that the wave function of the Cooper pair, (4.5.7), calculated at $T = 0$ is proportional to $\exp(-r/\pi\xi_0)$ for large $r = |r_1 - r_2|$, where $\xi_0 = (\hbar^2 k_F/\pi m\Delta_0)$. Since $\xi_0 k_F = \mu/\Delta_0 \gg 1$, the spatial extention of the Cooper pair is much bigger than the average distance between electrons in the metal. We cannot accept the intuitive picture that the Cooper pair is a "dielectronic" molecule.

If we assume that $\langle a_{k\uparrow} a_{l\downarrow} \rangle = 0$ unless $k + l = q (\neq 0)$, we obtain the state, in which all the Cooper pairs are superflowing with the same center of mass momentum $\hbar q$ and therefore with the superflow velocity $v_s = (\hbar q/2m)$. The excitation energy of the quasiparticle is $E_k + \hbar k \cdot v_s$, which is obtained by Galilei transformation from (4.9.17). It is positive for all k when $q < \text{Min}(2m|\Delta|/\hbar^2 k)$, which means $q \lesssim \xi_0^{-1}$ at $T = 0$. Near T_c, on the other hand, we introduce the power series expansion of the free energy, up to the fourth order of $|\Delta|$ and up to the second order of q, and thus reproduce GL theory (with $A = 0$) from the microscopic point of view. For instance, the coherence length of GL theory is now given a microscopic expression $\xi = 0.74\xi_0[T_c/(T_c - T)]^{1/2}$.

Despite the degeneracy of the BCS ground state against the gauge transformation, the excitation energy of the quasiparticle has the finite gap and fails to satisfy Goldstone's theorem. The situation is somewhat similar to the Stoner model; if we allow for Δ in (4.9.12) to oscillate with definite wave vector and frequency and apply RPA, we obtain a collective mode of motion (*Anderson mode*), whose frequency vanishes in the long-wavelength limit. This motion is essentially the rotation of $\hat{t}$ spins round the Z axis and therefore corresponds to the phase mode in the pseudospin model of superfluid ^{4}He described in Sect. 4.6.2. In fact the RPA calculation reduces, in the long-wavelength limit $q\xi_0 < 1$, to the description by the quantum hydrodynamical Hamiltonian (4.7.5) *without* the Coulomb repulsion term. We can include this term in the RPA calculation by retaining the long-range part of the Coulomb repulsion explicitly in (4.9.1) or (4.9.12). Then, as we have already seen in Sect. 4.7 Goldstone's theorem is broken again, because of the plasma oscillation due to the surface charge.

4.9.3 Excitonic States

According to usual band theory, the insulating crystal is characterized by a finite gap E_g between valence band (called the a band hereafter) and conduction band (the b band) in the one-electron energy spectrum. At $T = 0$, the a band is fully occupied and the b band is completely vacant (Sect. 2.3). Let us assume a simple model, in which the a band has a single maximum at $k = 0$ in the reduced wave vector space and the b band has a minimum at $k = w$ ($\neq 0$). Because of time-reversal symmetry, the b band energy should be degenerate at $k = \pm w$ and therefore $w = K/2$, where K is a reciprocal lattice vector.

When the lattice constant and therefore the gap E_g decrease under external pressure, for instance, band theory predicts that the transition from insulator to metal will take place at $E_g = 0$ (though it is not always possible actually to produce such high pressure in practice). In fact, for $E_g < 0$, the maximum of the a band is higher than the minimum of the b band, so that we have a number of positive holes near the former and the equal number of electrons near the latter. These free *carriers* can conduct the electric current. For small $|E_g|$, however, the carrier concentration $2n$ is still small and the crystal is called the *semimetal*.

When the Coulomb interaction between electrons is explicitly taken into consideration, however, the concentration n should show a finite jump at the metal–insulator transition, which is therefore the first-order transition and called the *Mott transition* in contradistinction to the continuous transition predicted by band theory.

There are two types of discontinuous transitions; one is associated with the so-called excitonic state and the other is not. Let us begin with the former type and first suppose that $E_g > 0$. When one electron is excited from a to b bands, the electron in the b band and the hole in the a band thus produced exert the Coulomb attraction on one another and can form the exciton, a bound state similar to the hydrogen atom. Here we assume that the "Bohr" radius of the exciton given by (2.3.16) is much bigger than the lattice constant. When E_g

decreases and gets smaller than the binding energy given by (2.3.15), the "excitation" energy of the exciton, $E_g - R$ is actually negative, so that the state in which a macroscopic number of excitons exist can have a lower energy than the band theoretical ground state. The new ordered state is called the *excitonic state* and characterized mathematically by the nonvanishing expectation value of the type $\langle a_{k\sigma}^\dagger b_{k+Q\tau} \rangle$, which means the exciton condensation similar to the Cooper pair condensation in superconductors. The wave vector Q is either equal to w or close to it and represents new periods different from those represented by reciprocal lattice vectors of the original crystal.

Similarly, when $E_g < 0$ and $|E_g|$ decreases, we can also infer the possibility of the excitonic state. The Coulomb interaction between electron and hole is screened by other carriers and has a finite range of the order of the average carrier distance $n^{-1/3}$. When this range gets bigger than the exciton radius, the pair consisting of an electron and a hole will form an exciton. This takes place when $|E_g|$ is of the order of R.

A simple description of the excitonic state is given by the BCS-like mean field approximation, in which the Coulomb interaction is truncated by retaining only the long-range part

$$\mathcal{H}_{\text{int}} \cong (2V)^{-1} \sum v(q) \, (a_{k\sigma}^\dagger a_{k'\sigma'}^\dagger a_{k'-q\sigma'} a_{k+q\sigma}$$
$$+ \, b_{w+k\sigma}^\dagger b_{w+k'\sigma'}^\dagger b_{w+k'-q\sigma'} b_{w+k+q\sigma} + 2a_{k\sigma}^\dagger b_{w+k'\sigma'}^\dagger b_{w+k'-q\sigma'} a_{k+q\sigma}) \; .$$

$$(4.9.19)$$

Here $v(q) = (4\pi e^2/\varepsilon q^2)$ is the Fourier transform of the Coulomb potential in the medium of the dielectric constant ε, and we have to exclude the term with $q = 0$ because of electrical neutrality. Terms we have omitted in (4.9.19) have q comparable to w, so that $v(q)$ is small. We apply the HF approximation to (4.9.19) and only retain expectation values of the type $\langle a_{k\uparrow}^\dagger b_{w+k\downarrow} \rangle$ and their complex conjugate (up to the approximation we are applying here, the result is the same for triplet pairs). For simplicity, let us assume the same mass m for electron and hole and write $\xi_k = (\hbar^2 k^2/2m) + (E_g/2)$. The equation to determine the order parameter $\Delta_k = V^{-1} \sum v(k - k') \langle a_{k'\uparrow}^\dagger b_{w+k'\downarrow} \rangle$ self-consistently takes the same form as the BCS gap equation and the quasiparticle energy is also given by $E_k = (\xi_k^2 + |\Delta_k|^2)^{1/2}$. Writing $\Delta_k = 2E_k\varphi_k$, we have

$$2E_k\varphi_k - V^{-1} \sum_{k'} v(k - k') \, \varphi_{k'} = 0 \; . \tag{4.9.20}$$

When $E_g > 0$ and in the limit of $\Delta \to 0$, it reduces to the Schrödinger equation of a particle with mass $m/2$ moving in the Coulomb field $- e^2/\varepsilon r$ and therefore gives the energy of a single exciton.

Whether this kind of HF approximation is actually adequate or not, the above mentioned general definition applies also to CDW and SDW states in metals; they may be regarded as excitonic states. The SDW state of metallic Cr is such an example. Paramagnetic Cr has a bcc structure; the Fermi surface partly consists of an octahedral electron surface centered at $k = 0$ and the hole surface centered at $w = (2\pi/d)(1,0,0) = K/2$ with same shape and slightly smaller volume. If they had exactly the same volume, two surfaces would coincide with each other by the translation of $K/2$ in k space (*nesting*). The electron system would take advantage of the mutual interaction to produce self-perturbation, which has the periodicity characterized by the wave vector $K/2$, and therefore lower its translational symmetry from body-centered lattice to simple cubic. In actuality, electron and hole Fermi surfaces enclose slightly different volumes, so that the wave vector Q characterizing the SDW state is also slightly different from $K/2$.

As regards nesting, the simplest example would be the so-called *quasi-one-dimensional conductor* (e.g., TTF-TCNQ). The crystal is composed of parallel chains of molecules and the conduction electron moves overwhelmingly along the chain. As an idealized model, we may think of a one-dimensional Fermi liquid, which possesses the one-dimensional k space and two "Fermi points" $\pm k_F$. They coincide with each other by the translation of $\pm 2k_F$, so that we expect CDW or SDW states with the fundamental period characterized by $Q = 2k_F$. In an ideal one-dimensional system, however, the long-range order cannot be established at finite temperature. Even if the interchain coupling is taken into account, the SDW long-range order is impossible as far as the electron exchange between chains is negligible. The CDW long-range order, on the other hand, can be stabilized by the interchain Coulomb coupling. The three-dimensional structural change which organic conductor TTF-TCNQ shows below 60 K seems to result from such a CDW long-range order coupled with the lattice through the electron–phonon interaction.

4.9.4 Electron–Hole Metals

There have been a number of experimental attempts to observe the excitonic transition by controlling band parameters of typical semimetals As, Sb, Bi in adequate ways, but none seems to confirm the excitonic state unambiguously. It is not sufficient simply to observe a metal–insulator transition. The electron–hole system may undergo a first-order transition without entering in the excitonic state and therefore without changing the translational symmetry of the crystal.

In fact, by irradiating a semiconductor such as Ge, Si with laser light, we can excite a macroscopic number of electrons and holes and observe that this man-made electron–hole matter undergoes the above-mentioned first-order transition. Since the electron–hole recombination is a rather slow process, we can, under steady conditions of irradiation, regard the electron–hole system as being in thermal equilibrium with definite carrier number $2N$ and definite volume V. When the density $n = N/V$ is small and the temperature is not too high, our system is a classical gas of excitons or exciton molecules similar to hydrogen molecules. When n reaches a certain critical value, the system separates into two phases; a metallic electron–hole drop with a high density n_0 appears in the exciton gas. When the average density n is equal to n_0, the whole system is an electron–hole metal. The situation is similar to the usual first-order transition from gas to liquid under isothermal compression.

Theoretically we need to calculate the energy $E_0 = N\varepsilon(n)$ of the electron–hole Fermi liquid as a function of n, or $r_s = (3/4\pi na^3)^{1/2}$ by the method described in Sect. 3.3. We do not include the creation energy E_g of the electron–hole pair in E_0, so that $\varepsilon(n) \to 0$ as $n \to 0$, while $\varepsilon(n)$ diverges like $n^{2/3}$ in the high density limit where the zero-point kinetic energy is predominant. For a certain intermediate density n_0, $\varepsilon(n)$ will take a negative minimum value ε_0. If $|\varepsilon_0|$ is bigger than the exciton binding energy R, the electron–hole metal with density n_0 will be formed. In the case of Ge, n_0 and ε_0 ($r_s \cong 0.6$, $\varepsilon_0 \cong -2R$) calculated by RPA are in very good agreement with observed values. Here RPA means substituting the RPA expression in the dielectric function (3.3.12) and gives the Gell-Mann-Brueckner energy (3.3.14) in the case of the electron gas. In actual calculations, it is essential to take into account the energy band structure of Ge, in particular, the fact that the conduction band has four valleys. Thanks to this fourfold degeneracy, perturbation terms to be summed up by RPA have four times as much weight as terms to be discarded. Anyway, the calculation of the electron–hole metal energy in Ge is one of most successful cases, in which many-electron theory is applied to practical problems.

A similar first-order transition might be observed also in usual semimetals. Adding E_g, we need in this case to consider $\omega(n) = nE_g + n\varepsilon(n)$ as the function of n. When E_g is positive and sufficiently large, ω will be a minimum at $n = 0$ (insulator). When E_g is sufficiently small, a minimum can appear at $n \neq 0$. When $E_g = |\varepsilon_0|$, ω takes the minimum value at $n = n_0$ and $\omega(n_0) = \omega(0)$. Thus the first-order transition is possible from insulator to metal with density n_0.

4.10 Fluctuations

When we know the existence of the phase transition and the type of resulting order, we can make use of mean field approximation to establish a qualitative (sometimes even quantitative), but useful model. There are a number of cases, however, for which we know that the mean field approximation is definitely inadequate.

4.10.1 Low-Dimensional Systems

For example, the mean field approximation gives finite transition temperatures even to low-dimensional systems, which should not undergo phase transitions at finite temperature. As we have already seen, these low-dimensional systems are characterized by large *fluctuations*. In the case of the one-dimensional Heisenberg ferrogmagnet, for instance, all spins are aligned in the ground state, but close to this there exist disordered states with such overwhelming weight that the longrange order of spins is destroyed by thermal fluctuation as soon as the temperature becomes finite. The mean field approximation cannot take into account the fluctuation properly and gives a finite Curie temperature.

The *Kondo effect* shown by dilute magnetic alloys is similar in the sense that it is concerned with a single magnetic impurity and shows no singularity such as phase transitions. Suppose that the impurity spin S of the magnitude $1/2$ is placed in the Fermi gas of conduction electrons and interacting with electrons through the so-called sd interaction $\mathscr{H}_{sd} = - JS \cdot A_0^\dagger \sigma A_0$. Here A_j is the spinor with components (4.9.2) and $j = 0$ means the lattice point where the impurity spin is placed. In the case of weak coupling, $\rho_F|J| \ll 1$, the impurity spin behaves approximately like a free spin showing the Curie susceptibility $\chi_0(T) \propto T^{-1}$ in the temperature range $|J|/k_B < T < T_F$, while electrons show the Pauli susceptibility. The correction to the impurity spin susceptibility, calculated by the perturbation expansion into powers of $\mathscr{H}_{sd}$, exhibits a logarithmic singularity:

$$\Delta\chi/\chi_0(T) = \rho_F J[1 - \rho_F \log(T_F/T) + \ldots]$$
$$\cong \rho_F J[1 + \rho_F J \log(T_F/T)] .$$

The signularity is because the scatterer S of our problem possesses the internal degree of freedom and brings about many-body effects in the scattering. The expression on the second line is the sum of most divergent terms and, in the case of antiferromagnetic coupling ($J < 0$), diverges as T approaches the Kondo temperature $T_K = \exp[(\rho_F J)^{-1}]$. The divergence merely implies inadequacy of the approximation, since the effect of the single impurity spin (zero-dimensional perturbation) cannot result in singularities of thermodynamic functions at finite temperature.

In fact, theory at $T = 0$ (Yosida–Yoshimori theory) shows that the impurity spin, in the case of $J < 0$, forms a singlet bound state with conduction electrons and has a finite susceptibility of the order of $\chi_0(T_K)$. The impurity spin at $T = 0$ is thus a nonmagnetic scatterer, but still shows a (local) Fermi-liquid effect (Sect. 3.5) in the sense that the phase shift is a function of the electron distribution. Anyway the rapid, but continuous changeover from magnetic to nonmagnetic behavior of the impurity spin occurs around the Kondo temperature T_K. This situation can be seen more clearly by applying the so-called Anderson model to describe the impurity, but we will not enter into it in detail.

4.10.2 Critical Phenomena

Similarly, even in the case of the second-order phase transition with a finite transition temperature, the mean field approximation fails to describe properly large fluctuations in the so-called critical region. Thus, even above the transition temperature, the ordered state can appear as the thermal fluctuation within finite distance ξ in space and finite life τ in time. Such a fluctuation is ignored by the mean field approximation, which tries to describe the transition in terms of the average long-range order. The situation is similar below the transition temperature. As we have seen, RPA takes account of space–time fluctuations of the order parameter, but still is a dynamical mean field approximation in the sense that it ignores the *mode–mode coupling*, i.e., the interaction between modes of fluctuations with different wave vectors and frequencies.

As a result of this defect, the mean field approximation fails to give not only the correct value of the transition temperature T_C, but also correct critical indices of physical quantities. When the temperature T approaches T_c from above, specific heat diverges as $(T - T_C)^{-\alpha}$, susceptibility as $(T - T_C)^{-\gamma}$, and correlation distance ξ as $(T - T_C)^{-\nu}$. When T approaches T_C from below, the order parameter vanishes as $(T_c - T)^{\beta}$. In contrast to T_C, which is sensitive to the interaction between particles, *critical indices $\alpha, \beta, \gamma, \nu$* depend only upon rather general features of the system such as dimensionality of space, number of components of the order parameter, etc. Such universality of fluctuations in the vicinity of the phase transition point can be studied by the powerful method called theory of *renormalization group*, the detailed description of which is beyond the scope of the present volume. We may add, however, that the method is not restricted to phase transitions, but can be applied to such problems as the Kondo effect. In this case, as we have mentioned, the role of T_C is played by $T = 0$ and the critical region is defined by $T < T_K$. As T approaches $T = 0$, the effect of J upon the impurity spin gets more and more similar to the strong coupling limit of large $|J|$.

4.10.3 Superconductor and Superfluid ^{3}He

Mean field approximation and RPA are exceptionally successful in the case of superconductors. The GL theory in Sect. 4.5 and the corresponding microscopic BCS model in Sect. 4.9 can quantitatively describe not only the superconducting state below T_C, but also the effects of Cooper pairs appearing as thermal fluctuations above T_C. For instance, the observed electronic specific heat shows a finite jump at T_C in good agreement with mean field theory which entirely ignores the fluctuation of the order parameter. The reason is that the spatial extension ξ_0 of the Cooper pair is much larger than the average spacing $r_0 \sim k_F^{-1}$ of electrons, the ratio of the two being of the order of T_F/T_C. The number of other pairs existing within the extension of one pair is so large that fluctuations become very small, as we know well in statistics. The critical region should exist, but it is usually restricted to an exceedingly narrow temperature range around T_C and is difficult to observe except in special circumstances such as thin films and Josephson junctions.

Here, on the basis of GL free energy (4.5.10), we shall show that the shift of the transition temperature is small even if fluctuations are taken into account. In doing so, we assume that $A = 0$ and also make use of BCS expressions $a \cong (\hbar^2/2m\xi_0^2)t$, $b \cong (\hbar^2/2mn\xi_0^2)$, where $t = (T - T_C)/T_C$ with the mean field theoretical

T_C (Sect. 4.9.2). The transition point is determined by $t = 0$, as we have seen in Sect. 4.5, when we assume a spatially uniform order parameter Ψ from the outset in seeking for the minimum of the free energy (4.5.10). We then have $\Psi = 0$ for $t > 0$. In reality, the fluctuation of Ψ appears with the probability proportional to $\exp(- F/k_B T)$ and Ψ may vary in space. Let its Fourier expansion be $\Psi = V^{-1/2} \sum_q \Psi_q \exp [i q \cdot r]$. We restrict ourselves to long-wavelength modes, $q\xi < 1$, to which GL theory can reasonably be applied. Here ξ is the GL coherence length given by $[\hbar^2/4ma]^{1/2} \cong 0.7\xi_0 t^{-1/2}$. Ignoring the $|\Psi|^4$ term, we have $F = \sum_q a[1 + (\xi q)^2]| \Psi_q|^2$, so that each mode obeys the Gaussian probability distribution independent of others and $\langle | \Psi_q|^2 \rangle_0 = k_B T[a(1 + (\xi q)^2)]^{-1}$.

The $|\Psi|^4$ term gives the mode–mode coupling proportional to the product of the four Ψ_q. Let us again apply the mean field approximation and replace the product of Ψ_q and Ψ_q^* among the four by its expectation value. The second-order term is thus renormalized by including the effect of the fourth-order term, and the GL coefficient of the $q = 0$ mode is now given by

$$a^* = a + 2bV^{-1} \sum_q \langle | \Psi_q|^2 \rangle . \tag{4.10.1}$$

Hence the transition temperature determined by $a^* = 0$ is lower than the mean-field theoretical one given by $a = 0$. Provided that the fluctuation effect is small, we replace the expectation value on the right of (4.10.1) by the above mentioned Gaussian value. Apart from numerical factors, we then have $(a^* - a)a^{-1} \sim (bk_B T_C/a^2\xi^3) \sim (r_0/\xi_0)^2 t^{-1/2}$, so that we can ignore the difference between a^* and a in so far as t is bigger than $(r_0/\xi_0)^4$. The transition temperature is practically determined by $a = 0$.

When the dimensionality of space is d, a similar calculation gives $(a^* - a)a^{-1} \sim (bk_B T_C/a^2\xi^d) \sim [r_0/\xi_0]^{d-1}/t^{(4-d)/2}$.

Now, in the case of liquid ^{3}He, the Fermi temperature T_F is of the order of 1 K and the transition to the superfluid state takes place in the milli-Kelvin range, so that we have the same condition $\xi_0 \gg r_0$ as in the case of superconductors. In fact, the observed specific heat shows a jump at the second-order transition from normal to superfluid states. The critical region should be very narrow. Upon the second-order transition (λ transition) of liquid ^{4}He to the superfluid state, on the other hand, the logarithmic divergence of specific heat in the critical region has been observed. It must be because ξ_0 of liquid ^{4}He is close to the acoustic "Compton" wavelength ($\hbar/m_4 c_s$), which is comparable to the average atomic spacing r_0. Here m_4 is the atomic mass of ^{4}He and c_s is the velocity of sound in liquid ^{4}He.

In the case of superconductors, the collective mode of motion whose excitation energy is lower than that of the individual electron motion is the vibration of heavy ions, i.e., the phonon, which gives the attraction for electrons to form Cooper pairs. The phonon itself remains almost unchanged upon the superconducting transition, since the static dielectric polarizability of the electron system, the main factor to affect the phonon, remains almost unchanged. Hence we can assume the same phonon-mediated interaction between electrons in the superconducting state as well as in the normal state.

In the case of superfluid ^{3}He, the situation is somewhat different. Liquid ^{3}He is a single component system and has nothing to correspond to the ionic vibration in a metal. Since the order parameter of superfluid ^{3}He has degrees of freedom of orbital and spin rotations in addition to the gauge freedom, a variety of collective modes of motion are possible. Furthermore it should be remembered that liquid ^{3}He in the normal state shows the nuclear spin susceptibility much enhanced in comparison with the Pauli susceptibility of the Fermi gas of the same density (Sect. 3.5.3). We slightly exaggerate the situation by saying that there exist in the liquid strong spin fluctuations of long wavelength as the precursor of ferromagnetic instability. Unlike spin waves in a real ferromagnet, they are not stable collective modes of motion, but their lifetimes must be rather long. Generalizing the concept of elementary excitations, we sometimes call them *paramagnons* (which may also be applied to metals with large spin susceptibilities, e.g. Pd). In analogy with the phonon-mediated interaction in metals, we can think of the paramagnon-mediated interaction in liquid ^{3}He, though the paramagnon is one mode of motion of liquid atoms themselves in contrast to the phonon mode which is external to metallic electrons.

We should remember that the bare interaction potential between two He atoms has the form shown by Fig. 1.1 (Sect. 1.2). It gives a hard-core-like repulsion at short distances, so that even the perturbation expansion into powers of the bare potential cannot be cut off at a finite order. To say the least, we should take into account multiple scatterings of two particles to infinite order, so that the bare interaction potential is replaced by the so-called K matrix, the well-known concept in nuclear physics. Since the K matrix no longer has the hard-core singularity, we can take it as the effective interaction to start with in the BCS model of Cooper-pair condensation. In the case of weak coupling, we are concerned with the K matrix element $\langle k', -k' | K | k, -k \rangle \equiv K(n \cdot n')$ corresponding to the scattering of pairs from $k, -k$ states to $k', -k'$ states on the Fermi surface, where n is the unit vector in the direction of k ($k = k_F n$). We introduce the Legendre (partial wave) expansion of $K (n \cdot n')$ and write the coefficient of the lth-order term as $[(2l + 1)/4\pi] K_l$. Then $- K_l$ plays the role of g in Sect. 4.9.2. The most favorable value of l is such that K_l is negative and its absolute value is the maximum; the Cooper-pair condensation with the relative orbital angular momentum l should then take place at $T_c = 1.14 \, T_0 \exp [(\rho_F K_l)^{-1}]$ where $k_B T_0$ is the cut off energy corresponding to $\hbar \omega_0$ in Sect. 4.9.2, and T_0 is of the order of T_F.

Parameters K_l calculated for the density of liquid ^{3}He under saturation vapor pressure were found to be positive for $l = 0, 1$ and negative with a large absolute value for $l = 2$, so that the ^1D pair was first predicted theoretically. Experimentally the condensation of ^3P pairs has been observed. One possible explanation is that the paramagnon-mediated interaction, i.e. the sum of those perturbation terms which would produce the ferromagnetic instability, is equally important as, or even more important than the above mentioned K matrix. In fact, the paramagnon-mediated interaction for the scattering of the pair from $(k'\sigma', -k'\tau')$ to $(k\sigma, -k\tau)$ is proportional to $-\chi(k_F n - k_F n') \times S_{\sigma\sigma'} \cdot S_{\tau\tau'}$ where $\chi(q)$ is the spin susceptibility against the magnetic field varying in space with the wave vector q. In RPA, $\chi(q) = \chi_0(q)[1 - \rho_F U \chi_0(q)]^{-1}$, where $\chi_0(q)$ is the susceptibility of the Fermi gas and U is an effective repulsion between atoms, approximately given by K_0. We thus see that the paramagnon-mediated interaction is attractive for the triplet pair and repulsive for the singlet pair. The difference gets more pronounced with increasing $\chi(0)/\chi_0(0)$ and therefore spin fluctuations. This is perhaps the simplest mechanism to explain the observed ^3P pair condensation.

In the case of weak coupling, the macroscopic amplitude of the Cooper pair, $\langle a_{k\sigma} a_{-k\tau} \rangle$, is finite only in the vicinity of the Fermi surface, so that it is more convenient to take the sum over magnitudes of the wave vector k with its direction n being fixed. We write the sum as $\Psi_{\sigma\tau}(n)$. In the case of the triplet pair, it is symmetric with respect to the spin ($\Psi_{\sigma\tau} = \Psi_{\tau\sigma}$) and a linear combination of n_x, n_y, n_z. According to spinor analysis, $d_x(n)$, $d_y(n)$, $d_z(n)$ defined by $\Psi_{\downarrow\downarrow}(n) - \Psi_{\uparrow\uparrow}(n) = 2\Psi d_x(n)$, $\Psi_{\downarrow\downarrow}(n) + \Psi_{\uparrow\uparrow}(n) = 2i\Psi d_y(n)$, $\Psi_{\uparrow\downarrow}(n) = \Psi d_z(n)$ are transformed as vector components under spin rotation, where Ψ is the constant to normalize the average of $|d(n)|^2$ over the Fermi surface to unity. The simplest example is the spherically symmetric distribution $d(n) = n$, which defines the BW (Balian–Werthamer) state and characterizes the B phase of two superfluid phases of ^{3}He. We should remember, however, that the Hamiltonian is invariant against independent rotations of spin and orbit. Hence the order parameter $d(Rn)$ gives the same free energy as $d(n)$ for arbitrary rotation R.

The A phase of two superfluid phases, on the other hand, is characterized by $d(n) = (3/2)^{1/2} [(e_1 + ie_2) \cdot l] d$, where e_1, e_2, l are constant unit vectors forming a triad and d is another constant unit vector. The order parameter of this form defines the ABM (Anderson–Brinkman–Morel) state, in which the Cooper pair has the relative orbital angular momentum $\hbar$ in the direction of l. When both d and l are parallel to the x axis, we have $\Psi_{\uparrow\uparrow} \propto -n_y + in_z$, $\Psi_{\downarrow\downarrow} \propto n_y + in_z$, $\Psi_{\uparrow\downarrow} = 0$ (Note that the dipole–dipole interaction between nuclear magnetic moments favors $d//l$).

Now, if we assume the same effective interaction in the superfluid state as in the normal state, we reach the conclusion that the BCS model takes the minimum free energy for the BW state among all ^3P substates and thus excludes the possibility of A phase. Here again the paramagnon-mediated interaction can account, at least qualitatively, for the fact that the A phase does exist under higher pressure where the spin susceptibility is also higher. In fact the paramagnon-mediated interaction depends on the choice of $d(n)$, since the susceptibility in the superfluid state depends on the energy gap of the quasiparticle spectrum. The spin polarization is produced by the shift of the up-spin Fermi surface relative to the down-spin Fermi surface and therefore results in the reduction of condensation energy of pairs described by $\Psi_{\uparrow\downarrow}$. These pairs suppress the spin polarization if the magnetic field is weak. This is the reason why the electron spin susceptibility of the superconductor goes down to zero with decreasing temperature. As we can guess from the above mentioned form of Ψ, the spin susceptibility in the ABM state keeps the normal value when the magnetic field is perpendicular to d and decreases only in the direction of d. Near the transition point, the decrease is proportional to $|\Psi|^2$. The susceptbility in the BW state, on the other hand, decreases in all directions with decreasing temperature. Hence the paramagnon-mediated attraction is stronger in the ABM state than in the BW state.

These qualitative arguments of paramagnon effects may be put in a mathematical scheme by adding RPA terms of spin fluctuations to the BCS model. It seems difficult, however, to develop a theory a consistent and quantitative theory.

Finally, if the paramagnon effect is really responsible to the ^3P pairing, we might expect the same type of electron pairing in strongly spin-paramagnetic metals such as Pd.

4.10.4 Ferromagnetic Metals

Let us go back to the Stoner model in Sect. 4.9.1. and try to see how the transition from ferromagnetic to paramagnetic states proceeds when the temperature T is increased. We assume the case where $\rho_F U$ slightly exceeds unity and the spontaneous spin polarization ζ at $T = 0$ is small.

Let us first examine the analogy with the superconducting transition and suppose that the transition is determined by disordering of the individual mode of electron motion, i.e., by smoothing out the jump of the Fermi distribution function round the Fermi level. We thus calculate the free energy $F(T, \zeta)$ on the basis of the Hartree–Fock Hamiltonian (4.9.7) and expand it into powers of ζ. We obtain the same form as (4.9.8), but coefficients now depend on T. For $T \ll T_F$, we have $a = \chi_0^{-1} - (U/2)$, where $\chi_0(T) = 2\rho_F[1 - (\pi^2/12) \, (T/T_F)^2]$. This is a kind of GL theory and the Curie temperature is determined by $a = 0$:

$$T_C = [12(\rho_F U - 1)/\pi^2 \rho_F U]^{1/2} T_F \; . \tag{4.10.2}$$

We are assuming that this T_C is much lower than T_F. For $T > T_C$, the susceptibility is given by $[\partial^2 F/\partial \zeta^2]_{\zeta=0}$, so that

$$\chi(T) \sim 12\chi_0(0) \, [T_F/\pi T]^2[T_C^2/(T^2 - T_C^2)] \; . \tag{4.10.3}$$

Now, as we see from the relation between Stoner and spin wave excitations in Sect. 4.9.1, the fluctuation of the order parameter itself is lower in energy and therefore more important than the individual motion at least in the long-wavelength region. However, the spin wave theory in Sect. 4.9.1 is RPA, in which the coupling between modes with different wave vectors q is neglected, so that the

$q = 0$ mode (which determines the Curie temperature) is *not* renormalized by the coupling with other modes. To include the mode–mode coupling, let us follow the method by which we have obtained the shift of the superconducting transition temperature. Instead of the GL expansion of the free energy, however, we will start here with the expansion (4.9.8), in which coefficients are evaluated at $T = 0$. Furthermore we allow for the order parameter to fluctuate gradually in space. Thus

$$E = \text{const} + \int n d\mathbf{r} \left[\frac{1}{2} a\zeta^2 + \frac{1}{4} b\zeta^4 + \frac{1}{2} c(\nabla\zeta)^2 \right] \tag{4.10.4}$$

where $n = NV^{-1}$ and the new coefficient c evaluated by the same model as a, b is given by $- \chi_0^{-2}(0)[d^2\chi(0)/dq^2] = [12\rho_F k_F^2]^{-1}$ in terms of the q-dependent susceptibility of the electron gas, $\chi_0(q)$. The role of GL coherence length is played by $\zeta = [c/|a|]^{1/2} = [6(\rho_F U - 1)k_F^2]^{-1/2}$.

Again we introduce the Fourier expansion of ζ and renormalize the $q = 0$ mode by applying the mean field approximation to the mode–mode coupling arising from the ζ^4-term. Assuming that the fluctuation of ζ appears with the probability proportional to $\exp[- E/k_B T]$, which means classical statistics, we obtain $\langle |\zeta_q|^2 \rangle = k_B T[n(a^* + cq^2)]^{-1}$ and the following equation corresponding to (4.10.1):

$$X = - (T_C/T_0) + (T/T_0)(1 - X^{1/2} \tan^{-1} X^{1/2}) . \tag{4.10.5}$$

Here $X = (a^*/cq_m^2)$, $T_0 = (2\pi^2 c^2 n q_m/3b)$, $T_C = (|a|/cq_m^2)T_0$, and q_m is a cut-off wave number. The Curie temperature T_C is defined by $X = 0$. In the region $(4T_0/\pi^2 T_C) \gg (T - T_C)/T_C > 0$, we have $X \propto (T - T_C)$ and the Curie–Weiss susceptibility

$$\chi = [\chi_0(0)/(\rho_F U - 1)] [T_C/(T - T_C)] . \tag{4.10.6}$$

If we substitute electron-gas values in a, b, c, we have q_m as the parameter of our model. In terms of ξ and q_m, $T_C = (\rho_F U - 1)^{1/2} (3\sqrt{6} q_m \xi)^{-1} T_F$, $(4T_0/\pi^2 T_C) = 16(q_m\xi/2\pi)^2$. We should probably assume $q_m\xi \sim 1$. The advantage of the model (Murata–Doniach theory), however, is that we can see the important role played by the mode–mode coupling in a simple, qualitative way. Quantitatively, the application of classical statistics to (4.10.4) is useful only when spin fluctuations ζ_q with $q \ll k_F$ and excitation energy of the order of $k_B T$ play the decisive role. It is interesting to remember that a similar mechanism was conceived a long time ago concerning displacive ferroelectrics. In that case, normal coordinates of an optic branch of lattice vibrations correspond to our ζ_q and the parameter corresponding to our a is also negative if we take into account the dipole–dipole interaction between electric moments produced by the optic vibrations. At high temperature, however, the effect of the anharmonic coupling between normal modes is so large that the remormalized a^* is positive. This is akin also to the idea of self-consistent phonon described in Sect. 1.9.

For quantitative purposes, we need a more microscopic formulation. One method is to express the free energy in terms of the dynamical response function by the use of Feynman's theorem given in Sect. 3.3. In (4.9.3), the interaction term may be written as $\mathcal{H}_{int} (U) = (NU/2)[1 - N^{-1} \Sigma \{\sigma^{(+)}(q), \sigma^{(-)}(-q)\}]$, where $2\sigma^{(\pm)}(q) = \sigma_q^{(x)} \pm i\sigma_q^{(y)}$ with use of (4.9.3) and $\{a, b\} \equiv ab + ba$. Hence the free energy can be expressed in terms of the transverse spin susceptibility by applying Feynman's theorem, in which U is regarded as the coupling parameter to change from zero up to the actual value with spin polarization $M = N\zeta$, temperature and electron number being kept constant. Let us define the thermodynamic potential Ω by $\exp[-\Omega/k_B T] = \mathrm{tr}\{\exp[-(\mathcal{H} - B\sigma_0^{(z)})/k_B T]\}$, where the parameter B is determined by $M = -(\partial\Omega/\partial B)$ for the given value of M. Then the free energy is given by $F = \Omega + BM$. Remembering that B depends upon U and differentiating F with respect to U, we obtain Feynman's theorem in the form $U(\partial F/\partial U) = \langle\mathcal{H}_{int}(U)\rangle_M$, where the expectation value should be taken at constant M. For instance, if we substitute the expectation value for noninteracting electrons, $(U/N)N_\uparrow N_\downarrow = (U/4N)(N^2 - M^2)$, and integrate with respect to U, we obtain the free energy in the Hartree–Fock approximation

$$F_{HF}(M) = F_0(M) + (U/4N)(N^2 - M^2) . \tag{4.10.7}$$

Here $F_0(M)$ is the free energy at $U = 0$.

In general, writing $\Delta F = F(M) - F_0(M)$, we have

$$\Delta F = \frac{N}{2} U - \frac{1}{2} \sum_q \int_0^U dU \langle\{\sigma^{(+)}(q), \sigma^{(-)}(-q)\}\rangle . \tag{4.10.8}$$

To find the relation with the dynamic susceptibility, we first define a Green's function by replacing Φ_λ, $\Phi_{-\lambda}$ in (1.8.2) with $\sigma^{(+)}(q)$, $\sigma^{(-)}(-q)$, respectively and introduce the Fourier expansion of the form (1.8.3). We write the Fourier coefficient as $-\chi^{(t)}(q, iv_l)$ and its analytic continuation from iv_l to $\omega + i0^+$ as $-\chi^{(t)}(q, \omega)$. This $\chi^{(t)}$ is the transverse susceptibility and corresponds to the dynamical structure factor in Sect. 3.3. We should remember that we are concerned here with the susceptibility under the condition $M = \mathrm{const}$. With use of $\chi^{(t)}$, we can write

$$\frac{1}{2}\langle\{\sigma^{(+)}(q), \sigma^{(-)}(-q)\}\rangle = k_B T \sum_{l=-\infty}^{+\infty} \chi^{(t)}(q, iv_l)$$

$$= \int_{-\infty}^{\infty} \frac{d\omega}{2\pi} \coth\left(\frac{\omega}{2k_B T}\right) \mathrm{Im}\{\chi^{(t)}(q, \omega)\} . \tag{4.10.9}$$

In the Hartree–Fock approximation, we substitute the expression $\chi_0^{(t)}$ of noninteracting electrons. Hence we write (4.10.8) as the sum of the Hartree–Fock expression and the correction term:

$$\Delta F = \frac{1}{4} U(N^2 - M^2) + \Delta_2 F \tag{4.10.10}$$

$$\Delta_2 F = - k_B T \sum_{ql} \int_0^U dU [\chi^{(t)}(q, i\nu_l) - \chi_0^{(t)}(q; i\nu_l)] \; . \tag{4.10.11}$$

The static susceptibility given by $N\chi^{-1} = \partial^2 F/\partial M^2$ takes the form

$$\chi = \chi_0 \left(1 - \frac{1}{2} U\chi_0 + \lambda\right)^{-1} \; . \tag{4.10.12}$$

Here λ is the correction due to $\Delta_2 F$ and given as

$$\lambda = - k_B T \sum_l G(i\nu_l)$$

$$= - \chi_0 k_B T \sum \int_0^U dU \frac{\partial^2}{\partial \zeta^2} [\chi^{(t)}(q, i\nu_l) - \chi_0^{(t)}(q, i\nu_l)] \; . \tag{4.10.13}$$

These are all general expressions. A specific model is defined by an approximate expression of $\chi^{(t)}$. For instance, in analogy with the electron gas in Sect. 3.3, we might take the RPA expression $\chi_{RPA}^{(t)} = \chi_0^{(t)}[1 - U\chi_0^{(t)}]^{-1}$. But this does not reduce to (4.10.12) in the limit $\nu \to 0$, $q \to 0$ [Note that $\chi_0^{(t)} \to (1/2)\chi_0$]. In order to make theory self-consistent, we modify the RPA expression as

$$\chi^{(t)} = \chi_0^{(t)}[1 - U\chi_0^{(t)} + \lambda]^{-1} \; . \tag{4.10.14}$$

It is always possible to write $\chi^{(t)}$ in this form, but then λ depends upon U, M, q, ν. Our approximation consists in neglecting this dependence. Substituting (4.10.14) in (4.10.13), we have the equation to determine λ self-consistently. Hence λ depends on T and may result in the Curie–Weiss law, even if we assume the $T = 0$ expression for $\chi_0^{(t)}$. In fact (4.10.13) reduces to the form (4.10.5) if we retain only the term with $\nu_l = 0$ and expand $\chi_0^{(t)}$ up to q^2, exaggerating the importance of spin fluctuations of low energy and long wavelength. Thus (4.10.13) is the microscopic expression for renormalizing the $q = 0$ mode by the mode–mode coupling.

In RPA, as we have mentioned, the quantum mechanical equation of motion of the operator $a_{k\sigma}^\dagger a_{k+q\sigma}$ is linearized by replacing appropriate pairs of a operators in the product of the four with their expectation values. This is to decouple the hierarchy of equations of motion at the lowest stage. If we proceed one step further and decouple the hierarchy at the equation of motion for the product of four a operators, we can include the effect of the mode–mode coupling to the lowest order. If we further assume that fluctuations of low energy and long wavelength are important, we obtain (4.10.13, 14).

Thus, this formalism (Moriya–Kawabata theory) surpasses the traditional Hartree–Fock approximation and can deal with spin fluctuations in a self-consistent manner. In fact it has succeeded in accounting for all features of weak ferromagnets (with low Curie temperatures and small magnetizations at $T = 0$) such as $ZrZn_2$, $Sc_2 In$. The same method can also be applied to the spin ordering characterized by a nonvanishing wave vector Q, provided that spin flucutations with wave vectors close to Q are important. Furthermore the basic idea of this theory must be also useful in dealing with many other phase transitions shown by Fermi liquids.

Part II

Interaction Between Elementary Excitations

5. Linear Interactions and Coupled Modes

In the previous chapters, we have been mainly concerned with the elementary excitations in a system of like particles such as a many-atom system or a many-electron system. However, a realistic system is usually composed of different kinds of particles and displays a variety of properties or phenomena which are governed by the interaction between elementary excitations of different species. For instance, the electrical resistivity of a metal is usually ascribed to the electron–phonon interaction; the optical spectra of an insulator very often reflect the interplay of electrons and phonons. We shall now describe the dynamics of such interactions, starting this chapter with the simplest case of linear interaction between two kinds of elementary excitations.

5.1 Linear Interaction

The interactions between elementary excitations of different species can be classified into linear and nonlinear ones. Consider two interacting fields, x and y, which may be the (quasi) boson fields of collective motions (amenable to classical description) such as lattice vibrations, electronic polarizations, plasma oscillations and magnetization fluctuations, or the fermion fields (amenable to no classical description) such as creation–annihilation operators of electrons. Confining ourselves to the former for the convenience of explanation, we consider the classical equations of motion for them.

Denote by $U(x, y)$ the potential energy of the two interacting fields, and expand it in a Taylor series around the minimum point $x = y = 0$:

$$U(x, y) = \frac{\omega_1^2}{2} x^2 + \frac{\omega_2^2}{2} y^2 + \gamma\omega_1\omega_2\, xy + \text{(high-order terms)}. \qquad (5.1.1)$$

The kinetic energy can be written as $T = (\dot{x}^2 + \dot{y}^2)/2$ by appropriate normalizations of x and y. From the Lagrangian:

$$\mathscr{L} = T - U = \frac{1}{2}(\dot{x}^2 - \omega_1^2 x^2) + \frac{1}{2}(\dot{y}^2 - \omega_2^2 y^2) - \gamma\omega_1\omega_2 xy$$
$$+ \text{(high-order terms)}, \qquad (5.1.2)$$

one obtains the equations of motion:

$$(\ddot{x} + \omega_1^2 x) + \gamma\omega_1\omega_2 y + \text{(nonlinear terms)} = 0 , \qquad (5.1.3a)$$

$$(\ddot{y} + \omega_2^2 y) + \gamma\omega_1\omega_2 x + \text{(nonlinear terms)} = 0 \ . \tag{5.1.3b}$$

The small oscillations, for which nonlinear interaction terms can be neglected, consist of eigenvibrations in which x and y are coupled together through the linear interaction with the coupling constant γ (the bilinear term in $\mathscr{L}$). Equation (5.1.3) has a nontrivial solution only when the secular equation

$$\begin{vmatrix} \omega_1^2 - \omega^2 & \gamma\omega_1\omega_2 \\ \gamma\omega_1\omega_2 & \omega_2^2 - \omega^2 \end{vmatrix} = 0 \tag{5.1.4}$$

is satisfied, from which one obtains the two eigenfrequencies $\omega_\pm$ and the mixing ratios $x:y$ in the corresponding eigensolutions. In Fig. 5.1, one sees how $\omega_\pm$ vary as ω_1 for a fixed value of ω_2. The extent of mixing and the change in frequency are most significant at $\omega_1 \sim \omega_2$. In the extreme case of $\omega_1 \gg \omega_2$ (or $\omega_1 \ll \omega_2$), the slow mode is subject to a finite influence of the other mode as regards mixing and frequency, whereas the rapid mode is subject to almost none. In fact, for slow eigenvibration with $\omega = \omega_-$, the inertia term $\ddot{x}(= -\omega^2 x)$ can be neglected in (5.1.3a) since $\omega < \omega_2 \ll \omega_1$, and x follows instantaneously the adiabatic change of y: $x = -(\gamma\omega_2/\omega_1)y$. Inserting this into (5.1.3b), one finds that the restoring force $-\omega_2^2 y$ for y is reduced by a factor $(1 - \gamma^2)$, which effect is sometimes called the *screening* of the y field by the x field. Only the rapid motion can follow and screen the slow motion. This was implicitly assumed in Sect. 2.1, where the electronic polarization was considered to instantaneously follow the displacement polarization.

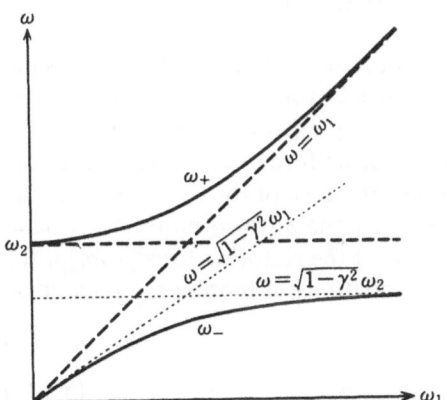

Fig. 5.1. Coupled oscillations ($\omega = \omega_\pm$) of two harmonic oscillators ($\omega = \omega_{1,2}$)

Solving the secular equation (5.1.4) for the eigenvibrations is equivalent to diagonalizing the quadratic form of (5.1.2) by an orthogonal transformation. The system consists of two independent harmonic oscillators in the new coordinate system. Such a complete decoupling is impossible, however, for nonlinear interactions, namely when we have higher-order terms of (5.1.2). There take place not only modes-mixing, but also a variety of nonlinear effects as will be described in later chapters. Confining ourselves to linear interactions in this chapter, we present a few typical examples of mixed modes.

5.2 Carrier Plasma Coupled to the Optical Mode of Lattice Vibrations in Polar Semiconductors

Consider a compound semiconductor, such as GaAs, which has infrared-active optical modes with transverse and longitudinal frequencies ω_t and ω_l. If one puts in an impurity atom with one valency greater than the host atom for which it is substituted, an extra electron not participating in the valence bond is liberated into the conduction band as a *carrier* of electric current. If the number density n of such carriers is sufficiently large, they will make a collective plasma oscillation with frequency $\omega_p = (4\pi ne^2/\epsilon_\infty m)^{1/2}$ [see (2.2.29)] where ϵ_∞ is the high-frequency dielectric constant of the host crystal and m is the effective mass of the electron in the conduction band. It is a merit of semiconductors that one can vary n and ω_p over a wide range by controlling the impurity concentration. Since the latter is much smaller than the host-atom concentration, the frequency ω_p of carrier plasma in the semiconductor is much smaller than that of host electrons in metals, and can be of the same order as the frequency ω_l of the longitudinal optical mode of lattice vibrations for moderate concentration n, in which case the two modes of collective oscillations, one electronic and other atomic, will be strongly coupled with each other through the depolarizing fields they produce.

Confining ourselves to the limit of vanishing wavevector $\mathbf{q}$, let us consider the equation of motion: $\ddot{\rho}_2 + \omega_l^2 \rho_2 = 0$ for the polarization charge $\rho_2 = -\operatorname{div} \mathbf{P}$ of optical lattice vibration. If it is written as $\ddot{\rho}_2 + \omega_t^2 \rho_2 = -(\omega_l^2 - \omega_t^2)\rho_2$, the right-hand side represents that part of the restoring force which comes from the depolarizing field, as is evident from (2.1.34). Because of the coexistence of the carrier plasma, one must add its polarization charge ρ_1 to ρ_2 on the right hand side. At the same time, one must add ρ_2 to ρ_1 of the restoring force $-\omega_p^2 \rho_1$ for the carrier plasma oscillation. The coupled equations of motion are then given by

$$\ddot{\rho}_1 + \omega_p^2 \rho_1 + \omega_p^2 \rho_2 = 0 \ , \tag{5.2.1a}$$

$$\ddot{\rho}_2 + \omega_l^2 \rho_2 + (\omega_l^2 - \omega_t^2)\rho_1 = 0 \ . \tag{5.2.1b}$$

The eigenfrequencies of the coupled modes, as functions of the variable $\omega_1 = \omega_p \propto \sqrt{n}$, are shown by solid lines in Fig. 5.2a, being similar to Fig. 5.1. The

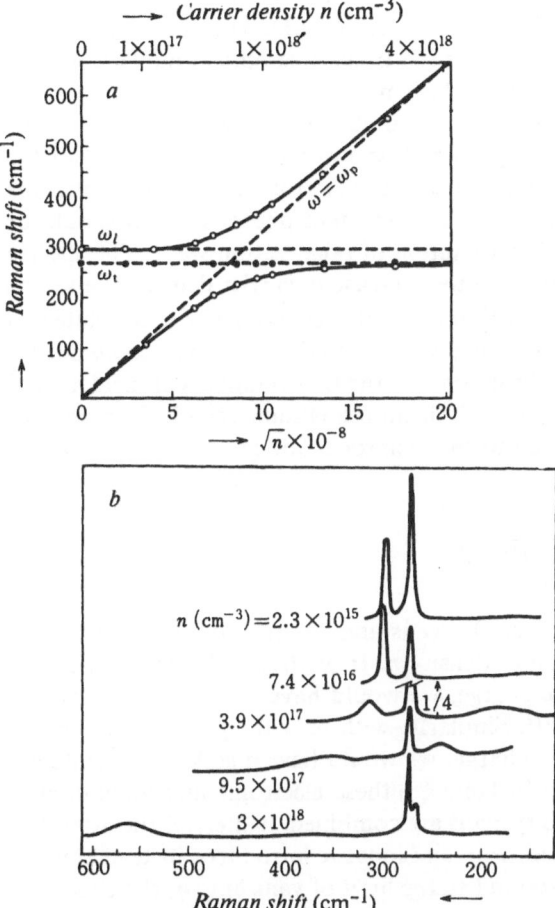

Fig. 5.2a,b. Coupled oscillations of carrier plasma (ω_p) and optical modes of lattice vibrations (ω_l, ω_t) in doped semiconductor GaAs. (a) Raman shifts versus carrier concentration n. (b) Raman spectra of GaAs with various carrier concentrations. [A. Mooradian, G.B. Wright: Phys. Rev. Lett. **16**, 999 (1966) A. Mooradian, A.L. McWhorter: In *Light Scattering Spectra of Solids*, Proceedings of the International Conference, 1968 at New York University, ed. by G.B. Wright (Springer, Berlin, Heidelberg, New York 1969) p.297]

low-frequency branch ω_- represents, when $\omega_p \ll \omega_l$, the carrier plasma screened by static dielectric constant ϵ_0 instead of high-frequency dielectric constant $\epsilon_\infty (\omega_-^2 \approx 4\pi n e^2/\epsilon_0 m^*)$, and when $\omega_p \gg \omega_l$, the longitudinal optical mode of lattice vibrations with depolarizing field completely screened by carrier plasma (hence, $\omega_- \approx \omega_t$).

As an alternative approach to the same problem, let us calculate the contributions of the two modes of polarization to the dielectric constant $\epsilon(\omega)$, with the use of the E method described in (II) of Sect. 2.2.2. According to this picture, the contributions should be additive if one takes the effective polarizability α_{eff}. With the use of (2.1.33′, 35, 2.2.27, 29), one obtains

$$\epsilon(\omega) = \epsilon_\infty \left[1 + \frac{\omega_l^2 - \omega_t^2}{\omega_t^2 - \omega^2} + \frac{\omega_p^2}{-\omega^2} \right] . \tag{5.2.2}$$

The equation $\epsilon(\omega) = 0$ for the longitudinal eigenvibrations of this system turns out to be the same as the secular equation for (5.2.1). The fact that $\epsilon(\omega)$ of (5.2.2) has a pole at $\omega = \omega_t$ means that the transverse optical mode is not influenced by the longitudinal plasma oscillation.

All branches of eigenvibrations mentioned above have been observed in the Raman spectra of n-type GaAs. Figure 5.2b shows the anti-Stokes part of the Raman spectra, where, the abscissa (increasing towards the left) represents the energy difference between the scattered and incident photons which should be equal to the energy of the absorbed quantum of polarization wave. The spectra consist of two peaks which move with the increase in carrier density n, and one peak which does not move. Their positions are plotted agsinst $\sqrt{n}$ by white and black points in Fig. 5.2a. The agreement with theoretical curves calculated from (5.2.1 or 2) is excellent. It should be noted that the wave numbers of incident and scattered photons, as well as that of polarization quantum absorbed or emitted, are negligilby small in comparison to the reciprocal lattice.

5.3 The Plasma Model of Metal

Let us regard the metal as a gas of electrons and positive ions with respective masses m and M and with number density n. If we had only electrons in the background of uniform positive charge, we would have a collective oscillation with frequency $\omega_p = (4\pi n e^2/m)^{1/2}$. Similarly, with only the positive ions in the background of uniform negative charge, we would have a collective oscillation with frequency $\omega_P = (4\pi n e^2/M)^{1/2}$. How are these electronic and ionic oscillations coupled together if the two systems are combined, namely, if the attractive Coulomb interactions between electrons and ions are introduced instead of interactions with the charged background? In the limit of vanishing $\boldsymbol{q}$, the equations of motion consist of (5.2.1a) for the electrons and its counterpart for the ions. The eigenvalues turn out to be $\omega = 0$ and $\omega = (\omega_p^2 + \omega_P^2)^{1/2} \approx \omega_p$. The electrons and ions oscillate completely out of phase ($\rho_1 + \rho_2 \neq 0$) in the eigenvibration with $\omega = 0$. For finite $\boldsymbol{q}$, however, the screening of the ionic plasma by the electron gas is no more complete ($\rho_1 + \rho_2 \neq 0$) leading to finite restoring force. To take this into account appropriately, we start from the coupled equations for a given wavevector $\boldsymbol{q}$:

$$\ddot{\rho}_1 + \omega_p^2 \rho_1 = -\lambda_1 \omega_p^2 \rho_2 \tag{5.3.1a}$$

$$\ddot{\rho}_2 + \omega_P^2 \rho_2 = -\lambda_2 \omega_P^2 \rho_1 \tag{5.3.1b}$$

where λ_1 and λ_2 are to be determined in the following argument.

Taking (5.3.1a) as the equation for the forced oscillation of the electron gas with ρ_1 under the external field $\rho_2 \propto \exp(i\boldsymbol{q} \cdot \boldsymbol{r} - i\omega t)$, one can write

$$\frac{\rho_1}{\rho_2} = \frac{-\lambda_1 \omega_p^2}{\omega_p^2 - \omega^2} = -\left[1 - \frac{1}{\epsilon(q, \omega)}\right] \tag{5.3.2}$$

with the use of the dielectric function $\epsilon(q, \omega)$ of the electron gas, because the ionic charge density ρ_2 is to be screened, by the electronic charge density ρ_1, into $\rho_2 + \rho_1 = \rho_2/\epsilon$.

The dielectric function of the degenerate electron gas with Fermi velocity v_F is given, in the limits of $\omega \gtrless qv_F$, by

$$\epsilon(q, \omega) \approx \begin{cases} \epsilon(0, \omega) = 1 - \dfrac{\omega_p^2}{\omega^2}, & (5.3.3a) \\[2mm] \epsilon(q, \omega) = 1 + \dfrac{q_0^2}{q^2}, & (5.3.3b) \end{cases}$$

respectively, as was derived in Sect. 2.2 [see (2.2.27, 30)]. In the case of $\omega \gg qv_F$, one obtains $\lambda_1 = 1$ from (5.3.2, 3a) in agreement with (5.2.1a), while $\lambda_1 \neq 1$ in the case of $\omega \ll qv_F$. In contrast, the ionic velocity V is so small (due to large M) that $\omega \gg qV$ always holds; namely, one can put $\lambda_2 = 1$ in (5.3.1b). Inserting (5.3.2) into (5.3.1b) to eliminate ρ_1, one obtains $\ddot{\rho}_2 + [\omega_p^2/\epsilon(q, \omega)]\rho_2 = 0$, whence the eigenfrequencies are given by the roots of

$$\omega = \frac{\omega_p}{\sqrt{\epsilon(q, \omega)}}. \tag{5.3.4}$$

Making use of (5.3.3), we find one branch in the region $\omega \ll qv_F$ with

$$\omega = \frac{\omega_p q}{\sqrt{q_0^2 + q^2}}, \tag{5.3.5}$$

and another branch in the region $\omega \gg qv_F$ with

$$\omega = \sqrt{\omega_p^2 + \omega_P^2} \approx \omega_p. \tag{5.3.6}$$

While the collective motion of electrons is almost unaffected by that of ions, the latter is strongly screened by the former. The screening becomes complete as $q \to 0$, resulting in the dispersion $\omega = cq$, as shown in Fig. 5.3. The proportionality constant:

$$c = \frac{\omega_P}{q_0} = \sqrt{\frac{m}{3M}} v_F \tag{5.3.7}$$

represents the velocity of longitudinal sound wave in the metal. The inequality $\omega \ll qv_F$ assumed in deriving (5.3.5) is in fact well satisfied since $c \ll v_F$ according to (5.3.7).

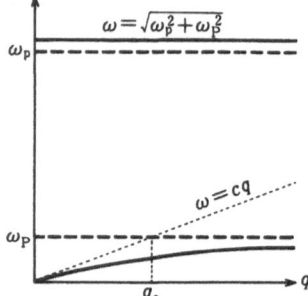

Fig. 5.3. Coupled oscillations of electron plasma (ω_p) and ion plasma (ω_p) in metals. The lower frequency branch represents the longitudinal acoustic mode of lattice vibrations

In spite of drastic simplification, the plasma model is not very far from reality in simple metals such as alkalis, where the expression (5.3.7) for the longitudinal sound velocity gives the right order of magnitude.

If one takes into account the spatial dispersion (q dependence) of the dielectric function of the electron gas in the coupled system described in Sect. 5.2, one finds that under the condition $\omega_p \gg \omega_l$ the frequency ω_- of the lower branch ($\sim$ longitudinal optical mode of lattice vibrations) varies from ω_t at $q \ll q_0$ to ω_l at $q \gg q_0$ because the screening by electrons becomes ineffective in the latter regime.

5.4 Polariton

Coupled modes may be formed not only between elementary excitations of a material system but also between one of them and an external field put in as a probe. A typical example is the polariton which is the coupled mode of electric polarization and electromagnetic wave.

5.4.1 Polariton and Dielectric Dispersion

Let us consider isotropic matter with dielectric constant $\epsilon(\omega)$ and magnetic permeability $\mu = 1$. The equation for the transverse part of the electromagnetic field within the matter:

$$c^2 \nabla^2 E = \frac{\partial^2 D}{\partial t^2}$$

(5.4.1)

has a plane-wave solution: $\exp(i\boldsymbol{k} \cdot \boldsymbol{r} - i\omega t)$, with the dispersion given by

$$\frac{c^2 k^2}{\omega^2} = \epsilon(\omega) = n(\omega)^2 ,$$

(5.4.2)

where $n(\omega)$ is the index of refraction.

The electric polarization of matter consists of contributions from various modes or oscillators, which may be atomic (optical modes of lattice vibrations) or electronic (excitons). Out of them, we consider a particular oscillator explicitly, denoting its contribution by P. Describing the contributions from all the remaining oscillators phenomenologically by the residual dielectric constant ϵ' (assumed to be positive), one can write

$$D = \epsilon'E + 4\pi P \; . \tag{5.4.3}$$

The equation of motion for this particular oscillator with its own frequency ω_t must be coupled to the electric field through the (static) polarizability α_0: the coupled equations now read:

$$\begin{cases} \ddot{P} + \omega_t^2 P = \omega_t^2 \alpha_0 E \; , & (5.4.4) \\[2mm] \epsilon'\ddot{E} - c^2\nabla^2 E = -4\pi\ddot{P} \; . & (5.4.5) \end{cases}$$

Since $P \| E$ due to isotropy, the plane-wave solutions of (5.4.4, 5) satisfy

$$\begin{bmatrix} -\omega^2 + \omega_t^2 & -\alpha_0\omega_t^2 \\ -4\pi\omega^2 & -\epsilon'\omega^2 + c^2 k^2 \end{bmatrix} \begin{bmatrix} P \\ E \end{bmatrix} = 0 \; . \tag{5.4.6}$$

By equating the determinant of coefficients to zero, one obtains the eigenfrequencies ω of the P-E coupled modes as functions of wave vector k. Or solving for k as a function of ω, one simply obtains

$$\frac{c^2 k^2}{\omega^2} = \epsilon(\omega) = \epsilon' + \frac{4\pi\alpha_0\omega_t^2}{\omega_t^2 - \omega^2} = \epsilon'\frac{\omega_l^2 - \omega^2}{\omega_t^2 - \omega^2} \; , \tag{5.4.7}$$

where

$$\omega_l = \omega_t + \Delta, \qquad \Delta = \left[\left(1 + \frac{4\pi\alpha_0}{\epsilon'}\right)^{1/2} - 1\right]\omega_t \; . \tag{5.4.8}$$

ϵ' may be assumed to be ω independent if the eigenfrequencies of other oscillators are far enough from the region $\omega \sim \omega_t$ with which we shall be concerned.

Equation (5.4.7) gives the wave vector k and dielectric constant ϵ as functions of ω, which are schematically shown in Fig. 5.4a, b. Namely, the electromagnetic wave propagating through matter with dielectric dispersion as given in Fig. 5.4b has $\omega(k)$ dispersion as shown in Fig. 5.4a. The quantum of this coupled wave is called the *polariton*. It is not essential whether the electromagnetic field of the polariton comes from the external source or not. The polarization P is always accompanied by the electric field E and vice versa, through the coupled equations (5.4.4, 5), irrespective of the origin of E. As is easily seen from (5.4.6),

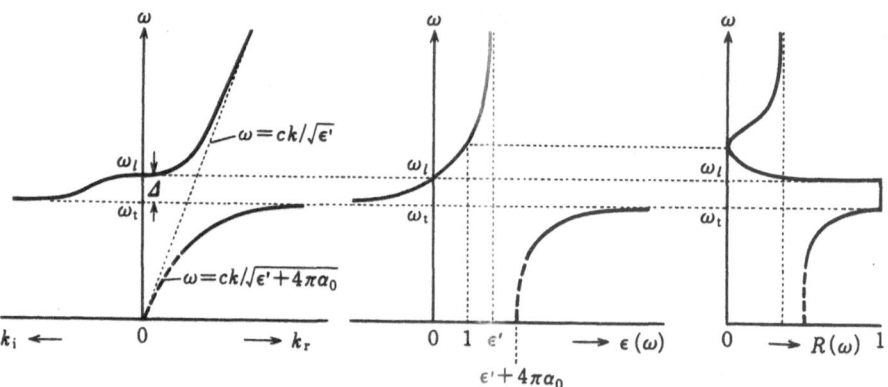

Fig. 5.4. (a) Dispersion of polariton: $\omega(k)$. (b) Dielectric dispersion: $\varepsilon(\omega)$. (c) Reflectivity: $R(\omega)$

the upper polariton branch reduces to pure E mode (with index of refraction $n = \sqrt{\epsilon'}$) when (I) $\omega - \omega_t \gg \Delta$, while the lower polariton branch reduces to pure P mode when (II) $0 < \omega_t - \omega \ll \Delta$, but in general they are E-P mixed modes. In the region (III) $\omega_t - \omega \gg \Delta$, however, the polariton is essentially an electromagnetic wave with constant index of refraction $n = \sqrt{\epsilon' + 4\pi\alpha_0}$, the difference from the region (I) being the participation of the particular oscillator of our concern. In the bottom region of the upper branch such that $k \ll \sqrt{\epsilon'}\omega/c$, the polariton is subject to the depolarizing field $E = -4\pi P/\epsilon'$ since the second term on the left side of (5.4.5) is negligible. This is the reason why the frequency of the upper polariton tends to the longitudinal frequency ω_l with vanishing k, in spite of its transverse nature. This is in contrast to the region $k \gg \sqrt{\epsilon'}\omega/c$ of the lower branch, namely, the region (II) mentioned above, where the depolarizing field is negligibly small. It is in this "off-polariton" region that we can neglect the retardation effect of the electromagnetic field in lattice vibrations and excitons as we did in Chap. 2.

The spectroscopic observation of the polarization wave now reduces to the problem of a boundary condition through which the polariton inside the matter is connected to the electromagnetic wave outside. An electromagnetic wave of the form: $\exp(ik_0 z - i\omega t)$ incident perpendicularly upon the surface of matter will give rise to the polariton wave of the form: $\exp[ik(\omega)z - i\omega t]$ inside the matter, where k_0 and $k(\omega)$ are given by ω/c and $\sqrt{\epsilon(\omega)}\,\omega/c$, respectively [see (5.4.7)]. The latter is shown schematically in Fig. 5.4a. The fraction R of incident electromagnetic energy which is reflected by the surface and the fraction $1 - R$ which penetrates into the matter as a polariton are determined by the boundary conditions associated with Maxwell's equation—the continuity, across the surface, of tangential components of E and H and normal components of D and B. For normal incidence, the reflectivity is given by

$$R(\omega) = \left| \frac{n(\omega) - 1}{n(\omega) + 1} \right|^2 = \frac{(n_r - 1)^2 + n_i^2}{(n_r + 1)^2 + n_i^2} , \tag{5.4.9}$$

where $n(\omega) = n_r(\omega) + in_i(\omega) = \sqrt{\epsilon(\omega)}$ is the complex refractive index. $R(\omega)$ corresponding to Fig. 5.4a, b is shown in Fig. 5.4c.

In the region $\omega_t < \omega < \omega_l$, the dielectric constant is negative while the refractive index and the wave vector are purely imaginary, as shown in the left halves of Fig. 5.4a, b. There is no propagating mode in this ω region, instead, E and P have the form $\exp[- k_i(\omega)z - i\omega t]$, and hence, the energy density decreases exponentially with the attenuation constant

$$A(\omega) = 2k_i(\omega) . \tag{5.4.10}$$

This does not mean, however, that the incident electromagnetic energy is absorbed by the matter. Instead, it is totally reflected by the surface without energy loss as shown in Fig. 5.4c. The imaginary part of $\epsilon(\omega)$ which is related to the absorption [see (2.2.35)] is zero in the present case, although $n(\omega)$ and $k(\omega)$ are purely imaginary.

However, the electromagnetic wave with $\omega = \omega_t$ should be subject to resonance absorption: $\epsilon(\omega)$ with real part (5.4.7) should have the imaginary part of $\delta(\omega - \omega_t)$ type according to the dispersion relation (2.2.12). To be more realistic, one can take into account a damping of the oscillator phenomenologically by adding the term $- \gamma \dot{P}$ to the right hand side of (5.4.4). This modifies the dielectric constant (5.4.7) into

$$\epsilon(\omega) = \epsilon' + \frac{4\pi\alpha_0\omega_t^2}{\omega_t^2 - \omega^2 - i\gamma\omega} . \tag{5.4.11}$$

The situation varies depending on the relative magnitude of the damping constant γ and the l–t splitting Δ given by (5.4.8). In the "dissipative" case where

$$\Delta \ll \frac{\gamma}{2} \ll \omega_t , \tag{5.4.12}$$

one finds from (5.4.8) that $2\pi\alpha_0/\epsilon' \approx \Delta/\omega_t < 1$, and hence, that the second term of (5.4.11) is small compared to the first term by factor $\Delta/(\gamma/2)$ even at resonance: $\omega = \omega_t$. On then obtains, from (5.4.2, 10),

$$A(\omega) = 2k_i(\omega) \approx \frac{\omega}{c\sqrt{\epsilon'}} \, \mathrm{Im}\{\epsilon(\omega)\} \approx \frac{\omega\sqrt{\epsilon'}}{c} \Delta \frac{\gamma/2}{(\omega_t - \omega)^2 + (\gamma/2)^2} . \tag{5.4.13}$$

Namely, the spatial attenuation of an electromagnetic wave is almost due to the absorption of its energy by matter. The attenuation constant $A(\omega)$ can then be called the absorption constant. Within the narrow resonance region, it is proportional to $\mathrm{Im}\{\epsilon(\omega)\}$. The absorption spectrum is a Lorentzian, with full width at half-maximum being given by the damping rate γ. The oscillator makes an important contribution to the absorption but almost none to the dispersion $[\epsilon(\omega) \approx \epsilon' = \text{const.}]$.

It is in the "dispersive" case:

$$\frac{\gamma}{2} \ll \varDelta \tag{5.4.14}$$

that the polariton effect manisfests itself clearly. In the very small region: $|\omega - \omega_t| \lesssim \gamma/2$ is $A(\omega)$ still related with absorption through $A(\omega) \approx (\sqrt{2}\omega/c) [\text{Im}\{\epsilon(\omega)\}]^{1/2}$ (instead of being proportional). In the region $\gamma/2 < \omega - \omega_t < \varDelta$, namely, in the greatest part of the l–t gap ($\omega_t < \omega < \omega_l$), the spatial attenuation $A(\omega)$ is due to the total reflection and is not associated with the energy absorption by matter.

As was mentioned before, the polariton is a coupled mode of a probe (electromagnetic wave) and an object (polarization wave), and the measurement of dielectric dispersion $\epsilon(\omega)$ (Fig. 5.4b is equivalent to the observation of polariton dispersion $\omega(k)$ (Fig. 5.4a). However, the polariton itself as an object can be observed by other electromagnetic waves, as were in fact performed in GaP. Being a zinc-blende-type crystal, it is isotropic but has no inversion symmetry, and hence, its optical mode is Raman active as well as infrared active. The polariton can therefore be excited by Raman scattering. Denote the energy and wave vector (inside the matter) of the incident and the scattered photons respectively by (ω_1, k_1) and (ω_2, k_2). One can obtain the polariton dispersion $\omega(k)$ by measuring the Raman shift $\omega_1 - \omega_2$ as a function of the scattering angle ($\theta = \angle\ k_1, k_2$), with the use of energy wave vector conservation rule: $\omega(k) = \omega_1 - \omega_2$, $k = k_1 - k_2$, The results obtained in this way are shown in Fig. 5.5, for the polariton and the (non polaritonic) longitudinal optical phonon. They are in good agreement with the dispersion curves (solid lines) calculated from known values of ω_t, ϵ_0 and ϵ_∞.

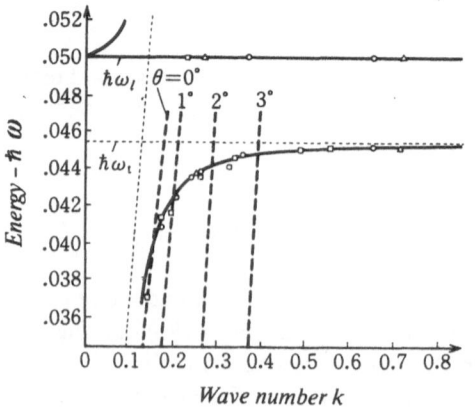

Fig. 5.5. Dispersion of phonon–polariton in GaP crystal as observed by Raman scattering. The abscissa denotes the wave vector k represented by the corresponding photon energy $\hbar ck$ in eV.
[C.H. Henry, J.J. Hopfield: Phys. Rev. Lett. **15**, 964 (1965)]

5.4.2 Spatial Dispersion and Optical Processes

If we take into account the k dependence of the frequency ω_t of transverse polarization wave which has so far been assumed to be constant, the dielectric constant depends not only on ω but also on k:

$$\frac{c^2k^2}{\omega^2} = \epsilon(\omega, k) = \epsilon' + \frac{4\pi\alpha_0(k)\omega_t(k)^2}{\omega_t(k)^2 - \omega^2} \ . \tag{5.4.15}$$

With this *spatial dispersion*, the solution $k(\omega)^2$ of (5.4.15) and hence $n(\omega)^2 = c^2k(\omega)^2/\omega^2$ can be a multivalued function of ω. A monochromatic wave will then excite more than one (say, ν) polariton wave with different k's and n's, which phenomenon is called birefringence. To relate the phases and the amplitudes of these polariton waves (and the reflected waves) to those of incident wave, $(\nu - 1)$ *additional boundary conditions* (sometimes abbreviated as ABC) are required besides the electromagnetic boundary condition. In contrast to the latter which is essentially macroscopic, the ABC is related to the microscopic nature of the polarization waves and the surface. Without going into the details of the ABC, we will give here qualitative arguments on the optical processes of the exciton-polariton for which the spatial dispersion is more important than for the phonon-polariton because of much larger k values of the polariton region concerned.

Assuming the translational mass M of the exciton to be positive, one finds the polariton dispersion as shown schematically in Fig. 5.6. The wave number K_p of the intersecting point of the dispersion curves for the noninteracting exciton and electromagnetic wave is of the order of 10^6 cm^{-1} even for the exciton of 10 eV, being much smaller than the reciprocal lattice ($\sim 10^8$ cm^{-1}). The exciton dispersion becomes significant much beyond K_p. Still, a part of the incident light with frequency ω_1 within the l–t gap ($\varepsilon_t(0)/\hbar < \omega_1 < \varepsilon_l(0)/\hbar$) can be converted into the lower branch polariton with ω_1 and K_1 (see Fig. 5.6); the reflectivity R is no longer total in the l–t gap. Unless ω_1 is very close to $\varepsilon_t(0)/\hbar$,

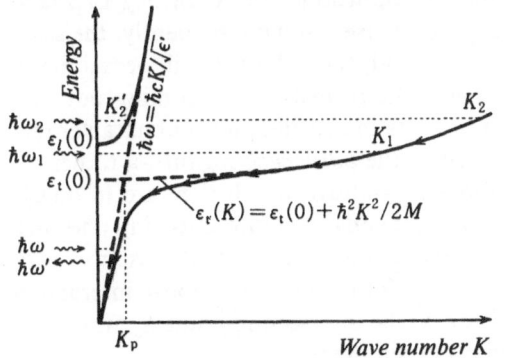

Fig. 5.6. The conversion between the incident or emitted electromagnetic waves (*wavy arrows*) and the polaritons (*solid lines*), and the energy degradation of the latter (*solid arrows*) due to inelastic scattering.

however, K_1 is much larger than K_p; this means that the electromagnetic component of the polariton is small, and hence, that the conversion ratio $1 - R$ of the incident light into the polariton is also small. For frequency ω_2 larger than $\varepsilon_l(0)/\hbar$, a significant part of the incident wave can get into the matter, of which by far the greater part is converted to the photon-like upper polariton with K_2' and the smaller to the exciton-like lower polariton with K_2 (see Fig. 5.6). This is the birefringence. The polariton wave of each branch will propagate through the matter with group velocity given by the gradient of its dispersion curve, and at another surface, a part of it will get out of the matter as transmitted light. However, the conversion ratio is small for the exciton-like lower branch (K_2 and K_2').

In deriving (5.4.11), we introduced the damping term $- \gamma \dot{\boldsymbol{P}}$ in the equation of motion (5.4.4), to describe phenomenologically the relaxation of the polarization wave. In the case of the exciton-polariton, the most important microscopic process responsible for the relaxation is the exciton–phonon interaction. The interaction Hamiltonian consists of matrix elements corresponding to the scattering of the exciton from $\boldsymbol{K}$ to $\boldsymbol{K} \mp \boldsymbol{q}$ with emission or absorption of a phonon with wave vector $\boldsymbol{q}$. The scattering rate γ_K of the exciton with wave vector $\boldsymbol{K}$ corresponds to the damping constant γ mentioned above. Because of the uncertainty principle, the exciton has the level width $\hbar\gamma_K$, which directly manifests itself as the spectral width of the absorption spectra [see (5.4.13)] in the case (5.4.12) of negligible polariton effect. Leaving the detailed study of this case to Chap. 7, we will consider here the opposite case (5.4.14), assuming H_{eL} to be small enough. Namely, we first diagonalize the energy of the exciton–photon system to get the polariton states as shown in Fig. 5.6, and then introduce H_{eL} as a perturbation. The polariton is now subject to interbranch as well as intrabranch scattering (shown by arrows in Fig. 5.6). At low temperature such that $k_B T \ll \Delta$, the thermalized polaritons will mostly be distributed in the lower branch, which however extends down to the photon-like region. Since the exciton energy $\varepsilon(K)$ is much larger than the lattice temperature $k_B T$, the polaritons would like to cascade down this branch, emitting (rather than absorbing) phonons. As its wave number decreases, the polariton becomes less exciton-like (hence, less interacting with phonons), and its density of states [proportional to $K^2 dK/d\varepsilon_p(K)$ where $\varepsilon_p(K)$ denotes the lower polariton energy] decreases, and consequently, the scattering rate decreases rapidly. On the other hand, the polariton will occasionally come to the surface and be reflected back, the probability of coming out as a photon increases as K decreases because of its increasing photon character. As a result, the polaritons will be distributed in the *bottleneck* region—a rather restricted region of K usually several times K_p—above which they are in quasithermal equilibrium and below which they are converted into outgoing photons. This is the polariton picture for the radiative annihilation of the exciton. According to a less accurate perturbation theory for the exciton–photon interaction H_{eR}, the exciton is converted to a photon at the intersecting point K_p where the energy–momentum conservation holds between them.

An incident photon with energy $\hbar\omega$ smaller than $\varepsilon_t(0)$ (see Fig. 5.6) is converted, with a certain probability, into a polariton, which in turn is scattered, emitting a phonon through its small exciton component, into a polariton with smaller energy $\hbar\omega'$, and then comes out of the matter as a photon. This is the polariton picture for the first-order Raman scattering. It is the first-order polariton–phonon scattering, as far as one is concerned with that part of the process which takes place within the matter.

If one takes H_{eR}, as well as H_{eL}, as the perturbation, the first-order Raman scattering is the third-order process of the type $H_{eR}H_{eL}H_{eR}$: a photon (inside the matter) with energy $\hbar\omega = \hbar cK/\epsilon'$ is converted through H_{eR} into an exciton with energy $\varepsilon(K)$, which is then scattered into $\varepsilon(K - q)$ by emitting a phonon with wave vector q through H_{eL} and finally converted through H_{eR} to a photon with energy $\hbar\omega' = \hbar c|K - q|/\epsilon'$. The energy need not be conserved in each step; the energy differences $[\varepsilon(K) - \hbar\omega]$ and $[\varepsilon(K - q) + \hbar\omega_q$ (phonon energy) $- \hbar\omega]$ appear as denominators of the third-order matrix element.

As $\hbar\omega$ approaches the energy of either intermediate state, the simple perturbation theory mentioned above diverges because of the vanishing energy denominator. This divergence, characteristic of *resonance Raman scattering*, is in fact removed by the finite level width of the intermediate state (imaginary part of energy). Does this imply that the absorption or emission of a photon takes place as a real (instead of virtual) process in this resonance situation? Putting it in a more general way, one may well ask whether one can conceptually distinguish between the second-order optical process (virtual absorption of photon followed by virtual emission of another photon) on the one hand and the successive but separable steps of the first-order optical processes (real absorption followed by real emission) on the other hand, under the resonance condition. This kind of question, which is relevant to the limitation of the concept of "elementary process," arises occasionally in systems with a great number of degrees of freedom. The general answer is that one cannot distinguish in principle. The separability of the two steps depends on the extent to which the relaxation has been attained in the intermediate states, as well as on the ways of excitation (incident light) and observation (emitted light). In certain idealistic situations can one decompose the emitted light spectrally into Raman and luminescence components corresponding to unrelaxed and relaxed intermediate states.

It should be emphasized that this problem arises only as the result of perturbation expansion. For instance, according to the polariton picture in which the exciton–photon interaction is exactly incorporated into the nonperturbed system, the (first order) Raman scattering is the first-order scattering of the polariton by a phonon, and its cross section does not diverge as the incident photon energy $\hbar\omega$ approaches the exciton energy ε_t. In fact, however, the cross section is subject to the resonance enhancement which corresponds to the divergence in perturbation theory: as $\hbar\omega$ approaches ε_t from below, the exciton component of the polariton with energy $\hbar\omega$, and hence, the polariton–phonon interaction, increases rapidly; as $\hbar\omega$ approaches $\varepsilon_l \sim \varepsilon_t$ from above, the amplitude of exciton-like lower polariton relative to that of photon-like upper polariton, as they are excited by incident photon, increases rapidly.

As is evident from the arguments, "elementary excitation" and "elementary process" are the relative concepts whose meanings vary depending on the non-perturbed system to be chosen for reference. If all the eigenstates of the total Hamiltonian H were known, these concepts would be neither necessary nor useful. In real macroscopic systems, however, the exact eigenstates of H are usually so complicated that they are almost of no use in understanding physical meaning and in calculating observable quantities. Still, it happens very often that H can be decomposed into H_0 and H_1 with the following properties: the eigenstates of nonperturbed system H_0 are simple enough to be accessible to physical intuitions and mathematical manipulations, while H_1 is small enough to be treated by perturbation theory or its modifications. It is in this situation that the concepts of elementary excitation and elementary process are really useful and powerful. However, the decomposition of H into $H_0 + H_1$ is not unique, as has been mentioned above on the exciton (e)–phonon(L)–photon (R) system with Hamiltonian $H = \mathscr{H}_e + H_L + H_R + H_{eL} + H_{eR}$. A simple decomposition is to put: $H_0 \equiv \mathscr{H}_e + H_L + H_R$ and $H_1 \equiv H_{eR} + H_{eL}$, as was done in the third-order perturbation theory $(H_{eR}H_{eL}H_{eR})$ of the first-order Raman scattering. A more sophisticated decomposition is to put $H_0 \equiv (\mathscr{H}_e + H_R + H_{eR}) + H_L$ and $H_1 \equiv H_{eL}$, where $(\mathscr{H}_e + H_R + H_{eR})$ can be diagonalized by introducing the polariton as mentioned above. Another decomposition is to put $H_0 = (\mathscr{H}_e + H_L + H_{eL}) + H_R$ and $H_1 = H_{eR}$; the primary interest here is the interaction dynamics within the matter, while the electromagnetic field plays the role of a probe for that. As will be described in Chap. 6, the electron (or exciton) in the phonon field becomes dressed by virtual phonons, the quasiparticle being called "polaron." The dynamics of exciton–phonon interaction is reflected in the optical spectra through H_{eR}, as will be shown in Chap. 7. How to decompose H, and which reference system H_0 to start from, depend not only on the smallness of H_1 to be left as perturbation but also on the aspect in which we are interested.

6. Renormalization and Damping—Centering Around Electron-Phonon Interaction

When the interaction between elementary excitations is written as a bilinear form, the problem can be solved exactly by means of suitable linear transformations as was shown in a previous chapter. However, for a general interaction not having such a simple form, it is very difficult to find an exact solution. As an effect of physical interest in a general case, we first mention that physical quantities associated with elementary excitations e.g., mass and sound velocity, etc., change because of interactions. This phenomenon is called "renormalization". Secondly, the elementary excitation has a finite lifetime. This is called "damping". If there exist interactions between elementary excitations, a given state of elementary excitation makes transitions to other quantum states and they cause the damping. In this chapter we will discuss renormalization and damping, concentrating on electron–phonon interaction.

6.1 Electron–Phonon Interaction in an Ionic Crystal

An electron in a solid moves in the potential field from the lattice points. If the lattice points compose a regular crystal structure, this potential has the same periodicity as the crystal itself and represents the so-called periodic potential. But, if lattice vibration takes place, this potential changes so that there appears an interaction between electron and phonon. This is called "electron–phonon interaction". Its explicit form depends on the type of solid; in this section we will discuss the electron–phonon interaction taking an ionic crystal as an example.

6.1.1 Optical Lattice Vibration in the Presence of an Electron

Although an ionic crystal is an insulator in the usual state, it is possible to make conduction electrons by a suitable method. For example, if we expose a crystal to light with an appropriate wavelength, some electrons in the filled band are excited to the conduction band and they behave as conduction electrons. According to recent studies, in some ionic crystals, the conduction band is isotropic, and if the magnitude of electron momentum p is small, the energy of the conduction electron is expressed as $p^2/2m$. Here, m stands for the band mass which has in general a value different from the value m_0 in a vacuum (see the next section). In the following, we assume that the energy of the conduction electron is given by the above equation.

In ionic crystals, positive and negative ions compose lattice points, taking alternate positions. If these lattice points vibrate, the electric field acting on the electron changes. In the case where the change in relative position between positive and negative ions is small, the change in electric field is small. On the contrary, if the former is large the latter is large. Therefore, in dealing with electron-phonon interaction in an ionic crystal, we may take into account only optical modes, neglecting acoustic modes. Since we discussed optical lattice vibration in Sect. 2.1, we will cite suitable equations in Sect. 2.1. As in there, we assume that the wavelength under consideration is much longer than the lattice spacing.

If there exist electrons in the crystal, they give rise to a certain charge distribution. Let the charge density at the position r be $\rho(r)$. These electrons produce an electric field that influences the lattice vibration. In order to study the relation between electric field and lattice vibration, let us observe the Maxwell equation div $D = 4\pi\rho$. If we substitute (2.1.1) into $D = E + 4\pi P$, we have

$$\text{div}(\epsilon_\infty E + 4\pi N_0 \sum_\nu e_\nu \xi_\nu) = 4\pi\rho \ .$$

This equation is interpreted in such a way that the electric field in the parentheses is produced by the presence of charge density ρ. Therefore, noting that only the longitudinal component remains if we take the divergence, we find

$$\epsilon_\infty E + 4\pi N_0 \sum_\nu (e_\nu \xi_\nu)_{\parallel} = -\ \text{grad} \int \frac{\rho(r')}{|r - r'|}\ dr' \equiv E' \ . \tag{6.1.1}$$

Or, we can write

$$E = -\frac{4\pi N_0}{\epsilon_\infty} \sum_\nu (e_\nu \xi_\nu)_{\parallel} + \frac{1}{\epsilon_\infty} E' \ . \tag{6.1.2}$$

The first and the second terms represent the electric fields produced by longitudinal waves and electrons, respectively.

It is convenient to use the law of energy conservation in order to derive electron–phonon interaction. For this purpose, taking account of the current density j arising from the charge density ρ of electrons, we consider the following Maxwell equations (we put $B = H$):

$$\left. \begin{array}{ll} \text{div } D = 4\pi\rho, & \text{div } H = 0 \\[2mm] \text{rot } E = -\dfrac{1}{c}\ \dot{H}, & \text{rot } H = \dfrac{1}{c}\ (\dot{E} + 4\pi\dot{P} + 4\pi j) \end{array} \right\} . \tag{6.1.3}$$

The Poynting vector S which describes the flow of electromagnetic energy is given by $S = (c/4\pi)\ (E \times H)$. Taking its divergence and using (6.1.3), we have

$$\text{div } S = -\left[\frac{1}{4\pi}\ (E \cdot \dot{E} + H \cdot \dot{H}) + E \cdot \dot{P} + E \cdot j \right] \ .$$

Integrating the above equation in a small region (strictly speaking in a region where $\boldsymbol{\xi}_{n\nu}$ do not depend on n) and applying the Gaussian theorem, we obtain

$$-\int S_n dS = \int \left[\frac{1}{4\pi}(\boldsymbol{E}\cdot\dot{\boldsymbol{E}} + \boldsymbol{H}\cdot\dot{\boldsymbol{H}}) + \boldsymbol{E}\cdot\dot{\boldsymbol{P}} + \boldsymbol{E}\cdot\boldsymbol{j}\right]dv \ .$$

The left-hand side in the above equation represents energy flowing into the region through the surface per unit time. According to the law of energy conservation, the right-hand side should be equal to the energy produced in the same region per unit time. Since the term $\boldsymbol{E}\cdot\boldsymbol{j}$ represents the Joule heat, we do not need to consider it in the present problem. Thus, if we let the energy density of the electromagnetic field and the lattice vibration be U, the following relation holds

$$\dot{U} = \frac{1}{4\pi}(\boldsymbol{E}\cdot\dot{\boldsymbol{E}} + \boldsymbol{H}\cdot\dot{\boldsymbol{H}}) + \boldsymbol{E}\cdot\dot{\boldsymbol{P}} \ .$$

In order to integrate this equation, we add two terms which are identically zero by virtue of (2.1.1, 10), and write

$$\dot{U} = \frac{1}{4\pi}(\boldsymbol{E}\cdot\dot{\boldsymbol{E}} + \boldsymbol{H}\cdot\dot{\boldsymbol{H}}) + \boldsymbol{E}\cdot\dot{\boldsymbol{P}} + \dot{\boldsymbol{E}}\cdot\left(\boldsymbol{P} - \frac{\epsilon_\infty - 1}{4\pi}\boldsymbol{E} - N_0 \sum_\nu e_\nu \boldsymbol{\xi}_\nu\right)$$
$$+ N_0 \sum_\nu \dot{\boldsymbol{\xi}}_\nu \cdot (M_\nu \ddot{\boldsymbol{\xi}}_\nu + \sum_\mu U'_{\nu\mu}\boldsymbol{\xi}_\mu - e_\nu \boldsymbol{E}) \ .$$

Integrating this equation and eliminating $\boldsymbol{P}$ by the use of (2.1.1), we find

$$U = \frac{1}{8\pi}(\epsilon_\infty E^2 + H^2) + N_0 \sum_\nu \left(\frac{1}{2}M_\nu \dot{\boldsymbol{\xi}}_\nu^2 + \frac{1}{2}\sum_\mu \boldsymbol{\xi}_\nu U'_{\nu\mu}\boldsymbol{\xi}_\mu\right)$$

where the integration constant is put to zero. Here the term H^2 is not related to electron–phonon interaction so that we will neglect it in the following discussion.

Integrating the above equation over the crystal with volume V and taking into account the energy of the conduction electrons, we find that the Hamiltonian of the total system is expressed as

$$\mathcal{H} = \sum_i \frac{p_i^2}{2m} + \mathcal{H}_L + \frac{\epsilon_\infty}{8\pi}\int E^2 dv \ . \tag{6.1.4}$$

Here $\mathcal{H}_L$ is the Hamiltonian of lattice vibration in the case of no electric field and is defined by

$$\mathcal{H}_L = N_0 \int \sum_\nu \left(\frac{1}{2}M_\nu \dot{\boldsymbol{\xi}}_\nu^2 + \frac{1}{2}\sum_\mu \boldsymbol{\xi}_\nu U'_{\nu\mu}\boldsymbol{\xi}_\mu\right)dv \ .$$

In what follows, we consider the crystal composed of two atoms such as NaCl type or CsCl type and observe only the optical mode. As was done in Chap. 1, we introduce an operator for the phonon and expand:

$$\xi_\nu(r) = \sum_{q,s} \sqrt{\frac{\hbar}{2VN_0\Omega_s}}\, \xi_\nu^{(s)}(\beta_{qs}e^{iq\cdot r} + \beta_{qs}^\dagger e^{-iq\cdot r})\ , \tag{6.1.5a}$$

$$\dot{\xi}_\nu(r) = \sum_{q,s} i\sqrt{\frac{\hbar\Omega_s}{2VN_0}}\, \xi_\nu^{(s)}(-\beta_{qs}e^{iq\cdot r} + \beta_{qs}^\dagger e^{-iq\cdot r})\ . \tag{6.1.5b}$$

For the transverse wave, the above β and $\beta^\dagger$ are annihilation and creation operators for the phonon, respectively, but for the longitudinal wave β and $\beta^\dagger$ are not operators for the phonon because of the presence of the restoring force due to the electric field. We will refer to this point later on. Substituting (6.1.5a, b) into the equation for $\mathscr{H}_L$ and using the orthonormal conditions (2.1.14) and (2.1.16) which represents the properties of diagonality:

$$\sum_\nu M_\nu \xi_\nu^{(s)} \cdot \xi_\nu^{(s')} = \delta_{ss'},\quad \sum_{\nu\mu} \xi_\nu^{(s)} U_{\nu\mu}' \xi_\mu^{(s')} = \delta_{ss'}\Omega_s^2\ ,$$

we obtain

$$\mathscr{H}_L = \sum_{q,s} \frac{\hbar\Omega_s}{2}(\beta_{qs}^\dagger\beta_{qs} + \beta_{qs}\beta_{qs}^\dagger)\ . \tag{6.1.6}$$

For an isotropic crystal, Ω_s is supposed to be independent of s and the relation $\Omega_s = \omega_t$ holds. In the following, we will consider such a case.

Returning now to (6.1.4), we substitute (6.1.2) into the third term in this equation. It then follows that

$$\frac{\epsilon_\infty}{8\pi}\int E^2 dv = \frac{2\pi N_0^2}{\epsilon_\infty}\sum_{\nu\nu'}\int (e_\nu\xi_\nu)_{||} \cdot (e_{\nu'}\xi_{\nu'})_{||}\, dv$$

$$- \frac{N_0}{\epsilon_\infty}\int \sum_\nu E' \cdot (e_\nu\xi_\nu)_{||} dv + \frac{1}{\epsilon_\infty}\int E'^2 dv\ . \tag{6.1.7}$$

Since the third term in the above equation represents the Coulomb interaction between electrons, we will omit it in the following discussion. Also, the first term describes the restoring force due to the electric field acting on the longitudinal wave and the second term is the electron–phonon interaction. By means of (6.1.5a) and (2.1.21), we have

$$\sum_\nu (e_\nu\xi_\nu)_{||} = \sum_q \sqrt{\frac{\hbar}{2VN_0\omega_t}}\, p_{||}(\beta_q e^{iq\cdot r} + \beta_q^\dagger e^{-iq\cdot r}) \tag{6.1.8}$$

(considering only the longitudinal wave). Furthermore, by substituting (6.1.8) into the first term in (6.1.7), this term is expressed as

$$\frac{\hbar\pi N_0}{\epsilon_\infty\omega_t}\,(p_{\parallel})^2 \sum_q (\beta_q\beta_q^\dagger + \beta_q^\dagger\beta_q + \beta_q\beta_{-q} + \beta_{-q}^\dagger\beta_q^\dagger) \; . \tag{6.1.9}$$

By using the relation $|p_{\parallel}| = e_+/\sqrt{M_r}$ which was mentioned above (2.1.33) together with (2.1.34) and by adding (6.1.9) to the longitudinal part in (6.1.6), the Hamiltonian $\mathscr{H}_l$ of the longitudinal wave is given by

$$\mathscr{H}_l = \sum_q \left[\frac{\hbar(\omega_l^2 + \omega_t^2)}{4\omega_t}\,(\beta_q^\dagger\beta_q + \beta_q\beta_q^\dagger) + \frac{\hbar(\omega_l^2 - \omega_t^2)}{4\omega_t}\,(\beta_q\beta_{-q} + \beta_{-q}^\dagger\beta_q^\dagger)\right] \; .$$

In order to diagonalize the above equation, we introduce the Bogoliubov transformation

$$\beta_q = \frac{1}{2\sqrt{\omega_l\omega_t}}\,[(\omega_l + \omega_t)\,b_q - (\omega_l - \omega_t)\,b_{-q}^\dagger] \; . \tag{6.1.10}$$

Then it turns out that $\mathscr{H}_l$ is expressed as

$$\mathscr{H}_l = \sum_q \frac{\hbar\omega_l}{2}\,(b_q^\dagger b_q + b_q b_q^\dagger) \; . \tag{6.1.11}$$

As can be seen from this, b and $b^\dagger$ are annihilation and creation operators, respectively, for the phonon of the longitudinal wave.

6.1.2 Electron–Phonon Interaction

The Hamiltonian $\mathscr{H}'$ describing electron–phonon interaction is given by the second term in (6.1.7), i.e.,

$$\mathscr{H}' = -\frac{N_0}{\epsilon_\infty}\int \sum_\nu \mathbf{E}'\cdot(e_\nu\boldsymbol{\xi}_\nu)_{\parallel}dv \; . \tag{6.1.12}$$

From this equation it is seen that the electrons interact with only the longitudinal wave. In order to derive an explicit form for (6.1.12), let us assume that there exists one conduction electron and denote its position by r_e. If we put the electronic charge to be $-e$, we have $\rho(r') = -e\delta(r' - r_e)$. Therefore, substituting this into (6.1.1) and using the Fourier expansion, we find

$$\mathbf{E}'(\mathbf{r}) = e\,\mathrm{grad}\,\frac{1}{|\mathbf{r} - \mathbf{r}_e|} = -\frac{4\pi ei}{V}\sum_k \frac{\mathbf{k}\exp[-i\mathbf{k}\cdot(\mathbf{r} - \mathbf{r}_e)]}{k^2} \; . \tag{6.1.13}$$

Also, putting (6.1.10) in (6.1.8), we obtain

$$\sum_\nu (e_\nu\boldsymbol{\xi}_\nu)_{\parallel} = \sum_q \sqrt{\frac{\hbar}{2VN_0\omega_l}}\,p_{\parallel}(b_q e^{i\mathbf{q}\cdot\mathbf{r}} + b_q^\dagger e^{-i\mathbf{q}\cdot\mathbf{r}}) \; . \tag{6.1.14}$$

Therefore, by substituting (6.1.13, 14) into (6.1.12) and by performing the volume integration, $\mathcal{H}'$ is calculated to be

$$\mathcal{H}' = \frac{4\pi e i N_0}{\epsilon_\infty}\sqrt{\frac{\hbar}{2VN_0\omega_l}}\sum_q \frac{p_{||}\cdot q}{q^2}[b_q \exp(iq\cdot r_e) - b_q^\dagger\exp(-iq\cdot r_e)] \ .$$

Since the vector $p_{||}$ is along the longitudinal direction, by means of the relation which was mentioned after (6.1.9), it follows that

$$p_{||}\cdot q = \frac{e_+}{\sqrt{M_r}}q \ .$$

By the use of this relation and (2.1.34), $\mathcal{H}'$ is expressed as

$$\mathcal{H}' = 4\pi e i\left(\frac{\hbar}{2V}\right)^{1/2}\left(\frac{\omega_l^2 - \omega_t^2}{4\pi\epsilon_\infty\omega_l}\right)^{1/2}\sum_q \frac{1}{q}[b_q \exp(iq\cdot r_e) - b_q^\dagger\exp(-iq\cdot r_e)] \ .$$

Furthermore, by means of the Lyddane-Sachs-Teller relation (2.1.35) which states that $\omega_l^2/\omega_t^2 = \epsilon_0/\epsilon_\infty$, $\mathcal{H}'$ is written as

$$\mathcal{H}' = \sum_q (V_q b_q e^{iq\cdot r} + V_q^* b_q^\dagger e^{-iq\cdot r}) \tag{6.1.15}$$

where we denote the position of the electron by r to simplify the notation. Also, V_q is given by

$$V_q = 4\pi e i\omega\left[\frac{1}{4\pi}\left(\frac{1}{\epsilon_\infty} - \frac{1}{\epsilon_0}\right)\right]^{1/2}\left(\frac{\hbar}{2V\omega}\right)^{1/2}\frac{1}{q}$$

where we put ω_l to be ω for simplicity. Or, introducing α defined by

$$\alpha = \frac{1}{2}\left(\frac{1}{\epsilon_\infty} - \frac{1}{\epsilon_0}\right)\frac{e^2}{\hbar\omega}\left(\frac{2m\omega}{\hbar}\right)^{1/2} \ , \tag{6.1.16}$$

we see that V_q is expressed as

$$V_q = \frac{i\hbar\omega}{q}\left(\frac{\hbar}{2m\omega}\right)^{1/4}\left(\frac{4\pi\alpha}{V}\right)^{1/2} \ . \tag{6.1.17}$$

The α is a nondimensional quantity and a coupling constant which describes the strength of interaction between the electron and the phonon. It plays an important role in the problem of the polaron which will be discussed in the next section.

6.2 Polaron

In a previous section, it was shown that an electron excited to the conduction band in an ionic crystal has an interaction with the longitudinal optical lattice vibration. The physical origin of this interaction is that the electric field arising from the electron induces displacements of positive and negative ions and consequently the restoring force acting on the longitudinal wave changes. On the other hand, electric polarization in the crystal induced by an electron affects conversely the motion of the electron. It may be understood intuitively that the electron moves accompanied by a phonon cloud around it. The situation is similar to the case that when a ball rolls on a rubber membrane a hollow in the membrane moves with the ball itself (Fig. 6.1). In this way, the electron and the phonon cloud compose a kind of composite particle. This is called "polaron". The polaron is one of the elementary excitations in a solid. Because of the presence of the phonon cloud, the polaron mass becomes larger than the band mass. In other words, renormalization of the mass takes place. In this section, we shall discuss the problem of the polaron on the basis of a simple perturbation procedure.

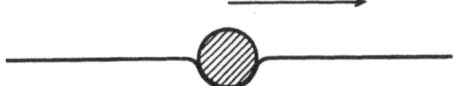

Fig. 6.1. When a ball (electron) rolls on a rubber membrane, a hollow (phonon) in the membrane moves with the ball itself

6.2.1 Renormalization of Mass (Perturbation Calculation of Second Order)

The Hamiltonian describing the polaron is written as

$$\mathcal{H} = \frac{p^2}{2m} + \sum_q \hbar\omega b_q^\dagger b_q + \sum_q (V_q b_q e^{i\boldsymbol{q}\cdot\boldsymbol{r}} + V_q^* b_q^\dagger e^{-i\boldsymbol{q}\cdot\boldsymbol{r}}) \tag{6.2.1}$$

where the zero-point lattice vibration is neglected. Taking the first and second terms as the unperturbed Hamiltonian $\mathcal{H}_0$ and the third term $\mathcal{H}'$ as the small perturbation, we will carry out perturbation calculation. The eigenfunction of $\mathcal{H}_0$ is in general given by $|k; n_1, \ldots, n_q, \ldots\rangle$ in terms of the electron wave vector $\boldsymbol{k}$ and the phonon occupation number n_q. We will here restrict ourselves to the phonon vacuum and consider $|k; 0_1, \ldots, 0_q, \ldots\rangle$ as the eigenfunction of $\mathcal{H}_0$. The eigenfunction $|k\rangle$ of the electron normalized in the volume V is a plane wave given by $|k\rangle = V^{-1/2} \exp(i\boldsymbol{k}\cdot\boldsymbol{r})$. The relation $(p^2/2m)|k\rangle = \varepsilon_k |k\rangle$ holds where $\varepsilon_k = \hbar^2 k^2/2m$ is the energy of the electron with wave vector $\boldsymbol{k}$. Needless to say, the unperturbed energy E_0 is expressed as $E_0 = \varepsilon_k$.

In (6.2.1) the $b^\dagger$ or b increases or decreases the phonon number by one so that $\mathcal{H}'$ has no diagonal element. Because of this the perturbation term of the first order for energy vanishes. Next, the second-order perturbation term E' is given by the formula in quantum mechanics:

$$E' = \sum_{n \neq 0} \frac{\langle 0 | \mathscr{H}' | n \rangle \langle n | \mathscr{H}' | 0 \rangle}{E_0 - E_n} \ .$$

Noting that $b_q | 0_q \rangle = 0$, $b_q^\dagger | 0_q \rangle = | 1_q \rangle$ and $\langle k - q | \exp(- iq \cdot r) | k \rangle = 1$, we see that in the present case the phonon with q is excited and the wave vector of the electron becomes $k - q$ at the intermediate state. Or, as depicted by the diagram, the second term in $\mathscr{H}'$ represents a process in which the electron with k is scattered to a state with $k - q$ emitting the phonon with q, as shown in Fig. 6.2a. In the same way, the first term is represented by Fig. 6.2b. As a result, E' is expressed as

$$E' = \sum_q \frac{|V_q|^2}{\varepsilon_k - \varepsilon_{k-q} - \hbar\omega} \ . \tag{6.2.2}$$

Substituting (6.1.17) into this equation and transforming the sum over q into the integral, we obtain

$$E' = \frac{\alpha \hbar \omega \gamma}{2\pi^2} \int \frac{1}{k^2 - (k - q)^2 - \gamma^2} \frac{dq}{q^2} \tag{6.2.3}$$

where γ is defined by $\gamma = (2m\omega/\hbar)^{1/2}$. The γ corresponds to the wave number of electron having the energy of phonon ($\hbar\omega = \hbar^2\gamma^2/2m$).

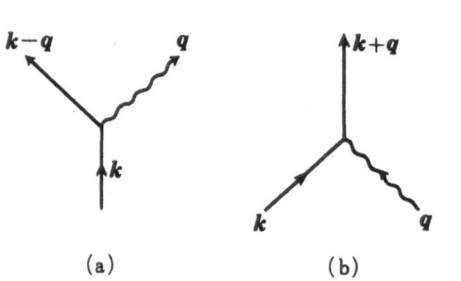

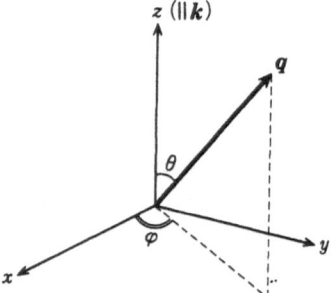

Fig. 6.2. Emission (a) and absorption (b) of the phonon by the electron

Fig. 6.3. Polar coordinate in q space

In order to carry out the integration in (6.2.3), we introduce a polar coordinate taking the k direction as the z axis (Fig. 6.3). Integrating by φ and θ, we are led to

$$E' = \frac{\alpha \hbar \omega \gamma}{4\pi k} \int_{-\infty}^{\infty} \frac{1}{q} \ln \left| \frac{2kq - q^2 - \gamma^2}{2kq + q^2 + \gamma^2} \right| dq \ . \tag{6.2.4}$$

Here, we have used the facts that the integrand is an even function of q and that the cutoff wave number for the phonon (e.g., Debye cutoff) can be made ∞ since the integrand tends to 0 rapidly as $q \to \infty$. Denoting the integral with respect to q in (6.2.4) by I and regarding it as a function of k, we have by differentiation

$$\frac{dI}{dk} = - \int_{-\infty}^{\infty} \left(\frac{2}{q^2 - 2kq + \gamma^2} + \frac{2}{q^2 + 2kq + \gamma^2} \right) dq \ .$$

By the use of the formula

$$\int_{-\infty}^{\infty} \frac{dx}{ax^2 + 2bx + c} = \frac{\pi}{\sqrt{ac - b^2}} \quad (a > 0, \ ac > b^2)$$

and from the condition that $I = 0$ at $k = 0$, I is calculated to be

$$I = - 4\pi \int_{0}^{k} \frac{dx}{\sqrt{\gamma^2 - x^2}} = - 4\pi \sin^{-1} \frac{k}{\gamma}$$

for $k < \gamma$. Thus, E' is expressed as

$$E' = - \alpha \hbar \omega \frac{\sin^{-1}(k/\gamma)}{k/\gamma} \ . \tag{6.2.5}$$

If we use the expansion $\sin^{-1} x = x + (1/6)x^3 + (3/40)x^5 + \ldots$ and take the unperturbed energy $\hbar^2 k^2 / 2m$ into account, the total energy E for $k \ll \gamma$ is given by

$$E = - \alpha \hbar \omega + \frac{\hbar^2 k^2}{2m} \left(1 - \frac{\alpha}{6} \right) + O(k^4) \ . \tag{6.2.6}$$

If we write the second term as $\hbar^2 k^2 / 2m^*$, the m^* represents the mass of the polaron. From (6.2.6) we obtain

$$m^* = \frac{m}{1 - (\alpha/6)} \quad \text{or} \quad \frac{m^*}{m} = 1 + \frac{\alpha}{6} \quad (\alpha \ll 1) \ . \tag{6.2.7}$$

6.2.2 Phonon Cloud

To study the phonon cloud existing around the polaron, we will discuss the expectation value of the total number of phonons on the basis of perturbation calculation as in Sect. 6.2.1. According to the formula in quantum mechanics, the perturbed wave function $| k ; 0 \rangle_1$ up to the first order is expressed as

$$| k ; 0 \rangle_1 = | k ; 0 \rangle + \sum_{q} \frac{\langle k - q ; 1_q | \mathcal{H}' | k ; 0_q \rangle}{\varepsilon_k - \varepsilon_{k-q} - \hbar \omega} | k - q ; 1_q \rangle \ .$$

Since the total number N of phonons is given by $N = \sum_q b_q^\dagger b_q$, the expectation value of N for the above state is written as

$$\langle N \rangle = {}_1\langle k; 0 | N | k; 0 \rangle_1 = \sum_q \frac{|\langle k - q; 1_q | \mathscr{H}' | k; 0_q \rangle|^2}{(\varepsilon_k - \varepsilon_{k-q} - \hbar\omega)^2}$$

$$= \sum_q \frac{|V_q|^2}{(\varepsilon_k - \varepsilon_{k-q} - \hbar\omega)^2} . \tag{6.2.8}$$

Substituting (6.1.17) into (6.2.8), we have

$$\langle N \rangle = \frac{\alpha\gamma^3}{2\pi^2} \int \frac{1}{[k^2 - (k - q)^2 - \gamma^2]^2} \frac{dq}{q^2} . \tag{6.2.9}$$

In order to calculate the above equation, we take a limit $k \to 0$ assuming that the electron is moving slowly. Then, $\langle N \rangle$ is calculated to be

$$\langle N \rangle = \frac{2\alpha\gamma^3}{\pi} \int_0^\infty \frac{dq}{(q^2 + \gamma^2)^2} = \frac{2\alpha\gamma^3}{\pi} \cdot \frac{\pi}{4\gamma^3} = \frac{\alpha}{2} . \tag{6.2.10}$$

In an actual ionic crystal α takes a value larger than 1 as will be explained in Sect. 6.2.4. Therefore, it is possible that $\langle N \rangle$ calculated from (6.2.10) exceeds 1. However, since we have been considering states where the number of phonons is at most one, this implies a contradiction. Thus, it turns out that the simple perturbation approach is insufficient to deal with the polaron quantitatively and an improved method is required. We will discuss this problem again in Sect. 6.3.

6.2.3 Damping

The above-mentioned wave function $| k; 0 \rangle_1$ includes various wave vectors of the electron other than k, so that the wave vector of the electron is no longer a constant of motion. Because of this situation, the electron which is in the state $| k \rangle$ initially makes a transition to other states and consequently the state $| k \rangle$ has a finite mean lifetime. In other words, damping takes place. In the following, we will study the mean lifetime τ of the electron assuming that the initial state is $| k; n_1, \ldots, n_q, \ldots \rangle$. According to quantum mechanics, the probability that the initial state i makes a transition to the final state f because of the perturbation Hamiltonian $\mathscr{H}'$ is given by $(2\pi/\hbar) |\mathscr{H}'_{fi}|^2 \delta(E_i - E_f)$ per unit time. Since τ is equal to the inverse of the transition probability W per unit time, considering two processes shown in Fig. 6.2, we have

$$\frac{1}{\tau} = W = \frac{2\pi}{\hbar} \sum_q |V_q|^2 [\delta(\varepsilon_k - \varepsilon_{k-q} - \hbar\omega)(n_q + 1)$$

$$+ \delta(\varepsilon_k - \varepsilon_{k+q} + \hbar\omega)n_q] . \tag{6.2.11}$$

In order to calculate this equation, we express the sum over q by the integral and substitute (6.1.17). Furthermore, we assume that n_q is given by its thermal equilibrium value at the absolute temperature T. That is, by using the Planck distribution, we put

$$\langle n_q \rangle = \frac{1}{e^{\beta\hbar\omega} - 1} \equiv \langle n \rangle$$

where $\beta = 1/k_B T$. As in Fig. 6.3, we introduce the polar coordinate and put $\cos\theta = t$. Then, $1/\tau$ is written as

$$\frac{1}{\tau} = 2\alpha\omega\gamma \int_0^\infty dq \int_{-1}^1 dt \, [\delta(2kqt - q^2 - \gamma^2)(\langle n \rangle + 1)$$
$$+ \delta(2kqt + q^2 - \gamma^2)\langle n \rangle] \, .$$

In order that the argument in the first δ function be zero, we should have $t = (q^2 + \gamma^2)/2kq$. From the inequalities $-1 \leq t \leq 1$, it follows that $\gamma \leq k$. Conversely, the first term vanishes if $k < \gamma$. In the same way, it is seen that the relation $\sqrt{k^2 + \gamma^2} - k < q < \sqrt{k^2 + \gamma^2} + k$ is required in order for the second term not to be zero. If this is satisfied, the integral by t leads to $1/2kq$ so that we obtain for $k < \gamma$

$$\frac{1}{\tau} = 2\alpha\omega\gamma\langle n \rangle \int_{\sqrt{k^2+\gamma^2}-k}^{\sqrt{k^2+\gamma^2}+k} \frac{dq}{2kq} = \frac{\alpha\omega\gamma\langle n \rangle}{k} \ln \frac{\sqrt{k^2 + \gamma^2} + k}{\sqrt{k^2 + \gamma^2} - k} \, .$$

In particular, for a slow electron ($k \ll \gamma$), $1/\tau$ is expressed as

$$\frac{1}{\tau} = \frac{2\alpha\omega}{e^{\beta\hbar\omega} - 1} \, . \tag{6.2.12}$$

The result (6.2.12) may be more easily derived if we put $k = 0$ in the equation for $1/\tau$ mentioned previously.

6.2.4 Numerical Values of α

Recently, it has become possible to measure experimentally the polaron mass m^* because of developments in experimental technique and sample preparation. The m^* is measured mainly by means of cyclotron resonance; the recent results are shown in Table 6.1. In the first column of this table, the values of m^*/m_e (m_e: electronic mass in vacuum) are mentioned for typical alkali halides and silver halides. On the other hand, the coupling constant α is expressed as $\alpha = (e^2/\hbar)(m/2\hbar\omega)^{1/2}[(\epsilon_0 - \epsilon_\infty)/\epsilon_0\epsilon_\infty]$ by virtue of (6.1.16). In this equation, ϵ_0, ϵ_∞ and ω are measured by the use of suitable experimental procedures. (In Table 6.1, the inverse of wavelength λ is shown instead of ω itself. The relation $\omega =$

$2\pi c/\lambda$ holds where c is the velocity of light.) Now, the numerical value of band mass m is necessary to calculate α by means of the above equation. For this purpose, the following two equations are frequently applied:

$$\frac{m^*}{m} = \frac{1 - 0.0008\alpha^2}{1 - (\alpha/6) + 0.0034\alpha^2} , \tag{6.2.13a}$$

$$\frac{m^*}{m} = 1 + \frac{1}{6}\alpha + 0.0236276\alpha^2 + 0.0014\alpha^3 . \tag{6.2.13b}$$

Equation (6.2.13a) is an interpolation formula derived on the basis of Feynman's theory which will be discussed in the next section and (6.2.13b) is the result of perturbation calculation up to the sixth order. If we put subscript a or b on m and α calculated from (6.2.13a) or (6.2.13b), their numerical values are as shown in Table 6.1. As is seen from this table, α takes values $\sim$ 2–4. Therefore, it is understood that α is a quantity of the order 1.

Table 6.1. Numerical values of α in ionic crystals

	$\dfrac{m^*}{m_e}$	ϵ_∞	ϵ_0	$\dfrac{1}{\lambda}$ (cm^{-1})	$\dfrac{m_a}{m_e}$	α_a	$\dfrac{m_b}{m_e}$	α_b
KCl	0.922 ± 0.04	2.20	4.49	212	0.432 ± 0.02	3.46	0.467 ± 0.02	3.60
KBr	0.700 ± 0.03	2.39	4.52	166	0.367 ± 0.02	3.07	0.388 ± 0.02	3.15
AgCl	0.431 ± 0.04	3.97	9.50	197	0.302 ± 0.03	1.90	0.305 ± 0.03	1.91
AgBr	0.289 ± 0.01	4.68	10.60	132	0.215 ± 0.01	1.59	0.217 ± 0.01	1.60

Cited from J.W. Hodby: J. Phys. C4, L8 (1971)

6.3 Intermediate Coupling Method and Method of Path Integral

Since the coupling constant α is a quantity of the order 1 in the problem of the polaron, simple perturbation calculation cannot lead to a satisfactory result. In order to derive a reliable result, a better mathematical procedure is necessary. In this section, we will introduce two methods along this line. Both are based on the variation principle.

6.3.1 Intermediate Coupling Method

This method is essentially an extension of Tomonaga's approach in the theory of the meson and is called "intermediate coupling method" in a sense that it is applicable to a case where the coupling constant takes an intermediate value. *Lee, Low* and *Pines* have applied this method to the polaron problem.

As was stated in Sect. 6.2.3, the momentum of the electron in the presence of electron–phonon interaction is no longer a constant of motion. However, if we take into account the momenta of phonons and introduce an operator

$P_{op} = p + \sum_q \hbar q b_q^\dagger b_q$ which represents the total momentum, it turns out the P_{op} commutes with the total Hamiltonian $\mathscr{H}$. Thus, P_{op} is a constant of motion and can be regarded as a c number. To clarify this point, let us consider the following transformations.

In general, if we make a unitary transformation $|\Phi\rangle = S|\Psi\rangle$ in the Schrödinger equation $\mathscr{H}|\Phi\rangle = E|\Phi\rangle$ it follows that $S^{-1}\mathscr{H}S|\Psi\rangle = E|\Psi\rangle$. Thus, the Hamiltonian is transformed as $\mathscr{H} \to \mathscr{\tilde{H}} = S^{-1}\mathscr{H}S$. The $\mathscr{\tilde{H}}$ describes the same physical system as the original one. In particular, if P is a c number vector and S is given by

$$S = \exp\left[\frac{i}{\hbar}(P - \sum_q \hbar q b_q^\dagger b_q) \cdot r\right],$$

by the use of the formula

$$e^{iX}Ae^{-iX} = A + i[X, A] + \frac{i^2}{2!}[X[X, A]] + \cdots,$$

we have

$$P_{op} \to S^{-1}P_{op}S = P + p,$$
$$p \to S^{-1}pS = P - \sum \hbar q b_q^\dagger b_q + p,$$
$$b_q \to S^{-1}b_q S = b_q \exp(-iq \cdot r).$$

Thus, as is seen from (6.2.1), the $\mathscr{\tilde{H}}$ does not depend on the position vector r of electron. For this reason, the momentum p of electron can be considered as a c number. In the following, we will put it to zero. As a result, $\mathscr{\tilde{H}}$ is expressed as

$$\mathscr{\tilde{H}} = \frac{1}{2m}(P - \sum_q \hbar q b_q^\dagger b_q)^2 + \sum_q \hbar \omega b_q^\dagger b_q + \sum_q (V_q b_q + V_q^* b_q^\dagger). \tag{6.3.1}$$

Since $\mathscr{\tilde{H}}$ includes the total momentum P as a parameter, its energy eigenvalue is a function of P. When $E(P)$ is expanded as $E(P) \simeq E_0 + (P^2/2m^*) + O(P^4)$, the m^* represents the mass of the polaron.

Let us make once more a unitary transformation to the equation $\mathscr{\tilde{H}}|\Psi\rangle = E|\Psi\rangle$, i.e.

$$|\Psi\rangle = U|0\rangle, \quad U = \exp[\sum_q (f_q b_q^\dagger - f_q^* b_q)] \tag{6.3.2}$$

where $|0\rangle$ represents the phonon vacuum ($\langle 0|0\rangle = 1$) and f_q is a variational parameter to be determined later on. By a simple calculation, it is seen that

$$U^{-1}b_q U = b_q + f_q, \quad U^{-1}b_q^\dagger U = b_q^\dagger + f_q^*. \tag{6.3.3}$$

From (6.3.2) it follows that the expectation value of the energy is written as

$E = \langle \Psi | \hat{\mathscr{X}}' | \Psi \rangle = \langle 0 | U^{-1} \mathscr{H} U | 0 \rangle$. Furthermore, by means of (6.3.1) and (6.3.3), E is calculated to be

$$E = \frac{P^2}{2m} + \sum_q (V_q f_q + V_q^* f_q^*) + \frac{\hbar^2}{2m} \left(\sum_q q |f_q|^2 \right)^2$$

$$+ \sum_q \left(\hbar\omega - \frac{\hbar q \cdot P}{m} + \frac{\hbar^2 q^2}{2m} \right) | f_q |^2 . \tag{6.3.4}$$

Now, by means of the variation principle, we determine f_q so as to make E a minimum. The necessary conditions are $\partial E/\partial f_q = \partial E/\partial f_q^* = 0$, so that from (6.3.4) we obtain

$$V_q + f_q^* \left(\hbar\omega - \frac{\hbar q \cdot P}{m} + \frac{\hbar^2 q^2}{2m} \right) + \frac{\hbar^2}{m} \left(\sum_{q'} q' |f_{q'}|^2 \right) \cdot q f_q^* = 0 . \tag{6.3.5}$$

In order to solve this equation, noting that only P is a particular direction in the present problem, we put

$$\eta P = \sum_q \hbar q |f_q^2| . \tag{6.3.6}$$

Substituting (6.3.6) into (6.3.5) and solving for f_q^*, we have

$$f_q^* = \frac{- V_q}{\left[\hbar\omega - \dfrac{\hbar q \cdot P}{m} (1 - \eta) + \dfrac{\hbar^2 q^2}{2m} \right]} .$$

Therefore, from (6.3.6) the following equation for η

$$\eta P = \sum_q \frac{|V_q|^2 \hbar q}{\left[\hbar\omega - \dfrac{\hbar q \cdot P}{m} (1 - \eta) + \dfrac{\hbar^2 q^2}{2m} \right]^2} \tag{6.3.7}$$

is derived.

If we return to (6.3.4) for E, by the use of (6.3.5) and (6.3.6) it follows that $E = (P^2/2m)(1 - \eta^2) + \sum_q V_q^* f_q^*$, i.e., E is expressed as

$$E = \frac{P^2}{2m}(1 - \eta^2) + \sum_q \frac{|V_q|^2}{(\hbar q \cdot P/m)(1 - \eta) - (\hbar^2 q^2/2m) - \hbar\omega} .$$

The second term in the above equation is equal to (6.2.2) for E' with k replaced by $k = P(1 - \eta)/\hbar$. As a result, from (6.2.5) we have

$$E = \frac{P^2}{2m}(1 - \eta^2) - \alpha\hbar\omega \frac{\sin^{-1} Q}{Q} \tag{6.3.8}$$

where Q is defined by

$$Q = \frac{P(1-\eta)}{\hbar\gamma} = \frac{P(1-\eta)}{(2m\hbar\omega)^{1/2}} .$$

Expanding (6.3.8) in powers of P, we obtain

$$E = -\alpha\hbar\omega + \frac{P^2}{2m}\left[1 - \eta^2 - \frac{\alpha}{6}(1-\eta)^2\right] + O(P^4) . \tag{6.3.9}$$

As is seen from (6.3.9), in order to calculate m^* it is sufficient to derive η in the limit $P \to 0$. Thus, by expanding the right-hand side in (6.3.7) it is sufficient to assume that

$$\eta P = (1-\eta) \sum_q |V_q|^2 \hbar q \frac{2\hbar q \cdot P}{m} \bigg/ \left(\hbar\omega + \frac{\hbar^2 q^2}{2m}\right)^3 .$$

Introducing the polar coordinate in which the P direction is taken as z axis, substituting (6.1.17) and carrying out necessary integrations, we have

$$\frac{\eta}{1-\eta} = \frac{8\alpha}{3\pi} \int_0^\infty \frac{x^2}{(1+x^2)^3} dx = \frac{\alpha}{6} \quad \therefore \quad \eta = \frac{\alpha/6}{1+(\alpha/6)} .$$

Substituting this result into (6.3.9), we find

$$E = -\alpha\hbar\omega + \frac{P^2}{2m[1+(\alpha/6)]} + O(P^4)$$

whence it follows that the polaron mass m^* is expressed as

$$\frac{m^*}{m} = 1 + \frac{\alpha}{6} . \tag{6.3.10}$$

We have already shown in (6.2.7) that the above equation is valid for $\alpha \ll 1$. However, in the present method we do not assume particularly the condition $\alpha \ll 1$. Therefore, it may be understood that (6.3.10) holds even in the case $\alpha \simeq 1$.

We have been discussing so far the mathematical aspects of the intermediate coupling method. In closing we would like to refer to its physical meaning. For this purpose, we apply the following formula (Glauber's formula) for phonon operators:

$$\exp(fb^\dagger - f^*b) = \exp\left(-\frac{1}{2}|f|^2\right)\exp(fb^\dagger)\exp(-f^*b) . \tag{6.3.11}$$

To prove this formula, we put

$$\exp[x(fb^\dagger - f^*b)] = A(x)\exp(xfb^\dagger)\exp(-xf^*b)$$

and differentiate both sides by x. Then, by means of the relation $\exp(xf^*b)b^\dagger$ $\times \exp(-xf^*b) = b^\dagger + xf^*$, i.e., $b^\dagger\exp(-xf^*b) = \exp(-xf^*b)(b^\dagger + xf^*)$, we obtain a differential equation for $A(x)$:

$$A'(x) + x|f|^2 A = 0 .$$

Solving this equation under the condition $A = 1$ at $x = 0$ and putting $x = 1$, we are led to (6.3.11). Applying (6.3.11) to (6.3.2) and noting $\exp(-f^*b)|0\rangle = |0\rangle$, we have

$$|\Psi\rangle = \exp\left(-\frac{1}{2}\sum_q |f_q|^2\right) \prod_q \exp(f_q b_q^\dagger)|0\rangle . \tag{6.3.12}$$

If we use the relation

$$(b^\dagger)^n|0\rangle = \sqrt{n!}\,|n\rangle,$$

the probability that the phonon numbers are $n_1, n_2, \ldots, n_q, \ldots$ is given by

$$P(n_1, n_2, \ldots, n_q, \ldots) = \exp\left(-\sum_q |f_q|^2\right) \prod_q \frac{(|f_q|^2)^{n_q}}{n_q!}$$

for the present trial function. That is, P has a separable form such as $P = g_1(n_1)g_2(n_2) \ldots g_q(n_q) \ldots$. This implies that the phonon excitations are assumed to be statistically independent. In other words, the correlation between excited phonons is neglected in the intermediate coupling method.

6.3.2 Path Integral[1]

The intermediate coupling method discussed above has a relatively simple structure from both physical and mathematical points of view. On the contrary, the method of path integral is much more complicated. However, it leads to a better accuracy compensating the complexity. A characteristic feature of this method is to eliminate phonon coordinates and to reduce the polaron problem essentially to a one-electron problem. Unfortunately, in order to understand the details of this method some preliminary discussion is necessary. Let us begin with it.

It is known in quantum mechanics that $\langle r'|\exp(-i\mathcal{H}t/\hbar)|r\rangle$ represents a probability amplitude that a particle at the position r at time 0 will be found at the position r' at time t. Feynman has shown that it is expressed as an integral over paths which start from r and end at r'. In classical mechanics, the trajectory of a particle which moves from r to r' is uniquely determined. However, in quantum mechanics, all the trajectories from r to r' should be taken into account

---

[1] What follows in this section may involve something difficult for beginners to understand so that those who are not interested in details may skip this part and proceed to the next section. It will not prevent the understanding of the whole scope of this book.

from the standpoint of a particle. We observe that the operator $\exp(-i\mathcal{H}t/\hbar)$ takes a form $\exp(-\beta\mathcal{H})$ which is the density matrix in statistical mechanics, if we assume that t is a pure imaginary number. For this reason, the concept of path integral is applicable to problems in statistical mechanics.

If we consider the density matrix $\exp(-\beta\mathcal{H})$ for the canonical ensemble and if the Hamiltonian $\mathcal{H}$ has a form $\mathcal{H} = \mathcal{H}_0 + \mathcal{H}'$, in general the following relation holds

$$\exp[-\beta(\mathcal{H}_0 + \mathcal{H}')] = \lim_{N\to\infty}\left[1 - \frac{\beta}{N}(\mathcal{H}_0 + \mathcal{H}')\right]^N = \lim_{N\to\infty}[1 - \varepsilon(\mathcal{H}_0 + \mathcal{H}')]^N$$

where ε is defined by $\varepsilon = \beta/N$. Even if there are correction terms of the order of ε^2 in the terms $1 - \varepsilon(\mathcal{H}_0 + \mathcal{H}')$, they will lead to a correction term of the order of $\varepsilon^2 \times N = \varepsilon\beta$ altogether and thus can be neglected in the limit $\varepsilon \to 0$. Because of this, the relation

$$\exp[-\beta(\mathcal{H}_0 + \mathcal{H}')] = \lim_{N\to\infty}[\exp(-\varepsilon\mathcal{H}_0)\exp(-\varepsilon\mathcal{H}')]^N \qquad (6.3.13)$$

is derived. This relation is valid even when $\mathcal{H}_0$ does not commute with $\mathcal{H}'$ and is sometimes called Trotter's formula.

In applying (6.3.13) to the problem of the polaron, it is quite natural to take the electron system and free-phonon system as $\mathcal{H}_0$ and electron–phonon interaction as $\mathcal{H}'$. Also, to simplify notation, we write the wave function for the phonon as $|n\rangle = |n_1, n_2, \ldots, n_q \ldots\rangle$. Then, taking the matrix element of (6.3.13), we have

$$\langle r'; n' | \exp[-\beta(\mathcal{H}_0 + \mathcal{H}')] | r; n\rangle$$
$$= \lim \sum_{n_1, \cdots n_{N-1}} \int \langle r'; n' | \exp(-\varepsilon\mathcal{H}_0)\exp(-\varepsilon\mathcal{H}') | r_{N-1}; n_{N-1}\rangle \cdots$$
$$\cdot \langle r_2; n_2 | \exp(-\varepsilon\mathcal{H}_0)\exp(-\varepsilon\mathcal{H}') | r_1; n_1\rangle$$
$$\cdot \langle r_1; n_1 | \exp(-\varepsilon\mathcal{H}_0)\exp(-\varepsilon\mathcal{H}') | r; n\rangle \, dr_1 \, dr_2 \ldots dr_{N-1} . \qquad (6.3.14)$$

In order to understand this equation intuitively, we denote both r and n simply by x and consider that β is a fictitious time. Then, as is shown conceptually in Fig. 6.4, if we divide the region between 0 and β in N equal parts and assign x at each time, the integrand in (6.3.14) is determined. In other words, if the path from x to x' is determined, the integrand is determined. Equation (6.3.14) represents an integral over all these paths, so it is called "path integral".

As we stated previously, correction terms of the order of ε^2 can be neglected. Therefore, noting that $\mathcal{H}'$ does not include the momentum of the electron and that r_j is an eigenfunction of r, i.e., $r|r_j\rangle = r_j|r_j\rangle$, we are allowed to put

$$\exp(-\varepsilon\mathcal{H}')|r_j; n_j\rangle \approx (1 - \varepsilon\mathcal{H}')|r_j; n_j\rangle = |r_j\rangle(1 - \varepsilon\mathcal{H}'_j)|n_j\rangle$$
$$\approx |r_j\rangle\exp(-\varepsilon\mathcal{H}'_j)|n_j\rangle$$

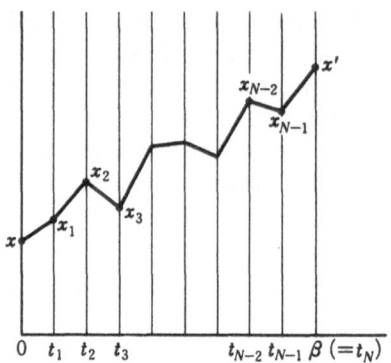

Fig. 6.4. A possible path contributing to the path integral

where $\mathscr{H}'_j$ implies that the position r of electron in $\mathscr{H}'$ is replaced by r_j. On the contrary, if we notice that $\mathscr{H}_0$ is a sum of electron part $\mathscr{H}_e$ and phonon part $\mathscr{H}_p$ ($\mathscr{H}_0 = \mathscr{H}_e + \mathscr{H}_p$), the right-hand side in (6.3.14) is expressed as

$$\lim \int \langle r'|\exp(-\varepsilon\mathscr{H}_e)|r_{N-1}\rangle \ldots \langle r_2|\exp(-\varepsilon\mathscr{H}_e)|r_1\rangle$$
$$\cdot \langle r_1|\exp(-\varepsilon\mathscr{H}_e)|r\rangle \, dr_1 \ldots dr_{N-1}$$
$$\cdot \sum_{n_1 \cdots n_{N-1}} \langle n'|\exp(-\varepsilon\mathscr{H}_p)\exp(-\varepsilon\mathscr{H}'_{N-1})|n_{N-1}\rangle \ldots$$
$$\cdot \langle n_2|\exp(-\varepsilon\mathscr{H}_p)\exp(-\varepsilon\mathscr{H}'_1)|n_1\rangle\langle n_1|\exp(-\varepsilon\mathscr{H}_p)\exp(-\varepsilon\mathscr{H}'')|n\rangle. \quad (6.3.15)$$

Here, by using the relation $\mathscr{H}_e = p^2/2m$, we have

$$\langle r_j|\exp(-\varepsilon\mathscr{H}_e)|r_{j-1}\rangle = \langle r_j|\exp\left(-\frac{\varepsilon p^2}{2m}\right)|r_{j-1}\rangle$$
$$= \int \langle r_j|p'\rangle \exp\left(-\frac{\varepsilon p'^2}{2m}\right)\langle p'|r_{j-1}\rangle \, dp' \ .$$

Furthermore, noting that $\langle r_j|p'\rangle = h^{-3/2}\exp(ip'\cdot r_j/\hbar)$, we obtain

$$\left.\begin{array}{l} \langle r_j|\exp(-\varepsilon\mathscr{H}_e)|r_{j-1}\rangle = \dfrac{1}{A}\exp\left[-\dfrac{m}{2\varepsilon\hbar^2}(r_j - r_{j-1})^2\right] , \\[3mm] \dfrac{1}{A} \equiv \dfrac{1}{h^3}\left(\dfrac{2m\pi}{\varepsilon}\right)^{3/2} . \end{array}\right\} \qquad (6.3.16)$$

6.3.3 Elimination of Phonon Variables

Let us now take a sum over all phonon states and introduce the following reduced density matrix:

$$\rho(r', r) = \sum_n \langle r'; n|\exp[-\beta(\mathscr{H}_0 + \mathscr{H}')]|r; n\rangle \ . \qquad (6.3.17)$$

By means of (6.3.14–16), $\rho(r', r)$ is expressed as

$$\rho(r', r) = \lim \int \exp\left[-\sum_{j=0}^{N-1} \frac{m}{2\varepsilon\hbar^2}(r_{j+1} - r_j)^2\right] \frac{1}{A} dr_1 \frac{1}{A} \cdots dr_{N-1} \frac{1}{A}$$

$$\cdot \sum_n \langle n | \exp(-\varepsilon\mathcal{H}_p) \exp(-\varepsilon\mathcal{H}_{N-1}') \cdots$$

$$\cdot \exp(-\varepsilon\mathcal{H}_p) \exp(-\varepsilon\mathcal{H}_1') \exp(-\varepsilon\mathcal{H}_p) \exp(-\varepsilon\mathcal{H}') | n \rangle . \qquad (6.3.18)$$

In the following, to simplify the notation, we write

$$\lim \frac{1}{A} dr_1 \frac{1}{A} \cdots dr_{N-1} \frac{1}{A} \to \mathscr{D}r(t) .$$

In (6.3.18) the sum over n means to take the trace for phonons. Let us denote it by Ψ, i.e.,

$$\Psi = \lim \mathrm{tr}\{\exp(-\varepsilon\mathcal{H}_p) \exp(-\varepsilon\mathcal{H}_{N-1}') \cdots$$

$$\cdot \exp(-\varepsilon\mathcal{H}_p) \exp(-\varepsilon\mathcal{H}_1') \exp(-\varepsilon\mathcal{H}_p) \exp(-\varepsilon\mathcal{H}')\} .$$

Observing the last four terms in the above equation, let us rewrite them as $\exp(-2\varepsilon\mathcal{H}_p) [\exp(\varepsilon\mathcal{H}_p) \exp(-\varepsilon\mathcal{H}_1') \exp(-\varepsilon\mathcal{H}_p)] \exp(-\varepsilon\mathcal{H}')$. Repeating the same procedures from right to left and putting $\exp(j\varepsilon\mathcal{H}_p) \mathcal{H}_j' \exp(-j\varepsilon\mathcal{H}_p) = \mathcal{H}'(j)$, we have

$$\Psi = \lim \mathrm{tr}\{\exp(-\beta\mathcal{H}_p) \exp[-\varepsilon\mathcal{H}'(N-1)] \cdots$$

$$\cdot \exp[-\varepsilon\mathcal{H}'(1)] \exp[-\varepsilon\mathcal{H}'(0)]\} . \qquad (6.3.19)$$

If we consider $j\varepsilon$ as a general time t, it is seen that operators in (6.3.19) are arranged from right to left with increasing time. Therefore, we can insert Wick's operator T_τ just after $\exp(-\beta\mathcal{H}_p)$. Since we are dealing with operators for phonons, all operators are supposed to be commutable under the T_τ operator. For example, $T_\tau[b(t)b^\dagger(s)]$ is equal to $b(t)b^\dagger(s)$ for $t > s$ and to $b^\dagger(s)b(t)$ for $t < s$, but $T_\tau[b^\dagger(s)b(t)]$ leads to exactly the same result. In this way, we obtain

$$\Psi = \lim \mathrm{tr}\{\exp(-\beta\mathcal{H}_p) T_\tau \exp[-\varepsilon \sum_j \mathcal{H}'(j)]\}$$

$$= \mathrm{tr}\{\exp(-\beta\mathcal{H}_p) T_\tau \exp[-\int_0^\beta \mathcal{H}'(t)\, dt]\} .$$

Here, $\mathcal{H}'(t)$ has a form $\mathcal{H}'(t) = \sum_q \varphi_q(t)$ with $\varphi_q(t)$ given by

$$\varphi_q(t) = V_q b_q(t)\exp[i\mathbf{q}\cdot\mathbf{r}(t)] + V_q^* b_q^\dagger(t)\exp[-i\mathbf{q}\cdot\mathbf{r}(t)] \qquad (6.3.20)$$

where $b_q(t)$ and $b_q^\dagger(t)$ are operators in the interaction representation defined by $b_q(t) = \exp(t\mathcal{H}_p)b_q\exp(-t\mathcal{H}_p)$ and $b_q^\dagger(t) = \exp(t\mathcal{H}_p) b_q^\dagger \exp(-t\mathcal{H}_p)$, respectively.

In order to calculate Ψ, we divide it by the partition function $Z_p = \text{tr}\{\exp(-\beta\mathcal{H}_p)\}$ for free phonons. Noting that $\varphi_q(t)$ with different q are commutable with each other, we have

$$\frac{\Psi}{Z_p} = \langle T_\tau \exp[-\int_0^\beta \mathcal{H}'(t)\,dt]\rangle_0 = \langle \prod_q T_\tau \exp[-\int_0^\beta \varphi_q(t)\,dt]\rangle_0 \qquad (6.3.21)$$

where $\langle\,\ldots\,\rangle_0$ indicates the statistical-mechanical average for the system of free phonons. Omitting subscript q for simplicity and expanding the exponential function in (6.3.21), we find

$$\langle T_\tau \exp[-\int_0^\beta \varphi(t)\,dt]\rangle_0$$

$$= \langle 1 - \int_0^\beta \varphi(t)\,dt + \frac{1}{2!}T_\tau \int_0^\beta \int_0^\beta \varphi(t_1)\,\varphi(t_2)\,dt_1dt_2 - \ldots\rangle_0 . \qquad (6.3.22)$$

The expectation values appearing in (6.3.22) are essentially the same as discussed in Sect. 1.5 for the case of the Mössbauer effect, so that a method used there is taken over to the present case. Thus, in (6.3.22) the terms of odd order vanish and the fourth-order term, for example, is expressed as

$$\langle T_\tau \varphi(1)\varphi(2)\varphi(3)\varphi(4)\rangle_0 = \langle T_\tau \varphi(1)\varphi(2)\rangle_0 \langle T_\tau \varphi(3)\varphi(4)\rangle_0$$

$$+ \langle T_\tau \varphi(1)\varphi(3)\rangle_0 \langle T_\tau \varphi(2)\varphi(4)\rangle_0 + \langle T_\tau \varphi(1)\varphi(4)\rangle_0 \langle T_\tau \varphi(2)\varphi(3)\rangle_0 \qquad (6.3.23)$$

where $t_1, t_2, \ldots$ are written simply as $1, 2, \ldots$ If we integrate this by t_1, t_2, t_3 and t_4, each term on the right-hand side leads to the same result so we have three identical terms. A similar situation exists for a general term in (6.3.22). If we consider the term of the order of $2m$, the number of identical terms is given by $(2m - 1) \cdot (2m - 3) \ldots 3 \cdot 1$. [If we observe one specified φ, the number of φ which make the pair with it is $(2m - 1)$. After these pairings, a remaining φ can make pairings with $(2m - 3)$ operators, and so on.] Taking into account the coefficient $1/(2m)!$ for the term of the order of $2m$ and using the relation $(2m - 1) \cdot (2m - 3) \ldots 3 \cdot 1/(2m)! = 1/2^m m!$, we obtain

$$\langle T_\tau \exp[-\int_0^\beta \varphi(t)\,dt]\rangle_0 = \sum_{m=0}^\infty \frac{1}{2^m m!} [T_\tau \int_0^\beta \int_0^\beta \langle\varphi(t_1)\varphi(t_2)\rangle_0\,dt_1dt_2]^m$$

$$= \exp\left[\frac{1}{2}\int_0^\beta \int_0^\beta D(t, s)\,dtds\right] . \qquad (6.3.24)$$

Here $D(t, s)$ corresponds to the temperature Green's function for the phonon. More precisely, if we restore the subscript q, we have

$$D_q(t, s) = \langle T_\tau \varphi_q(t)\varphi_q(s)\rangle_0 . \qquad (6.3.25)$$

To summarize above results, the $\rho(r', r)$ is expressed

$$\rho(r', r) = Z_p \int \exp\left[-\int_0^\beta \frac{m}{2\hbar^2}\left(\frac{dr}{dt}\right)^2 dt + \frac{1}{2}\int_0^\beta\int_0^\beta\sum_q D_q(t, s) \, dt ds\right] \mathscr{D}r(t) .$$

$$(6.3.26)$$

In the above equation, the definition of $\mathscr{D}r(t)$ on page 229 is employed. Although the phonon variables are completely eliminated in (6.3.26), there takes place an interaction for the electron between different times, i.e., a retardation effect appears. The $D_q(t, s)$ is a function describing this effect, and by the use of $b_q(t) = \exp(-\hbar\omega t)b_q$ and $b_q^\dagger(t) = \exp(\hbar\omega t)b_q^\dagger$ together with (6.3.20) and (6.3.25), it is calculated to be

$$D_q(t, s) = |V_q|^2 \left\{ e^{iq\cdot[r(t)-r(s)]-\hbar\omega(t-s)}\begin{bmatrix}\langle n_q\rangle + 1\\\langle n_q\rangle\end{bmatrix}\right.$$

$$\left. + e^{-iq\cdot[r(t)-r(s)]+\hbar\omega(t-s)}\begin{bmatrix}\langle n_q\rangle\\\langle n_q\rangle + 1\end{bmatrix}\right\}$$

$$(6.3.27)$$

where the upper or lower part in [...] corresponds to the case $t > s$ or $t < s$. Also, $\langle n_q\rangle$ represents the Planck distribution given by $\langle n_q\rangle = 1/[\exp(\beta\hbar\omega) - 1]$.

In applying the above-mentioned results to the problem of the polaron, it is convenient to introduce a unit system such as $\hbar = m = \omega = 1$. Then, we see from (6.1.17) that $|V_q|^2 = 4\pi\alpha/2^{1/2}q^2V$ and hence we have

$$\sum_q |V_q|^2 e^{iq\cdot[r(t)-r(s)]} = \frac{4\pi\alpha}{2^{1/2}V}\sum_q\frac{e^{iq\cdot[r(t)-r(s)]}}{q^2} = \frac{\alpha}{2^{1/2}|r(t) - r(s)|} .$$

Also, if we take a limit $\beta \to \infty$, we can put $\langle n_q\rangle = 0$ in (6.3.27). Thus, by introducing a step function $\theta(x)$ [$\theta(x) = 1$ for $x > 0$ and $\theta(x) = 0$ for $x < 0$], the following equation is derived.

$$\sum_q D_q(t, s) = \frac{\alpha}{2^{1/2}|r(t) - r(s)|}[e^{-(t-s)}\theta(t - s) + e^{(t-s)}\theta(s - t)]$$

$$= \frac{\alpha e^{-|t-s|}}{2^{1/2}|r(t) - r(s)|} .$$

On the other hand, since $Z_p = 1$ as $\beta \to \infty$, (6.3.26) in this limit is expressed as

$$\rho(r', r) = \int e^S \mathscr{D}r(t) , \qquad (6.3.28a)$$

$$S = -\frac{1}{2}\int\left(\frac{dr}{dt}\right)^2 dt + \frac{\alpha}{2^{3/2}}\int\int\frac{e^{-|t-s|}}{|r(t) - r(s)|} \, dt ds . \qquad (6.3.28b)$$

The quantity S corresponds to the action in mechanics; we will call it action in the following for convenience. Also, in (6.3.28b) the integral by t or s implies the one from 0 to β. We will hereafter use the same notation, unless otherwise specified.

6.3.4 Feynman's Variation Principle

The method of path integral mentioned above was applied to the study of the polaron by *Feynman*. In the following, we will review an outline of this theory. First of all, if we let the lth eigenvalue and eigenfunction of $\mathcal{H}_0 + \mathcal{H}'$ be E_l and $|\psi_l\rangle$, respectively, it follows that

$$\rho(r', r) = \sum_{n,l} e^{-\beta E_l} \langle r'; n | \psi_l \rangle \langle \psi_l | r; n \rangle .$$

Thus, if we put the ground state energy to be E_0, we have in the limit $\beta \to \infty$

$$\rho(r', r) \approx A e^{-\beta E_0} \tag{6.3.29}$$

where A is a function of only r and r', independent of β. Next, introducing a suitable trial action S_1, we rewrite (6.3.28a) as

$$\rho(r', r) = \int e^{S-S_1} e^{S_1} \mathscr{D}r(t) \tag{6.3.30}$$

and define the following average for an arbitrary function F:

$$\langle F \rangle = \int F e^{S_1} \mathscr{D}r(t) / \int e^{S_1} \mathscr{D}r(t) . \tag{6.3.31}$$

Then, (6.3.30) is written as

$$\rho(r', r) = \langle e^{S-S_1} \rangle \int e^{S_1} \mathscr{D}r(t) . \tag{6.3.32}$$

Noting the relation

$$e^F \doteq e^{\langle F \rangle} e^{F - \langle F \rangle} = e^{\langle F \rangle}(1 + F - \langle F \rangle + R), \quad R \geq 0$$

and taking the average of this equation, we are led to an inequality

$$\langle e^F \rangle \geq e^{\langle F \rangle} . \tag{6.3.33}$$

Also, in the limit $\beta \to \infty$, we assume that

$$e^{\langle S-S_1 \rangle} \int e^{S_1} \mathscr{D}r(t) \approx B e^{-\beta E} . \tag{6.3.34}$$

From (6.3.29, 32, 33) it follows that $A \exp(-\beta E_0) \geq B \exp(-\beta E)$. In order for this relation to hold in the limit $\beta \to \infty$, we should have an inequality $E_0 \leq E$. Namely, as in the variation principle in quantum mechanics, E yields an upper bound to the true energy eigenvalue. To actually calculate E from (6.3.34), it is convenient to express

$$\langle S - S_1 \rangle = \beta E', \quad \int e^{S_1} \mathscr{D}r(t) \propto e^{-\beta E_1} \tag{6.3.35}$$

in the limit $\beta \to \infty$. As a result, E is given by

$$E = E_1 - E' . \tag{6.3.36}$$

6.3.5   Application to the Polaron

In calculating the ground state energy by means of the above mentioned variation principle, it is desirable to choose a trial action as close as possible to the true action. However, there are only few cases where the path integral can be calculated rigorously. For this reason, there should be necessarily some limitation to an actual form of trial action. We will refer to the trial action which Feynman has introduced later on and for the time being we will discuss its particular case to simplify explanation. Nevertheless, the essential features of Feynman's theory will be sufficiently reflected in the following discussion.

Let us consider a trial action S_1 given by

$$S_1 = -\frac{1}{2} \int \left(\frac{dr}{dt}\right)^2 dt - \frac{C}{2} \int\int [r(t) - r(s)]^2 dtds \tag{6.3.37}$$

where C is a constant. This S_1 represents the action which will be obtained if the Coulomb potential is replaced by the potential for harmonic oscillator and the term $\exp(-|t - s|)$ is set to be 1 in the true action (6.3.28b). By the use of (6.3.28b, 35, 37), E' is expressed as

$$E' = A + B , \tag{6.3.38a}$$

$$A = \frac{\alpha}{2^{3/2}\beta} \int\int \left\langle \frac{1}{|r(t) - r(s)|} \right\rangle e^{-|t-s|} \, dtds, \tag{6.3.38b}$$

$$B = \frac{C}{2\beta} \int\int \langle [r(t) - r(s)]^2 \rangle \, dtds . \tag{6.3.38c}$$

To calculate (6.3.38b), it is convenient to use the Fourier transform:

$$\frac{1}{|r(t) - r(s)|} = \frac{1}{2\pi^2} \int \frac{e^{ik \cdot [r(t)-r(s)]}}{k^2} \, dk . \tag{6.3.39}$$

In order to avoid a possible confusion in notations, we write τ or σ instead of t or s. Then, from (6.3.31) we have

$$\langle e^{ik \cdot [r(\tau)-r(\sigma)]} \rangle = \int e^{ik \cdot [r(\tau)-r(\sigma)]} e^{S_1} \mathscr{D}r(t) / \int e^{S_1} \mathscr{D}r(t) . \tag{6.3.40}$$

Or, if we denote the numerator in the right-hand side of the above equation by I, I is written as

$$I = \int \exp\left[-\frac{1}{2}\int \left(\frac{dr}{dt}\right)^2 dt - \frac{C}{2}\int\int [r(t) - r(s)]^2 dtds\right.$$

$$\left. + \int f(t) \cdot r(t) dt \right] \mathscr{D}r(t) \tag{6.3.41}$$

where $f(t)$ is defined by

$$f(t) = ik[\delta(t - \tau) - \delta(t - \sigma)] \ . \tag{6.3.42}$$

The exponential function in (6.3.41) has a separable form with respect to x, y, and z components so I is expressed as $I = I_x I_y I_z$. Since I_y and I_z are calculated in the same way as I_x, we will restrict ourselves to discussion of I_x in the following. Then, the argument in the exponential function is a quadratic equation in $x(t)$. Let the $x(t)$ which makes the argument maximal be $X(t)$ and put $x = X + \delta x$. As a result, the path integral over δx is essentially the Gaussian integral and thus can be calculated rigorously. This is the reason why we employ the potential for the harmonic oscillator. That we could eliminate phonon variables in Sect. 6.3.3 is also due to the same reason. If we denote the path integral over δx by A_x, I_x is expressed as

$$I_x = A_x \exp\left[-\frac{1}{2} \int \left(\frac{dX}{dt}\right)^2 dt - \frac{C}{2} \int\int [X(t) - X(s)]^2 dtds \right.$$
$$\left. + \int f_x(t)X(t)dt \right] \ . \tag{6.3.43}$$

As will be seen later on, the factors A_x cancel out in the numerator and denominator of (6.3.40). Thus, we can assume that $A_x = 1$ without loss of generality. Also, $f_x(t)$ in (6.3.43) means the x component of $f(t)$ and is given by

$$f_x(t) = ik_x[\delta(t - \tau) - \delta(t - \sigma)] \ . \tag{6.3.44}$$

By our assumption, $X(t)$ is a function which maximizes the argument of the exponential function. Setting up the variation equation, we have

$$\frac{d^2 X}{dt^2} = 2C \int_0^\beta [X(t) - X(s)]ds - f_x(t) \ . \tag{6.3.45}$$

Also, by the use of the above equation, I_x is expressed as

$$I_x = \exp\left[\frac{1}{2} \int f_x(t)X(t)dt \right] \ . \tag{6.3.46}$$

In solving (6.3.45) some suitable boundary condition is necessary. Here we assume $X(0) = X(\beta) = 0$ for convenience. Arranging (6.3.45) slightly, we write

$$\frac{d^2 X}{dt^2} = v^2 X - 2CF - f_x(t), \quad F = \int_0^\beta X(s)ds, \quad v^2 = 2C\beta \ . \tag{6.3.47}$$

The v is essentially the same as C, but it is convenient to take v rather than C itself as a variation parameter. Putting $X = (2CF/v^2) + Y$ in (6.3.47), we have

$$\ddot{Y} = v^2 Y - f_x(t) \ .$$

If we note the following relation

$$\left(\frac{d^2}{dt^2} - v^2\right) e^{-v|t-s|} = -2v\delta(t-s) ,$$

the particular solution of the equation is obtained immediately. Thus, the general solution of (6.3.47) is obtained as follows

$$X(t) = \frac{2CF}{v^2} + Pe^{-vt} + Qe^{vt} + \frac{1}{2v}\int_0^\beta f_x(s)e^{-v|t-s|}ds \qquad (6.3.48)$$

where P and Q are arbitrary constants to be determined from the boundary condition. By means of (6.3.44), we have

$$P + Q = -\frac{2CF}{v^2} - \frac{ik_x}{2v}(e^{-v\tau} - e^{-v\sigma}) ,$$

$$Pe^{-v\beta} + Qe^{v\beta} = -\frac{2CF}{v^2} - \frac{ik_x}{2v}[e^{-v(\beta-\tau)} - e^{-v(\beta-\sigma)}]$$

whence Q is calculated to be

$$Q = -\frac{CF(1 - e^{-v\beta})}{v^2 \sinh v\beta} - \frac{ik_x e^{-v\beta}(\sinh v\tau - \sinh v\sigma)}{2v \sinh v\beta} .$$

If we take the limit $\beta \to \infty$ in the above equation, we see $Q \to 0$, so that we assume $Q = 0$ henceforth. Then, P is expressed as

$$P = -\frac{2CF}{v^2} - \frac{ik_x}{2v}(e^{-v\tau} - e^{-v\sigma}) . \qquad (6.3.49)$$

So far we have been considering F to be a known number. However, as a matter of fact, F should be determined from the defining equation (6.3.47). By a simple calculation, the necessary condition for it is shown to be

$$P + \frac{1}{2}\int_0^\infty dt \int_0^\infty ds\, ik_x[\delta(s-\tau) - \delta(s-\sigma)] e^{-v|t-s|} = 0$$

in the limit $\beta \to \infty$. Carrying out the above integration and using (6.3.49), we have

$$P = -\frac{ik_x}{2v}(e^{-v\sigma} - e^{-v\tau}), \quad \frac{2CF}{v^2} = \frac{ik_x}{v}(e^{-v\sigma} - e^{-v\tau}) .$$

As a result, $X(t)$ is given by

$$X(t) = \frac{ik_x}{v} (e^{-v\sigma} - e^{-v\tau}) - \frac{ik_x}{2v} (e^{-v\sigma} - e^{-v\tau}) e^{-vt}$$

$$+ \frac{1}{2v} \int_0^\infty f_x(s) e^{-v|t-s|} ds \ . \tag{6.3.50}$$

If we substitute the defining equation for $f_x(t)$ into (6.3.46), it follows that $I_x = \exp\{(ik_x/2) [X(\tau) - X(\sigma)]\}$. On the other hand, from (6.3.50) we have

$$X(\tau) - X(\sigma) = \frac{ik_x}{2v} (e^{-v\sigma} - e^{-v\tau})^2 + \frac{ik_x}{v} (1 - e^{-v|\tau-\sigma|}) \ .$$

The similar equations hold for I_y and I_z. In this way, we are led to

$$\langle e^{ik \cdot [r(\tau) - r(\sigma)]} \rangle = \exp\left[-\frac{k^2}{4v} (e^{-v\sigma} - e^{-v\tau})^2 - \frac{k^2}{2v} (1 - e^{-v|\tau-\sigma|}) \right] . \tag{6.3.51}$$

If we put $k \to 0$ in the above equation, we have $\langle 1 \rangle = 1$. This guarantees the validity of our assumption $A_x = 1$ which was introduced previously. Substituting (6.3.51) into (6.3.39) and performing the integration, we have

$$\left\langle \frac{1}{|r(t) - r(s)|} \right\rangle = \frac{1}{\pi^{1/2}} \left[\frac{1}{4v} (e^{-vs} - e^{-vt})^2 + \frac{1 - e^{-v|t-s|}}{2v} \right]^{-1/2} . \tag{6.3.52}$$

Furthermore, from (6.3.38b) A is expressed as

$$A = \frac{2\alpha}{2^{3/2}\pi^{1/2}\beta} \int_0^\beta dt \int_t^\beta ds \left[\frac{1}{4v} (e^{-vs} - e^{-vt})^2 + \frac{1 - e^{-v(s-t)}}{2v} \right]^{-1/2} e^{-(s-t)} \ .$$

Making a change of variable $s = t + x$, we see that the first term in [ ] can be neglected in the limit $\beta \to \infty$. Thus, A is given by

$$A = \frac{\alpha v^{1/2}}{\pi^{1/2}} \int_0^\infty \frac{e^{-x} dx}{(1 - e^{-vx})^{1/2}} \ . \tag{6.3.53}$$

Next, to calculate B we expand both sides in (6.3.51) and compare the coefficients of the terms of k^2. Using $\langle [x(\tau) - x(\sigma)]^2 \rangle = \langle [r(\tau) - r(\sigma)]^2 \rangle / 3$, we have

$$\frac{1}{6} \langle [r(\tau) - r(\sigma)]^2 \rangle = \frac{1}{4v} (e^{-v\sigma} - e^{-v\tau})^2 + \frac{1}{2v} (1 - e^{-v|\tau-\sigma|}) \ .$$

Substituting this into (6.3.38c), we obtain $B = 3v/4$. Furthermore, to calculate E_1 we take the logarithm of both sides in the right equation of (6.3.35) and differentiate it by C. It leads to

$$C \frac{dE_1}{dC} = B \ .$$

Solving this equation under the initial condition that $E_1 = 0$ at $C = 0$, we obtain $E_1 = 3v/2$. In this way, we find that $E = E_1 - E' = (3/2)v - (3/4)v - A$, i.e.,

$$E = \frac{3}{4} v - A \ . \tag{6.3.54}$$

As an application of (6.3.54), let us consider the case $v \gg 1$. In this limit, the term e^{-vx} in (6.3.53) can be neglected and A is calculated as

$$A \approx \frac{\alpha v^{1/2}}{\pi^{1/2}} \int_0^\infty e^{-x} dx = \frac{\alpha v^{1/2}}{\pi^{1/2}} \ .$$

Thus, E is given by

$$E = \frac{3}{4} v - \frac{\alpha v^{1/2}}{\pi^{1/2}} \ .$$

If we determine v so as to make this minimum, we have $v = 4\alpha^2/9\pi$. Also, the energy in this case is expressed as

$$E = -\frac{\alpha^2}{3\pi} \ . \tag{6.3.55}$$

From the above equation for v, we see that $v \gg 1$ implies that $\alpha \gg 1$. Although the perturbation procedure discussed in Sect. 6.2 leads to a correct result in the case of weak coupling ($\alpha \ll 1$), (6.3.55) is valid conversely in the case of strong coupling.

Now, the trial action which Feynman actually used is

$$S_1 = -\frac{1}{2} \int \left(\frac{d\mathbf{r}}{dt}\right)^2 dt - \frac{C}{2} \int\int [\mathbf{r}(t) - \mathbf{r}(s)]^2 e^{-w|t-s|} dt ds \ .$$

The trial action discussed here corresponds to the case $w = 0$. By the calculation similar to the present one, for the above S_1 it is shown that

$$E = \frac{3}{4v} (v - w)^2 - A \ ,$$

$$A = \frac{\alpha v}{\pi^{1/2}} \int_0^\infty \left[w^2 x + \frac{v^2 - w^2}{v} (1 - e^{-vx}) \right]^{-1/2} e^{-x} dx$$

where v and w are variation parameters. If we put $w = 0$, we can derive again (6.3.53, 54). Since A involves a rather complicated integral, numerical calculation is necessary to determine v and w which minimize E. Results of such calculations are shown in Table 6.2. In this table, E_p means the value calculated from above equations and E_i the value derived from the intermediate coupling method discussed in Sect. 6.3.1. Since both methods are based on the variation principle, the one leading to lower energy is supposed to yield a better approximation. As seen from this table, the method of path integral always leads to better results than the intermediate coupling method. In closing we would like to remark that the calculation of the polaron mass has been attempted by means of the method of path integral, though its mathematical aspect is not so clear as in the case of ground state energy. Furthermore, dynamical properties of the polaron, e.g., the mobility, the resistivity exerted on a moving electron in a crystal under an electric field, and so on, have been studied along the same line.

Table 6.2. Ground state energy of the polaron

α	3	5	7	9	11
E_p	-3.1333	-5.4401	-8.1127	-11.486	-15.710
E_i	-3.0000	-5.0000	-7.0000	-9.000	-11.000

Cited from T.D. Schultz: Phys. Rev. **116**, 526 (1959)

6.4 Electron–Phonon Interaction in Metals

We now turn to the method of Green's functions to describe elementary excitations and their interaction. General formulation and basic theorems of this method are given in a number of standard text books, but we need something more to apply the method to practical problems. Here we will take the electron–phonon interaction in metals as an example and show how to make the approximation and derive physical consequences. The method of Green's functions is perhaps more general than the methods of intermediate coupling and path integral described in preceding sections and can be applied to various types of problems. To show basic ideas, however, we restrict ourselves to the electron–phonon interaction in metals since this is a typical case where the method works very effectively. It seems better to give a physical picture of this interaction before going into mathematical details.

6.4.1 Hamiltonian

The difference from the ionic crystal discussed in Sect. 6.1 is that we have the Fermi distribution of conduction electrons with high density of the order of $n \approx 10^{22}$ cm^{-3} in a metal. We will think of simple metals, in which conduction electrons belong to s and p bands (alkali metals, aluminum, etc.). Then electronic parameters can be estimated by the free-electron model (or, more precisely, by the nearly-free-electron model): Fermi wave number is $k_F = (3\pi^2 n)^{1/3} \approx 10^8$ cm^{-1} and Fermi energy is $\varepsilon_F = \hbar^2 k_F^2/2m \approx 5$ eV. In contrast to ionic crystals, the electrostatic interaction in a metal is screened by the zero-point motion of electrons. Since ions move much more slowly than electrons, the screening effect upon ionic vibrations can be described in terms of the static, but q-dependent dielectric function $\varepsilon(q)$ discussed in Chap. 3. The effect is important for small q. RPA gives $\varepsilon(q) \approx 6\pi n e^2/\varepsilon_F q^2$ for small q, where $-e$ is the electronic charge. If the ion with charge Ze and mass M vibrates in the uniform negative charge of density $-ne$, the frequency is the ion plasma frequency $\Omega_p = (4\pi Z e^2 n/M)^{1/2}$. If we take into account the dielectric functions of the electron gas, however, the frequency of the ionic vibration with the wave number q will be renormalized as $\omega_q = \Omega_p/[\varepsilon(q)]^{1/2}$, which reduces to the acoustic spectrum $\omega_q = c_s q$, $c_s = (Zm/3M)^{1/2} v_F$ as $q \to 0$. Here $v_F = \hbar k_F/m$ is the electron Fermi velocity of the order of 10^8 cm · s^{-1}, so that the sound velocity c_s is of the observed order of magnitude 10^5 cm · s^{-1} (see Sect. 5.2).

As for the electron–phonon interaction, we can use the same expression as the one given in Sect. 6.1, provided that we take account of the q dependence of the ion frequency and dielectric constant. In quantum mechanics, the interaction Hamiltonian describes emission and absorption of the phonon by the electron, so that it is convenient to describe the electron system also in terms of destruction, creation operators as

$$\mathcal{H}' = \sum \frac{1}{\sqrt{N}} g_q a^\dagger_{k+q\sigma} a_{k\sigma} \phi_q \tag{6.4.1}$$

$$\phi_q = b_q + b^\dagger_{-q} . \tag{6.4.2}$$

The coupling constant is given by $g_q = (\hbar^2/2M\omega_q)^{1/2} [4\pi Z e^2 n/q\varepsilon(q)]$. Hereafter we write the phonon energy as ω_q instead of $\hbar\omega_q$. The above mentioned g_q is the coupling constant given by the continuum model in Sect. 6.1, but its order of magnitude will remain unchanged if we include the band structure. We will be concerned with short-wavelength phonons whose wave numbers are comparable to k_F or a^{-1}, where a is the average atomic spacing. The screening is not so important $[\varepsilon(q) \sim 1]$, and ω_q may be replaced approximately by some average $\omega_0 (\sim 10^{-2}$ eV). Thus

$$g \cong \left(\frac{\hbar^2}{2M\omega_0 a^2}\right)^{1/2} \left(\frac{Ze^2}{a}\right) . \tag{6.4.3}$$

The Coulomb energy Ze^2/a is comparable to the Fermi energy, but the square root on the right is the ratio of the zero-point amplitude of atomic vibrations to a. Hence g is much smaller than the Fermi energy and one might expect that perturbation theory can be applied in its primitive form.

6.4.2 Electron Self-Energy

The usual textbook description says that ions may be regarded as fixed centers of force when we deal with the electron motion (adiabatic approximation), since electrons move much faster than ions. But this does not apply to electrons near the Fermi surface. First of all, the approximation ignores the phonon-mediated attraction between electrons near the Fermi surface. The interaction arises, for instance, from the second-order scattering process shown in Fig. 6.5; the phonon emitted by one electron is absorbed by another electron. If all electronic states are close to the Fermi surface, the energy difference between initial and intermediate states is approximately equal to the energy of the intermediate phonon. The matrix element of this second-order scattering process is therefore of the order of $-(g^2/\omega_0)$. According to BCS theory, this attraction is responsible for the superconducting transition. The dimensionless coupling parameter is

$$\lambda = \frac{2}{\omega_0} g^2 \rho_F \tag{6.4.4}$$

where ρ_F is the density of one-electron states at the Fermi surface. In most super-conductors, λ is of the order of unity. In fact, substituting (6.4.3) and $\rho_F \cong \varepsilon_F^{-1}$ in (6.4.4), we have $\lambda \approx (Ze^2/a\varepsilon_F)^2 \approx 1$.

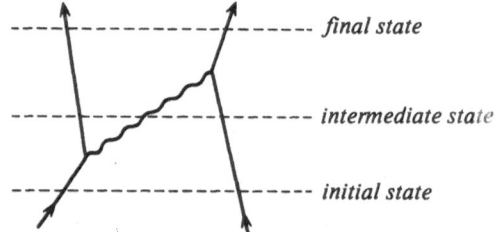

final state

intermediate state

initial state

Fig. 6.5. Phonon-mediated attraction between electrons

The superconducting transition is so dramatic that nobody would doubt the effect of ion vibrations. The same effect appears also on the electron energy spectrum in the normal state, however. To see it, let us calculate the second-order energy shift caused by perturbation (6.4.1). We simplify the calculation by assuming the Einstein model, in which the phonon energy and coupling constant are independent of q:

$$\omega_q = \omega_0, \quad g_q = g \ . \tag{6.4.5}$$

The neutron scattering experiment shows the phonon spectrum consisting of fairly sharp peaks, so that the Einstein model is not very far from reality.

As the unperturbed state, we take the state in which we have no phonons and the Fermi distribution of electrons at $T = 0$. Up to the second order of (6.4.1), one electron goes from inside to outside of the Fermi surface by emitting one phonon and gets back to the original state by reabsorbing the phonon (Fig. 6.6a). Let us write the energy shift caused by this process as ΔE_0. Suppose then that we add one extra electron with the wave vector k outside the Fermi surface ($k > k_F$) and write its unperturbed energy as ε_k ($= \hbar^2 k^2/2m$ in the free-electron model). The extra electron modifies the electron–phonon interaction energy in the following way. First, the extra electron can go to another state k' outside the Fermi surface by emitting one phonon and get back to the original state by reabsorbing the phonon (Fig. 6.6b). Secondly, emission and reabsorption of the phonon by the electron inside the Fermi surface are restricted by the presence of the extra electron through the Pauli principle. Hence E_0 is somewhat reduced. The sum of these two effects gives the self-energy, i.e., the energy shift from ε_k, of the extra electron:

$$\Delta\varepsilon_k = \frac{g^2}{N}\sum_{k'}\left[\frac{1 - n(k')}{\varepsilon_k - \varepsilon_{k'} - \omega_0} - \frac{n(k')}{\varepsilon_{k'} - \varepsilon_k - \omega_0}\right]. \qquad (6.4.6)$$

Here $n(k)$ is the Fermi distribution at zero temperature. It appears in (6.4.6) because the electron scatterer is not a static potential, but the dynamic phonon (the situation is similar in the case of the Kondo effect mentioned in Sect. 4.10). In fact the Fermi distribution disappears from (6.4.6) in the limit $\omega_0 = 0$.

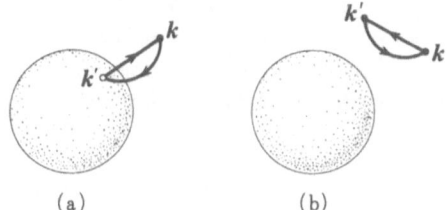

(a)   (b)

Fig. 6.6. Second-order perturbation by electron-phonon interaction

Let us now take the parameter D such that $\varepsilon_F \gg D \gg \omega_0$ and divide the sum over k' into two parts according to $|\xi_{k'}| \gtrless D$, where $\xi_k = \varepsilon_k - \varepsilon_F$. We assume that the density of states $\rho(\varepsilon_k)$ may be replaced by ρ_F in the region $|\xi_k| < D$. For $|\xi_{k'}| > D$, we ignore ω_0 in the denominator of (6.4.6), so that we have the contribution

$$-2g^2 \int_{|\xi'|>D}\frac{\rho(\xi')}{\xi'}\,d\xi' \cong -\lambda\omega_0 P\int\left[\frac{\rho(\xi')}{\rho_F}\right]\frac{d\xi'}{\xi'}. \qquad (6.4.7)$$

The symbol P means taking the principal part of the integral. The self-energy (6.4.7) is of the order of ω_0 and independent of k, so that it is not so important.

For $|\xi_{k'}| < D$, on the other hand, we ignore ω_0 and ξ against D and obtain the contribution

$$
\Delta_k = -g^2 \rho_F \left(\int_0^D \frac{d\xi'}{\xi' - \xi + \omega_0} + \int_{-D}^0 \frac{d\xi'}{\xi' - \xi - \omega_0} \right)
$$

$$
= -\frac{1}{2} \lambda \omega_0 \ln \left| \frac{\omega_0 + \xi}{\omega_0 - \xi} \right| \tag{6.4.8}
$$

where $\xi = \xi_k$. This self-energy is also of the order of ω_0, but its rate of change at the Fermi surface and therefore its effect upon the density of states are not small. In fact $\partial \Delta_k / \partial k = -\lambda (\partial \varepsilon_k / \partial k)$ at the Fermi surface. Since the density of states is proportional $\partial k / \partial \varepsilon_k$, ρ_F is renormalized to $\rho_F^* = (1 - \lambda)^{-1} \rho_F$ by including the self-energy (6.4.8). This expression, however, diverges when λ approaches unity and cannot be applied to actual metals. In other words, simple second-order perturbation theory is not adequate to deal with this problem and we need some more sophisticated formalism. The most adquate is the method of Green's functions.

6.5 Temperature Green's Function and Spectral Function

The formalism described in this and the next sections may be applied to many other systems, though we speak of the electron–phonon system for the sake of definiteness. We suppose that the system is in thermal equilibrium at temperature T, and describe it with use of the grand canonical distribution. We thus replace the Hamiltonian $\mathcal{H}$ by $\mathcal{H} - \mu \mathcal{N}$, which we again write $\mathcal{H}$ hereafter to save notation. Here μ is the chemical potential and $\mathcal{N}$ is the total number operator of the electrons. With use of $\xi_k = \varepsilon_k - \mu$, we write the "Hamiltonian" as

$$
\begin{aligned}
\mathcal{H} &= \mathcal{H}_0 + \mathcal{H}' \\
\mathcal{H}_0 &= \sum_{k\sigma} \xi_k a_{k\sigma}^\dagger a_{k\sigma} + \sum_q \omega_0 b_q^\dagger b_q .
\end{aligned} \tag{6.5.1}
$$

As we have mentioned in Chaps. 1, 3, we introduce the "imaginary" time $0 \leq \tau \leq \beta$ to define the temperature Green's functions, where $\beta = (k_B T)^{-1}$. The Heisenberg representation in this scheme is defined by $A_{k\sigma}(\tau) = U(\tau) a_{k\sigma} U(-\tau)$ with $U(\tau) = \exp(\tau \mathcal{H})$. Similarly the interaction representation is defined by $a_{k\sigma}(\tau) = U_0(\tau) a_{k\sigma} U_0(-\tau)$ with $U_0(\tau) = \exp(\tau \mathcal{H}_0)$ and represents the "motion" of the free electron. With use of anticommutation relations of a operators (see Sect. 3.4), we have

$$
\begin{aligned}
\frac{\partial}{\partial \tau} a_{k\sigma}(\tau) &= U_0(\tau) [\mathcal{H}_0, a_{k\sigma}] U_0(-\tau) \\
&= -\xi_k a_{k\sigma}(\tau) .
\end{aligned} \tag{6.5.2}
$$

The solution to satisfy the initial condition $a_{k\sigma}(0) = a_{k\sigma}$ is

$$a_{k\sigma}(\tau) = a_{k\sigma}\exp(-\xi_k\tau) \ . \tag{6.5.3}$$

Similarly

$$\frac{\partial}{\partial\tau} A_{k\sigma}(\tau) = -\ \xi_k A_{k\sigma}(\tau) - \frac{1}{\sqrt{N}} \sum_q g_q \Phi_q(\tau)\, A_{k-q\sigma}(\tau) \tag{6.5.4}$$

which should be coupled with the equation of motion for the Heisenberg representation $\Phi_q(\tau)$ of the phonon field (6.4.3). Both equations are difficult to solve rigorously, while the interaction representation of the phonon field is given as

$$\phi_q(\tau) = b_q \exp(-\tau\omega_0) + b^\dagger_{-q} \exp(\tau\omega_0) \ . \tag{6.5.5}$$

Now let us define the one-electron temperature Green's function by

$$\mathscr{G}(k;\tau - \tau') = -\ \langle T_\tau A_{k\sigma}(\tau)\, A^\dagger_{k\sigma}(\tau')\rangle \tag{6.5.6}$$

where $\langle Q \rangle = \mathrm{tr}\,\{Q \exp[\beta(\Omega - \mathscr{H})]\}$ and

$$\exp(-\beta\Omega) = \mathrm{tr}\,\{\exp(-\beta\mathscr{H})\} \ . \tag{6.5.7}$$

Wick's symbol T_τ means rearranging operators to its right according to the order of τ values and, when this is done, destruction and creation operators of Fermi particles should anticommute with one another. Thus, in the case of (6.5.6), we have

$$T_\tau A(\tau)A^\dagger(\tau') = \begin{cases} A(\tau)A^\dagger(\tau') & (\tau > \tau') \\ -A^\dagger(\tau')\, A(\tau) & (\tau < \tau') \ . \end{cases} \tag{6.5.8}$$

For $\tau > \tau'$, one extra electron is added to the system at "time" τ' and removed away at τ; the Green's function describes the electron propagation from τ' to τ. For $\tau < \tau'$, one electron is removed from the system at τ and added again at τ'; the Green's function describes the hole propagation from τ to τ'.

Now let us write down the matrix representation of (6.5.6) with the use of eigenvalues E_α and orthonormalized eigenfunctions $|\alpha\rangle$ of $\mathscr{H}$;

$$\mathscr{G}(k;\tau - \tau') = \sum_{\alpha\alpha'} \langle\alpha|a^\dagger_{k\sigma}|\alpha'\rangle \langle\alpha'|a_{k\sigma}|\alpha\rangle \exp[\beta(\Omega - E_\alpha)]$$
$$\times \exp[(E_{\alpha'} - E_\alpha)(\tau - \tau')]\,\{-\theta\,(\tau - \tau')\exp[\beta(E_\alpha - E_{\alpha'})] + \theta(\tau' - \tau)\} \ . \tag{6.5.9}$$

Here $\theta(x)$ is the step function equal to unity for $x > 0$ and equal to zero for $x < 0$. We define the one-electron spectral function by

$$S(k, E) = f^{-1}(E) \sum_{\alpha \alpha'} |\langle \alpha | a_{k\sigma}^\dagger | \alpha' \rangle|^2 \exp[\beta(\Omega - E_\alpha)] \, \delta(E_\alpha - E_{\alpha'} - E) \ .$$

$$(6.5.10)$$

Then

$$\mathscr{G}(k; \tau - \tau') = \int_{-\infty}^{\infty} dE \, S(k, E) \exp[-E(\tau - \tau')]$$

$$\times \ \{-\theta(\tau - \tau')[1 - f(E)] + \theta(\tau' - \tau)f(E)\}$$

$$(6.5.11)$$

where $f(z) = [\exp(\beta z) + 1]^{-1}$ is the Fermi distribution function. It reduces to the step function $\theta(-z)$ at zero temperature, so that the part $E > 0$ gives the electron spectrum and the part $E < 0$ the hole spectrum on the right hand side of (6.5.11).

From (6.5.9 or 11), we see that the Green's function depends only on the difference $\tau - \tau'$ and is antiperiodic

$$\mathscr{G}(k; \tau - \tau' + \beta) = - \ \mathscr{G}(k; \tau - \tau')$$

$$(6.5.12)$$

Hence it can be expressed as the Fourier series

$$\mathscr{G}(k; \tau) = \beta^{-1} \sum_{n=-\infty}^{+\infty} \mathscr{G}(k, i\varepsilon_n) \exp(-i\varepsilon_n \tau)$$

$$(6.5.13)$$

$$\varepsilon_n = \frac{\pi}{\beta} (2n + 1), \ n = 0, \pm 1, \pm 2, \ldots \ .$$

$$(6.5.14)$$

The Fourier coefficient is given by

$$\mathscr{G}(k, i\varepsilon_n) = \int_0^\beta d\tau \ \mathscr{G}(k; \tau) \exp(i\varepsilon_n \tau)$$

$$= \int_{-\infty}^{\infty} \frac{S(k, E)}{i\varepsilon_n - E} \, dE \ .$$

$$(6.5.15)$$

According to the theory of analytic functions,

$$G(k, z) = \int_{-\infty}^{\infty} \frac{S(k, E)}{z - E} \, dE$$

$$(6.5.16)$$

defines one analytic function on the upper half of the complex z plane and the other analytic function on the lower half-plane, provided that $S(k, E)$ is a continuous function of E. These two functions are analytic continuations of (6.5.15) given on the imaginary axis of the z plane. When z approaches the point E on the real axis from above, we obtain the retarded Green's function

$$G(k, E + i0^+) = P \int \frac{S(k, E')}{E - E'} \, dE' - i\pi S(k, E) \ .$$

$$(6.5.17)$$

In the case of free electrons, the motion is given by (6.5.3) and the expectation value $\langle a_{k\sigma}^{\dagger} a_{k\sigma} \rangle$ is equal to the Fermi distribution $f(\xi_k)$, so that

$$\mathscr{G}_0(k; \tau - \tau') = \exp[-\xi_k(\tau - \tau')] \{-\theta(\tau - \tau')[1 - f(\xi_k)]$$
$$+ \theta(\tau' - \tau) f(\xi_k)\} . \tag{6.5.18}$$

Comparing this with (6.5.11), we see that the spectral function in this case has an infinitely sharp peak at $E = \xi_k$:

$$S_0(k, E) = \delta(E - \xi_k) . \tag{6.5.19}$$

Substituting in (6.5.15), we obtain the Fourier coefficient of (6.5.18) as

$$\mathscr{G}_0(k, i\varepsilon_n) = \frac{1}{i\varepsilon_n - \xi_k} \tag{6.5.20}$$

In the presence of interaction, if $S(k, E)$ has a sharp peak at $E = E_k$, E_k defines the energy of the quasiparticle and the width Γ_k of the peak defines the life time $\hbar\Gamma_k^{-1}$ of the quasiparticle. When the temperature Green's function is known by some approximation, the approximate spectral function is obtained by analytic continuation (6.5.17) as $S = -\pi^{-1} \text{Im} \{G(k, E + i0^+)\}$.

a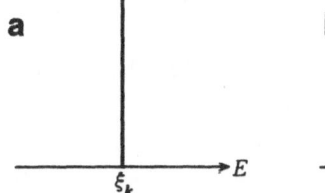

b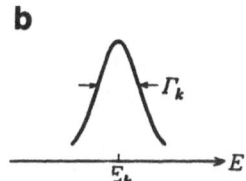

Fig. 6.7 a,b. One-electron spectral function. (a) free electron, (b) interaction electron

As in Chap. 1, the phonon Green's function is defined by

$$\mathscr{D}(q; \tau - \tau') = -\langle T_\tau \Phi_q(\tau) \Phi_{-q}(\tau') \rangle \tag{6.5.21}$$

where we need not worry about the change of sign in rearranging phonon operators according to T_τ since phonons are Bose particles. As a result, $\mathscr{D}$ is periodic and can be Fourier expanded as

$$\mathscr{D}(q: \tau) = \beta^{-1} \sum_{l=-\infty}^{+\infty} \mathscr{D}(q, i\nu_l) \exp(-i\nu_l\tau) \tag{6.5.22}$$

$$\nu_l = \frac{2\pi_l}{\beta} , \qquad l = 0, \pm 1, \pm 2, \dots . \tag{6.5.23}$$

With use of the spectral function $\sigma(q, E)$ and the Planck distribution $p(z) = [\exp(\beta z) - 1]^{-1}$, we can write

$$\mathscr{D}(q; \tau) = \int_{-\infty}^{\infty} dE\, \sigma(q, E) \exp(-E\tau)\, \{-\theta(\tau)\,[1 + p(E)] - \theta(-\tau)\,p(E)\} \ .$$

(6.5.24)

Hence

$$\mathscr{D}(q, i\nu_l) = \int_{-\infty}^{\infty} \frac{\sigma(q, E)}{i\nu_l - E}\, dE \ .$$

(6.5.25)

In the case of free phonons, the motion is given by (6.6.5) and the expectation value $\langle b_q^\dagger b_q \rangle$ by the Planck distribution $p(\omega_q)$, so that

$$\mathscr{D}_0(q; \tau) = -\theta(\tau)\,\{[1 + p(\omega_q)]\exp(-\omega_q\tau) + p(\omega_q)\exp(\omega_q\tau)\}$$
$$-\theta(-\tau)\,\{p(\omega_q)\exp(-\omega_q\tau) + [1 + p(\omega_q)]\exp(\omega_q\tau)\} \ .$$ (6.5.26)

The Fourier coefficient is

$$\mathscr{D}_0(q, i\nu_l) = \frac{1}{i\nu_l - \omega_q} - \frac{1}{i\nu_l + \omega_q} \ .$$

(6.5.27)

Thus the spectral function in this case is

$$\sigma_0(q, E) = \delta(E - \omega_q) - \delta(E + \omega_q) \ .$$

(6.5.28)

In general, with use of time-reversal symmetry, we can prove that the spectral function is odd in E and therefore write (6.5.25) as

$$\mathscr{D}(q, i\nu_l) = \int_{0}^{\infty} dE\, \sigma(q, E) \left(\frac{1}{i\nu_l - E} - \frac{1}{i\nu_l + E} \right) \ .$$

(6.5.29)

6.6 Perturbation Expansion and Partial Summation

6.6.1 Diagrams and Rules of Calculation

To deal with a many-body problem, a primitive, but systematic method is to apply the perturbation expansion. As we have emphasized in Chap. 1, the great advantage of using temperature Green's functions is that the perturbation expansion, i.e., the expansion into powers of $\mathscr{H}'$ in our case, takes a particularly simple form. In the case of the one-electron Green's function (6.5.6), for instance, we have

$$\mathscr{G}(\boldsymbol{k}\,; \tau - \tau') = - \sum_{n=0}^{\infty} \frac{(-1)^n}{n!} \int_0^\beta d\tau_1 \ldots \int_0^\beta d\tau_n$$

$$\langle T_\tau a_{\boldsymbol{k}\sigma}(\tau)\, \mathscr{H}'(\tau_1) \ldots \mathscr{H}'(\tau_n)\, a_{\boldsymbol{k}\sigma}^\dagger(\tau') \rangle_{0c}\,. \tag{6.6.1}$$

All operators on the right are defined in the interaction representation and the expectation value $\langle Q \rangle_0$ is defined with respect to the noninteracting system with the same T and μ as the interacting system. The expectation value on the right of (6.6.1) is thus factorized into the product of free-electron Green's functions and free–phonon Green's functions by applying the Bloch–Dominicis theorem mentioned in Chap. 1. In doing so, we should rearrange electron operators as if they anticommute. For instance, $\langle a_1^\dagger a_2^\dagger a_3 a_4 \rangle_0 = \langle a_1^\dagger a_4 \rangle_0\, \langle a_2^\dagger a_3 \rangle_0 - \langle a_1^\dagger a_3 \rangle_0\, \langle a_2^\dagger a_4 \rangle_0$ (see Sect. 3.4). We assume the normal state of the electron system, so that $\langle a_1^\dagger a_2^\dagger \rangle_0$, $\langle a_3 a_4 \rangle_0$ vanish. Since the product of an odd number of ϕ operators gives the vanishing expectation value, we retain terms of even n in (6.6.1).

Each term obtained by factorization can be represented by a Feynman diagram to show the way in which the term is composed of $\mathscr{G}_0$, $\mathscr{D}_0$ factors. For example, the term of $n = 2$ in (6.6.1) is factorized into the sum of four terms, which are represented by Fig. 6.8a, b and similar diagrams obtained by exchanging τ_1 with τ_2. The symbol c in (6.6.1) means disregarding disconnected diagrams such as Fig. 6.8b. The diagram obtained by exchanging τ_1 with τ_2 in Fig. 6.8a makes the same contribution as Fig. 6.8a when integrated over τ_1, τ_2, so that we retain Fig. 6.8a only and at the same time drop the factor 2! in the denominator of (6.6.1). In general, for the nth order term, we have $n!$ diagrams which differ from other only by a permutation of $\tau_1, \ldots, \tau_n$. We retain one of them and disregard the factor $n!$ in (6.6.1). This removal of $n!$ is essential to take a partial sum, which we will discuss later on.

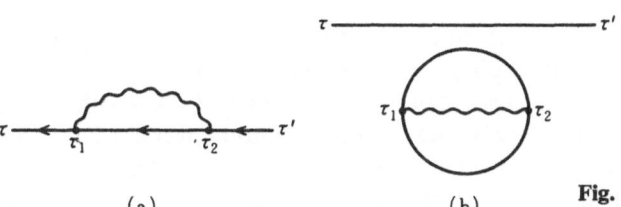

(a)   (b)

Fig. 6.8. Second-order thermal Green's function

Keeping these remarks in mind, we say that the second-order correction to the Green's function is represented by Fig. 6.8a, in which the full line connecting vertices τ, τ_1 represents the factor $\mathscr{G}_0\,(\boldsymbol{k}\,; \tau - \tau_1)$ and the wavy line represents the factor $\mathscr{D}_0\,(\boldsymbol{q}\,; \tau_1 - \tau_2)$. The latter is not directed because $\mathscr{D}_0(\boldsymbol{q}\,; \tau)$ is even in τ. The whole diagram represents

$$\mathcal{G}_2(k; \tau - \tau') = - \frac{1}{N} \sum_q g_q^2 \int_0^\beta d\tau_1 \int_0^\beta d\tau_2 \ \mathcal{G}_0(k; \tau - \tau_1)$$

$$\times \ \mathcal{G}_0(k - q; \tau_1 - \tau_2) \mathcal{D}_0(q; \tau_1 - \tau_2) \ \mathcal{G}_0(k; \tau_2 - \tau') \ . \qquad (6.6.2)$$

Substituting Fourier expansions, we obtain the second-order correction in the energy–momentum representation:

$$\mathcal{G}_2(k, i\varepsilon_n) = - \frac{1}{N\beta} \sum_{k'm} g_{k-k'}^2 \ \mathcal{G}_0(k, i\varepsilon_n)$$

$$\times \ \mathcal{G}_0(k', i\varepsilon_m) \mathcal{D}_0(k - k', i\varepsilon_n - i\varepsilon_m) \ \mathcal{G}_0(k, i\varepsilon_n) \qquad (6.6.3)$$

The right-hand side may be represented by the same Fig. 6.8a, in which the full line now represents $\mathcal{G}_0(k, i\varepsilon_n)$ and the wavy line $\mathcal{D}_0(q, i\nu_l)$. It is convenient to suppose that these lines carry energies ε_n, ν_l, respectively. Then we have the "energy" conservation law at each vertex (in the present model, we also have the "momentum" conservation law). In Fig. 6.8a, we call two full lines at the extreme ends external lines and others internal lines. By removing external lines, we obtain the diagram shown by Fig. 6.9, which gives the lowest-order approximation to the electron self-energy. We obtain its analytic expression by removing $\mathcal{G}_0$ factors at the extreme ends on the right of (6.6.3):

$$\Sigma_1 (k, i\varepsilon_n) = - \frac{1}{N\beta} \sum_{k'm} g_{kk'}^2 \ \mathcal{D}_0(k - k', i\varepsilon_n - i\varepsilon_m) \ \mathcal{G}_0(k', i\varepsilon_m) \qquad (6.6.4)$$

Fig. 6.9. The lowest-order electron self-energy Σ_1

In general, the term with $n = 2r$ in (6.6.1) is equal to the sum of terms represented by topologically distinct diagrams which we can draw in the following way. Take $2r$ points (vertices) on a sheet of paper and connect them with $2r + 1$ directed full lines (electron lines) and r wavy lines (phonon lines) in such a way that one phonon line and two electron lines (one coming in and the other going out) are connected at each vertex. We are thus left with two external electron lines. To obtain the analytic expression, we suppose as before that electron and phonon lines represent factors $\mathcal{G}_0$ and $\mathcal{D}_0$ respectively, and that the vertex represents the coupling constant g_q. We take the product of all these factors and further multiply it by $(- 1/N\beta)^r \times (- 1)^L$, where L is the number of closed loops ($L = 1$ in the case of Fig. 6.10c, for instance). Finally we take the sum over momenta and energies of internal lines with the momentum k and energy ε_n of external electron lines being fixed. Momentum and energy should be conserved at each vertex.

6.6.2 Self-Energy

The fourth-order correction to the Green's function is the sum of four diagrams Fig. 6.10a-d. The contribution of Fig. 6.10a is $\mathscr{G}_0 \Sigma_1 \mathscr{G}_0 \Sigma_1 \mathscr{G}_0$, as we can see by the use of the above rules and (6.6.4). All factors here carry the same energy-momentum ε_n, k. Hence it is a simple repetition of Fig. 6.8a. This type of repetition appears in all higher-order corrections. If we ignore all other correction terms, the electron Green's function is approximated by the partial sum of the perturbation series shown by Fig. 6.11. It is the geometric series whose sum is

$$\mathscr{G}(k, i\varepsilon_n) \cong \frac{1}{\mathscr{G}_0^{-1}(k, i\varepsilon_n) - \Sigma_1(k, i\varepsilon_n)}$$

$$= \frac{1}{i\varepsilon_n - \xi_k - \Sigma_1(k, i\varepsilon_n)} \ .$$

$$(6.6.5)$$

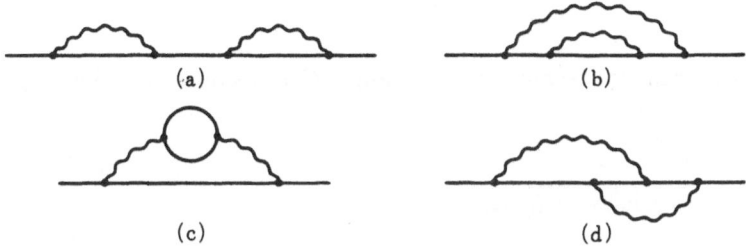

Fig. 6.10. a-d. The fourth-order correction

Fig. 6.11. An approximate Green's function ($\Sigma \approx \Sigma_1$)

In general we define the self-energy type diagram by removing two external lines from any diagram for the Green's function. It is called reducible when it can be decomposed into two disconnected pieces, like Fig. 6.10a, by cutting one electron (or phonon) line. Otherwise (Fig. 6.10b-d), it is called irreducible. Let the sum of all irreducible self-energy type diagrams be $\Sigma (k, i\varepsilon_n)$, which defines the self-energy of the electron. The lowest-order approximation is Σ_1 given by Fig. 6.9. The next-order correction is the sum Σ_2 of Fig. 6.12a-c. We now alter the order of taking the sum of the perturbation expansion of $\mathscr{G}$ and suppose that we first take the partial sum over irreducible self-energy type diagrams, so that the expansion of $\mathscr{G}$ has the form shown by Fig. 6.13. This may be regarded as replacing Σ_1 in (6.6.5) by the exact self-energy Σ. The analytic expression corresponding to Fig. 6.13 is called the *Dyson equation*, whose solution is

$$\mathscr{G}(k, i\varepsilon_n) = \frac{1}{i\varepsilon_n - \zeta_k - \Sigma(k, i\varepsilon_n)}.$$

(6.6.6)

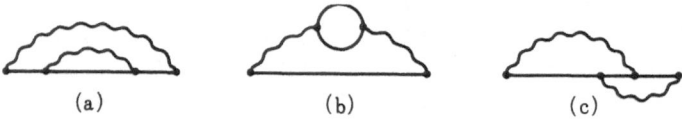

(a)   (b)   (c)

Fig. 6.12. a-c. The fourth-order electron self-energy Σ_2

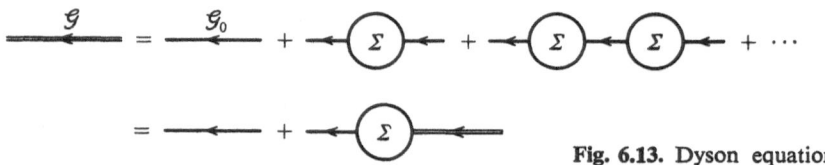

Fig. 6.13. Dyson equation

Similar analysis can also be made of the phonon Green's function, which can be written as

$$\mathscr{D}(q, i\nu_l) = \frac{1}{\mathscr{D}_0^{-1}(q, i\nu_l) - \Pi(q, i\nu_l)}$$

(6.6.7)

The phonon self-energy Π is defined by the sum of diagrams shown in Fig. 6.14.

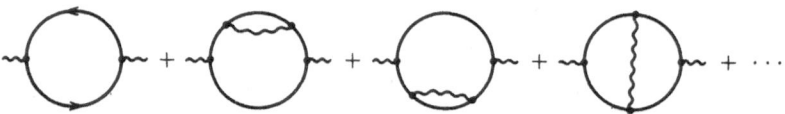

Fig. 6.14. The phonon self-energy

Now the fourth order self-energy diagrams Fig. 6.12a-c can be obtained from Fig. 6.9 for Σ_1 by adding the following corrections, respectively. Thus, Fig. 6.12a is obtained by adding the self-energy correction Σ_1 to the electron line of Fig. 6.9. Similarly Fig 6.12b is obtained by adding the lowest-order self-energy correction to the phonon line while Fig. 6.12c contains a new type of correction called the vertex correction. Thus Fig. 6.12c is obtained by adding the lowest-order correction to the electron–phonon interaction vertex in Fig. 6.9. This classification applies to all higher-order self-energy type diagrams. Supposing that we take the partial sum of each type of diagram, we see that the self-energy Σ should have the structure shown by Fig. 6.15, in which double full lines represent the exact electron Green's function $\mathscr{G}$ including all selfenergy corrections, double wavy lines represent the exact phonon Green's function $\mathscr{D}$, and shaded triangle defines the vertex function Γ. This is the sum of diagrams shown by Fig. 6.16. The analytic expression for Fig. 6.15 is

$$\Sigma(k) = -\frac{1}{N\beta} \sum_{k'} g_{k-k'} \, \mathscr{D}(k - k') \, \mathscr{G}(k') \, \Gamma(k', k, k - k') \tag{6.6.8}$$

where $k = (\mathbf{k}, i\varepsilon_n)$.

Similarly the phonon self-energy has the structure shown by Fig. 6.17, which means

$$\Pi(q) = \frac{2}{N\beta} g_q \sum_k \mathscr{G}(k + q) \, \mathscr{G}(k) \, \Gamma(k + q, k, q) \; . \tag{6.6.9}$$

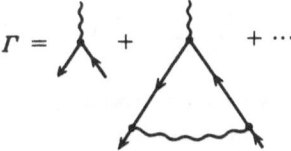

$\Gamma =$

Fig. 6.15. The structure of electron self-energy Σ

Fig. 6.16. Electron-phonon vertex

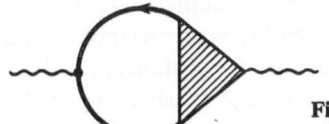

Fig. 6.17. The structure of phonon self-energy Π

6.7 Migdal Approximation and Electron Self-Energy

The general formalism described in the two preceding sections may be applied to many-body systems other than the electron–phonon one. In this section, we introduce the approximation specific to the electron–phonon system and seek for approximate, but explicit expressions for Green's functions.

6.7.1 Migdal Approximation

The electron–phonon interaction in the metal is characterized by the fact that the bare coupling constant g is already a good approximation of the vertex function Γ (Migdal approximation). The error committed by this approximation can be estimated by the ratio of the second-order term to the first in the perturbation expansion of Γ in Fig. 6.16. As we see below, the ratio is of the order of the average phonon energy ω_0 divided by the electron Fermi energy ε_F, i.e., of the order of 10^{-3}. In other words, the approximation is based *not* on the weakness of the interaction, but on the quickness of the electron zero-point motion in comparison with the phonon motion. It is not the usual Born–Oppenheimer approximation, which completely ignores the ion motion in determining the electron spectrum, and therefore the parameter (6.4.4) cannot affect the spectrum. The Migdal approximation claims that the dynamics of ions can give a subtle structure to the electron self-energy, but does not essentially affect the vertex function. It is an advantage of the Green's function method to make such analysis possible.

Now, for simplicity, let us take the Einstein model (6.4.5) to prove the Migdal approximation. The second-order term of Γ is given by Fig. 6.18, which we write as

$$\Gamma_1 = \frac{g^2}{N\beta} \sum_{k_1} \mathscr{G}_0(k_1 - q) \, \mathscr{G}_0(k_1) \, \mathscr{D}_0(k - k_1) \tag{6.7.1}$$

where we have introduced the four-dimensional notation $k = (\boldsymbol{k}, i\varepsilon_n)$ as before. We will be concerned with $\boldsymbol{k}$ near the Fermi surface and ε_n in the range $|\varepsilon_n| \lesssim \omega_0$. For $|\nu_l| \ll \omega_0$, (6.5.27) is equal to $-(2/\omega_0)$, and proportional to ν_l^{-2} for $|\nu_l| \gg \omega_0$. The upper limit of the phonon wave number q is of the order of the inverse lattice constant and therefore of the same order as the electron Fermi number k_F. Hence the range of $\boldsymbol{k}_1$ in (6.7.1) is restricted to $|\varepsilon_1 - \varepsilon| \lesssim \omega_0$, $|\boldsymbol{k}_1 - \boldsymbol{k}| \lesssim k_F$. To obtain a rough estimate, we therefore replace $\mathscr{D}_0$ in this region by $-(2/\omega_0)$. Furthermore the product of the two $\mathscr{G}_0$ is appreciable when at least one of $\boldsymbol{k}_1$ and $\boldsymbol{k}_1 - \boldsymbol{q}$ is close to the Fermi surface; for instance, $|\xi_{k_1}| \lesssim \omega_0$. Then the product is roughly equal to $[\omega_0(\omega_0 + \hbar v_F q)]^{-1}$, where v_F is the Fermi velocity and $\hbar v_F q \approx \varepsilon_F$ for q of the order of k_F. Remembering there are roughly $\beta\omega_0$ possible values of ε_1 and $N\rho_F\omega_0$ possible values of $\boldsymbol{k}_1$, we obtain the estimate

$$\Gamma_1 \approx g^2 \rho_F \omega_0^2 \frac{2}{\omega_0} \frac{1}{\omega_0 \varepsilon_F} = g\lambda \frac{\omega_0}{\varepsilon_F} . \tag{6.7.2}$$

Since λ given by (6.4.4) is at most of the order of unity,

$$\Gamma = g\left[1 + O\left(\frac{\omega_0}{\varepsilon_F}\right)\right] . \tag{6.7.3}$$

When the wave number q is less than $\omega_0/\hbar v_F$, the estimate (6.7.2) does not apply, but such a small q plays an essential role nowhere in the following analysis.

We go back to the general phonon model and admit the Migdal approximation, so that (6.6.8) takes the form

$$\Sigma(k, i\varepsilon_n) = -\frac{1}{N\beta} \sum_{k'm} g_{k-k'}^2 \mathscr{D}(k - k', i\varepsilon_n - i\varepsilon_m) \mathscr{G}(k', i\varepsilon_m) . \qquad (6.7.4)$$

This should be coupled with (6.6.6) to determine Σ, $\mathscr{G}$. Diagramatically Fig. 6.15 has been approximated by Fig. 6.19. If the electron Green's function represented by the double full lines is expanded, we obtain Fig. 6.20.

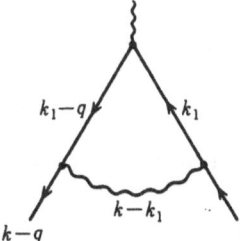

Fig. 6.18. The second-order vertex correction

Fig. 6.19. Migdal approximation for Σ

Fig. 6.20. The perturbational structure of the Migdal approximation

From the purely theoretical point of view, we should also set up the Dyson equation of the phonon Green's function. Instead of doing that, we assume here that we empirically know $\mathscr{D}$ or the spectral function σ in (6.5.29), say, from the neutron scattering experiment.

6.7.2 One-Electron Spectral Function

Substituting spectral representations (6.5.15, 29) in (6.7.4), we have

$$\Sigma(k, i\varepsilon_n) = -\frac{1}{N\beta} \sum_{k'} \sum_{m} g_{k-k'}^2 \int_{-\infty}^{\infty} dx \int_{0}^{\infty} dy\, S(k', x)\, \sigma(k - k', y)$$

$$\left(\frac{1}{i\varepsilon_m - i\varepsilon_n - y} - \frac{1}{i\varepsilon_m - i\varepsilon_n + y}\right) \frac{1}{i\varepsilon_m - x} . \qquad (6.7.5)$$

Let us first take the sum over $\varepsilon_m = (\pi/\beta)(2m + 1)$, $m = 0, \pm 1, \pm 2, \cdots$. It is convenient to sum $\beta^{-1} \Sigma\, \phi(i\varepsilon_m)$ with use of the theorem of residues. The Fermi function $f(z) = [\exp(\beta z) + 1]^{-1}$ possesses simple poles at $z = i\varepsilon_m$ with the residue $-\beta^{-1}$. Provided that $\phi(z)$ is analytic along the imaginary axis, we have

$$\frac{1}{\beta} \sum_m \phi(i\varepsilon_m) = \int \phi(z)\, f(z)\, \frac{dz}{2\pi i} \tag{6.7.6}$$

where the integral is taken along the path C enclosing the imaginary axis in Fig. 6.21. In the case of (6.7.5), $\phi(z)$ has simple poles at $z = i\varepsilon_n \pm y$, $z = x$, so that we deform C so as to enclose these poles and obtain

$$\frac{1}{\beta} \sum_m \left(\frac{1}{i\varepsilon_m - i\varepsilon_n - y} - \frac{1}{i\varepsilon_m - i\varepsilon_n + y} \right) \frac{1}{i\varepsilon_m - x}$$

$$= \frac{f(x)}{x - i\varepsilon_n - y} + \frac{1 - f(x)}{x - i\varepsilon_n + y} + p(y) \left(\frac{1}{x - i\varepsilon_n - y} + \frac{1}{x - i\varepsilon_n + y} \right).$$

$$\tag{6.7.7}$$

Here $p(y)$ is the Planck distribution. We will assume the zero temperature, so that $p(y) = 0$ and $f(x)$ is the step function $\theta(-x)$.

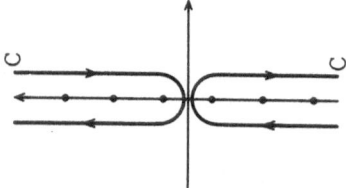

Fig. 6.21. The contour C enclosing $i\varepsilon_n$

Now, as was mentioned in Sect. 3.5, analytic continuation of Σ, like that of the Green's function, is possible with respect to $i\varepsilon_n$. We need only to replace $i\varepsilon_n$ by the complex z after taking the sum over energies of internal lines as we have done in (6.7.7). We are interested in the region $|z| \lesssim \omega_0$, where the z dependence of Σ is important compared with its dependence on k near the Fermi surface. We assume the free-electron model $\xi_k = \hbar^2 k^2/2m - \mu$ and take k at the Fermi surface $k = k_F$ as is shown by Fig. 6.22. So, we write the electron self-energy simply as $\Sigma(z)$.

Let $a(E)$, $b(E)$ be real and imaginary parts of the limit $\Sigma(E + i0^+)$ when z approaches a point E on the real axis from above. As we have mentioned in Chap. 3 (Sect. 3.5) concerning the Fermi-liquid theory and as we see from the explicit expression given below, we obtain the limit $a(E) + ib(E)$ when approaching E from below. The analytic continuation $G(k, z) = [z - \xi_k - \Sigma(z)]^{-1}$ of the temperature Green's function reduces to the retarded Green's function, i.e., the left-hand side of (6.5.17), as $z \to E + i0^+$. Thus

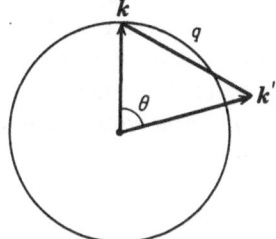

Fig. 6.22. Definitions of q, θ

$$G(k, E + i0^+) = \frac{1}{E - \xi_k - a(E) + ib(E)} . \tag{6.7.8}$$

The spectral function is thus given by

$$S(k, E) = \frac{1}{\pi} \frac{b(E)}{[E - \xi_k - a(E)]^2 + b(E)} . \tag{6.7.9}$$

From the general expression (6.5.10), $S \geq 0$ and therefore $b \geq 0$, which can be confirmed by the explicit expression of b given below. According to Fermi-liquid theory in Sect. 3.5, the parameter b represents the lifetime of the electron as the quasiparticle and should vanish at the Fermi level $E = 0$. This can also be confirmed later. Thus we have the *quasiparticle picture* in Landau's sense as far as the lower excited states of the electron system are concerned. In fact, as we will see soon, $a(E) \approx a(0) - \lambda E$ for $|E| \ll \omega_0$, where λ is defined by (6.4.4). Thus, for small $|E|$,

$$S(k, E) = \frac{1}{1 + \lambda} \delta(E - \zeta_k) \tag{6.7.10}$$

where $\zeta_k = (1 + \lambda)^{-1}[\xi_k + a(0)]$ is the energy of the quasiparticle measured from the Fermi level. The general formalism in Sect. 3.5 demands that the quasiparticle energy should vanish at the Fermi surface defined by $k = k_F$. Hence

$$\mu = \frac{\hbar^2 k_F^2}{2m} + a(0) \tag{6.7.11}$$

which gives the shift of the chemical potential for the constant electron density. We can then write

$$\zeta_k = \frac{\hbar^2}{2m^*} (k^2 - k_F^2) \tag{6.7.12}$$

$$m^* = m(1 + \lambda) . \tag{6.7.13}$$

The electron mass near the Fermi surface is enhanced by the electron–phonon interaction. The effect has been observed in such metals as K, Al, Pb, Hg through the electronic specific heat (see Sect. 3.5) and also through the cyclotron resonance whose resonance frequency is given by eH/m^*c in magnetic field H. The observed mass is sometimes 1.5–2 times as large as the band theoretical mass and the discrepancy can be accounted for by the electron–phonon enhancement factor $1 + \lambda$.

6.7.3 The Solution of Dyson's Equation

Writing $\Delta(z) = \Sigma(z) - \Sigma(0)$, we have from (6.7.5, 7)

$$\Delta(z) = -\frac{1}{N} \sum_{k'} g_{k-k'}^2 \int_{-\infty}^{\infty} dx \int_0^{\infty} dy \, S(k', x) \, \sigma(k - k', y)$$

$$\left[\theta(-x) \left(\frac{1}{x - z - y} - \frac{1}{x - y} \right) + \theta(x) \left(\frac{1}{x - z + y} - \frac{1}{x + y} \right) \right]. \qquad (6.7.14)$$

Since $|z|$ and y are of the order of ω_0, the region $|x| \lesssim \omega_0$ makes the main contribution. Since a and b are also of the order of ω_0, $S(k', x)$ is appreciable when $|\xi_{k'}| \lesssim \omega_0$. We replace the sum over k' by the integral over $\xi' = \xi_{k'}$ and q defined by Fig. 6.22. Note that $\sin\theta \, d\theta = (k_F k')^{-1} q \, dq$ since $q^2 = k^2 + k'^2 - 2kk'\cos\theta$. Note also that the spectral function (6.7.9) is a simple Lorentzian distribution with respect to ξ_k. Hence we first carry out the integral over ξ' in (6.7.14), in which q may be taken as connecting the fixed point k with another point on the Fermi surface. We define $I(y)$ and g^2 by

$$2g^2 I(y) = \frac{1}{k_F^2} \int g_q^2 \, \sigma(q,y) q \, dq \qquad (6.7.15)$$

$$\int_0^{\infty} I(y) \, dy = 1 \ . \qquad (6.7.16)$$

With use of $\rho_F = 3m/2\hbar^2 k_F^2$, we write

$$\Delta\Sigma(z) = -\rho_F g^2 \int_0^{\infty} dy I(y) \left[\int_{-\infty}^0 dx \left(\frac{1}{x - z - y} - \frac{1}{x - y} \right) + \right.$$

$$\left. + \int_0^{\infty} dx \left(\frac{1}{x - z + y} - \frac{1}{x + y} \right) \right] = -\rho_F g^2 \int_0^{\infty} dy \, I(y) \ln\left(\frac{y + z}{y - z} \right). \qquad (6.7.17)$$

As we have mentioned in Sect. 3.5, $\Sigma(z)$ is analytic on the complex z plane except on the real axis, which is its logarithmic branch line.

Substituting $z = E + i0^+$, we have real part $a(E) \equiv a(0) + a_1(E)$ and imaginary part $b(E)$ as

$$a_1(E) = -\rho_F g^2 \int_0^{\infty} dy \, I(y) \ln\left| \frac{E + y}{E - y} \right| \qquad (6.7.18)$$

$$b(E) = \pi \rho_F g^2 \int_0^{|E|} dy \, I(y) \tag{6.7.19}$$

If we define the average phonon energy ω_0 by

$$\frac{1}{\omega_0} = \int_0^\infty \frac{I(y)}{y} \, dy \tag{6.7.20}$$

we have $a_1(E) \approx -\lambda E$ for $|E| \ll \omega_0$ and λ takes the same form as (6.4.4).

If we ignore the shift of the chemical potential given by (6.7.11), the solution of Dyson's equation takes the same form as the second-order perturbation expression (6.6.4), provided that $\mathscr{D}_0$ in the latter is replaced by $\mathscr{D}$. With this replacement (6.6.4) reduces to the right hand side of (6.7.14), in which $S(k', x)$ is taken as the spectral function of the free electron $\delta(x - \eta_{k'})$ with

$$\eta_k = \frac{\hbar^2}{2m}(k^2 - k_F^2) \ . \tag{6.7.21}$$

6.7.4 Limitation of the Quasiparticle Picture

Qualitative features of the spectral function

$$S(k, E) = \frac{1}{\pi} \cdot \frac{b(E)}{[E - \eta_k - a_1(E)]^2 + b^2(E)} \tag{6.7.22}$$

can be seen by assuming the Einstein model $I(y) = \delta(y - \omega_0)$. Then,

$$\left.\begin{array}{l} a_1(E) = \dfrac{1}{2} \lambda \omega_0 \ln \left| \dfrac{E - \omega_0}{E + \omega_0} \right| \\[3mm] b(E) = \dfrac{\pi}{2} \lambda \omega_0 \theta(|E| - \omega_0) \end{array}\right\} . \tag{6.7.23}$$

Figure 6.23a-c show the results calculated for $\lambda = 1$, $\eta_k = 0$, $2\omega_0$, $5\omega_0$, respectively.

The infinitely sharp peak at $E = 0$ in Fig. 6.23a is obtained by putting $\zeta_k = 0$ in (6.7.10). As η increases, the peak moves to the right, and its position is $\zeta_k = (m/m^*) \eta_k$ as far as $\eta_k \ll \omega_0$. The one-electron excitation near the Fermi surface is determined by this quasiparticle peak. The peak approaches $E = \omega_0$ (Fig. 6.23b, c) as η_k exceeds ω_0 and, for $\eta_k \gg \omega_0$, the distance $\Delta\omega$ from ω_0 decreases like $2\omega_0 \exp[-2(\eta_k/\omega_0)]$ with the intensity proportional to $\Delta\omega$. Hence this peak becomes negligible since we have the sum rule that the integral of $S(k, x)$ over x is always equal to unity. The continuous spectrum in the region $E > \omega_0$, on the other hand, shows two peaks when $\eta_k > \omega_0$. As η_k increases, the peak at lower energy approaches ω_0, whereas the peak at higher energy not only approaches

η_k, but also takes up the most of spectral intensity. Hence this peak of highest energy is identified with the quasiparticle for $\eta_k \gg \omega_0$.

Thus the simple quasiparticle picture applies either for $|\eta_k| \ll \omega_0$ or for $|\eta_k| \gg \omega_0$. When $|\eta_k|$ is comparable with ω_0, the structure of the whole spectral function becomes important.

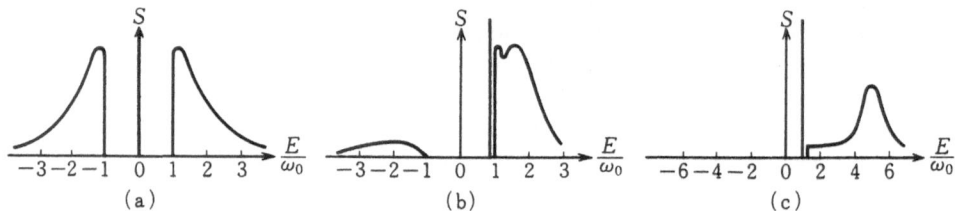

Fig. 6.23. a-c. One-electron spectral function (Einstein model). [J.R. Schrieffer: In *Many-Body Problems*, ed. by R. Kubo (W.A. Benjamin, New York 1966) p.98]

6.8 Electron-Phonon Interaction and Superconductivity

As we have mentioned in Chap. 4, the superconducting state of the metal is caused by the phonon-mediated interaction between electrons. With use of the terminology introduced in this chapter, we say that there appears a certain divergence in the electron–phonon vertex Γ at the transition temperature T_C from normal to superconducting states. The detail of the divergence and especially the value of T_C depend on our approximation to calculate Γ. Here we will apply the mean field approximation in accordance with the BCS model in Chap. 4.

6.8.1 Divergence of the Vertex Function

Among higher order terms in the perturbation expansion of Γ, we have a series of terms shown by Fig. 6.24a, where K means the sum of Feynman diagrams shown by Fig. 6.24a, b. We write the first term as K_0. For $q = 0$ and the Einstein model of phonons, we have

$$K_0(p, p') = \frac{1}{\beta N} g^4 \sum_{p_1} \mathscr{D}_0(p - p_1)\, \mathscr{G}_0(p_1)\, \mathscr{G}_0(-p_1)\, \mathscr{D}_0(p_1 - p') \ . \tag{6.8.1}$$

Let us replace $\mathscr{D}_0$ by $-(2/\omega_0)$ as we have done in the preceding section and restrict ε_1 in $p_1 = (\boldsymbol{p}_1, i\varepsilon_1)$ to the region $|\varepsilon_1| < \omega_0$. Assuming the constant density of states ρ_F in the integral over $\boldsymbol{p}_1$, we obtain

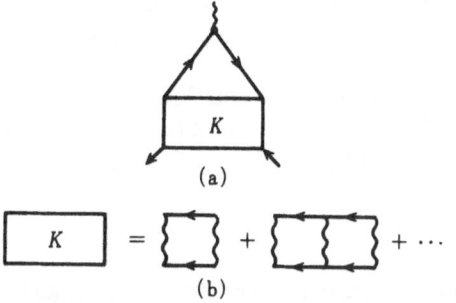

(a)

(b)

Fig. 6.24. a,b. Vertex function K

$$K_0 \approx \frac{2g^2}{\omega_0} \frac{\lambda}{\beta} \sum \int_{-\infty}^{\infty} \frac{d\xi}{(\xi + i\varepsilon_1)(\xi - i\varepsilon_1)}$$

$$= \frac{2g^2}{\omega_0} \cdot \frac{\pi\lambda}{\beta} \sum_{\varepsilon_1 > 0} \frac{1}{\varepsilon_1} \ . \tag{6.8.2}$$

When the maximum N of n is large, the asymptotic value of the sum $(\pi/\beta) \sum \varepsilon_1^{-1} = \sum (2n - 1)^{-1}$ is known to be $(1/2) [\ln(4N) + \gamma]$, where $\gamma = 0.577 \ldots$ is Euler's constant. Hence the correction arising from K_0 in Fig. 6.24b is of the order of $g\rho_F K_0$ and its ratio to the bare coupling constant g takes the form $\lambda\Lambda$, where

$$\Lambda = \lambda \left[\ln\left(\frac{2\beta\omega_0}{\pi}\right) + \gamma \right] . \tag{6.8.3}$$

Applying similar approximations to higher-order terms, we see that K in Fig. 6.24b is a geometric series whose sum is given by

$$\rho_F K = \lambda\Lambda(1 + \Lambda + \Lambda^2 + \ldots)$$

$$= \frac{\lambda\Lambda}{1 - \Lambda} \ . \tag{6.8.4}$$

This diverges at the critical temperature T_C defined by $\Lambda = 1$. Thus

$$k_B T_C = \frac{2C}{\pi} \omega_0 \exp\left(-\frac{1}{\lambda}\right) \tag{6.8.5}$$

where $C = \exp(\gamma) \approx 1.78$. When T approaches T_C from above, approximately $\Lambda \approx 1 - \lambda [(T - T_C)/T_C]$, so that

$$\rho_F K \cong \frac{T_C}{T - T_C} \ . \tag{6.8.6}$$

This is the "Curie–Weiss" law characteristic of the mean field approximation. As we have mentioned in Chap. 4, the approximation is exceptionally good in the case of superconductors.

Note that (6.8.3, 5) are applicable only in the weak coupling limit $k_B T_C \ll \omega_0$. Although λ is not very small, the Coulomb repulsion should actually be subtracted and most of superconductors belong to the weak-coupling case.

6.8.2 Nambu Representation

Below T_C, a macroscopic number of Cooper pairs are formed, or equivalently, the electron pair wave is in a coherent state. As we have mentioned in Sect. 4.9.2, it is then convenient to regard the destruction operator $a_{k\uparrow}$ and the creation operator $a_{-k\downarrow}^\dagger$ as two components of a two-dimensional spinor ψ_k (Nambu representation). In this section, we write Pauli matrices operating upon the spinor as

$$\hat{\rho}_1 = \begin{pmatrix} 0 & 1 \\ 1 & 0 \end{pmatrix}, \quad \hat{\rho}_2 = \begin{pmatrix} 0 & -i \\ i & 0 \end{pmatrix}, \quad \hat{\rho}_3 = \begin{pmatrix} 1 & 0 \\ 0 & -1 \end{pmatrix}. \tag{6.8.7}$$

The unperturbed Hamiltonian in (6.5.1), apart from a constant term, may be written as

$$\mathcal{H}_0 = \sum_q \psi_k^\dagger \xi_k \hat{\rho}_3 \psi_k + \sum_k \omega_0 b_q^\dagger b_q . \tag{6.8.8}$$

Similarly

$$\mathcal{H}' = \frac{g}{N} \sum_{kq} \psi_{k+q}^\dagger \hat{\rho}_3 \psi_k \phi_q . \tag{6.8.9}$$

The one-electron Green's function is defined by

$$\mathcal{G}_{rr'}(k; \tau - \tau') = - \langle T_\tau \psi_{kr}(\tau) \psi_{kr'}^\dagger(\tau') \rangle \quad (r, r' = 1, 2) \tag{6.8.10}$$

which may be regarded as forming a 2×2 matrix. Its diagonal elements $\mathcal{G}_{11}$, $\mathcal{G}_{22}$ are $- \langle T_\tau a_{k\uparrow}(\tau) a_{k\uparrow}^\dagger(\tau') \rangle$ and $- \langle T_\tau a_{-k\downarrow}^\dagger(\tau) a_{-k\downarrow}(\tau') \rangle$, respectively, and do not vanish in the normal state. Nondiagonal $\mathcal{G}_{12} = - \langle T_\tau a_{k\uparrow}(\tau) a_{-k\downarrow}(\tau') \rangle$ and $\mathcal{G}_{21}$ are characteristic of the superconducting state. General properties in Sect. 6.5, such as antiperiodicity (6.5.12), Fourier expansion (6.5.13), are applicable to $\mathcal{G}$, whose Fourier coefficients are of course 2×2 matrices. Similarly, with reference to the formal perturbation expansion, we can define electron self-energy Σ and vertex function Γ, and also derive Dyson's equation

$$\mathcal{G}(p) = \frac{1}{i\varepsilon_n - \xi_k \hat{\rho}_3 - \Sigma(p)} . \tag{6.8.11}$$

On the right hand side, $i\varepsilon_n$ means $i\varepsilon_n$ multiplied by the 2×2 unit matrix $\hat{i}$, which we will omit unless confusion may arise. Admitting the Migdal approximation $\Gamma \approx g\hat{\rho}_3$ again, we have

$$\Sigma(p) = - \frac{g^2}{N\beta} \sum_{p'} \mathscr{D}(p - p') \hat{\rho}_3 \mathscr{G}(p') \hat{\rho}_3 \tag{6.8.12}$$

to be coupled with (6.8.11).

Any 2×2 matrix is a linear combination of $\hat{i}$ and $\hat{\rho}_j (j = 1, 2, 3)$, so that we can assume

$$\Sigma(p) = \Sigma_0(p)\hat{i} + \Sigma_1(p)\hat{\rho}_1 + \Sigma_2(p)\hat{\rho}_2 + \Sigma_3(p)\hat{\rho}_3 . \tag{6.8.13}$$

For simplicity we neglect Σ_1, Σ_3. With use of $\hat{\rho}_j\hat{\rho}_l + \hat{\rho}_l\hat{\rho}_j = 2\delta_{jl}$, we obtain

$$\mathscr{G} = F^{-1}[i\varepsilon_n - \Sigma_0 + \Sigma_1\hat{\rho}_1 + \Sigma_3\hat{\rho}_3]$$
$$F = (i\varepsilon_n - \Sigma_0)^2 - \Sigma_1^2 - \Sigma_3^2 . \tag{6.8.14}$$

Substituting in (6.8.12), we first have

$$\Sigma_0(p) = - \frac{g^2}{N\beta} \sum_{p'} \mathscr{D}(p - p') \frac{1}{F(p')} [i\varepsilon' - \Sigma_0(p')] . \tag{6.8.15}$$

The right-hand side remains finite for $\Sigma_0 = \Sigma_1 = 0$ and gives the normal self-energy discussed in the preceding section. In other words, the perturbation expansion can be applied to Σ_0. On the contrary, Σ_1, which is characteristic of the superconducting state, satisfies

$$\Sigma_1(p) = \frac{g^2}{N\beta} \sum_{p'} \mathscr{D}(p - p') \frac{1}{F(p')} \Sigma_1(p') . \tag{6.8.16}$$

The perturbation expansion cannot give finite Σ_1 since the right-hand side vanishes for $\Sigma_0 = \Sigma_1 = 0$.

In the weak coupling limit, we may ignore Σ_0, too. As for Σ_1, we replace $\mathscr{D}$ by $- (2/\omega_0)$ in (6.8.16) and assume that Σ_1 is constant for $|\varepsilon| \leqq \omega_0$ and vanishes for $|\varepsilon| > \omega_0$. Then (6.8.16) reduces to

$$1 = \frac{\lambda}{\beta} \Sigma \int d\xi \frac{1}{\xi^2 + \varepsilon^2 + \Sigma_1^2} = \frac{\pi\lambda}{\beta} \Sigma \frac{1}{\sqrt{\varepsilon_1^2 + \Sigma_1^2}} . \tag{6.8.17}$$

For $\Sigma_1 = 0$, this gives $\Lambda = 1$, i.e., the equation to determine the transition temperature (6.8.5). At zero temperature, $\beta \to \infty$, on the other hand, we can replace the sum by integral:

$$1 = \lambda \int_0^{\omega_0} \frac{d\varepsilon}{\sqrt{\varepsilon^2 + \Sigma_1^2}} = \lambda \ln \left(\frac{2\omega_0}{\Sigma_1}\right) . \tag{6.8.18}$$

This gives $\Sigma_1 \approx 2\omega_0 \exp(-\lambda^{-1})$. The analytic continuation of the Green's function in this case is obtained by the replacement $i\varepsilon_n \to z$ in (6.8.14):

$$\hat{G}(z) = \frac{1}{2}\left[1 + \frac{1}{E_k}(\Sigma_1\hat{\rho}_1 + \xi_k\hat{\rho}_3)]\frac{1}{z - E_k}\right]$$
$$+ \frac{1}{2}\left[1 - \frac{1}{E_k}(\Sigma_1\hat{\rho}_1 + \xi_k\hat{\rho}_3)]\frac{1}{z + E_k}\right] \tag{6.8.19}$$

$$E_k = \sqrt{\xi_k^2 + \Sigma_1^2} . \tag{6.8.20}$$

Thus E_k is the one-electron excitation energy in the superconducting state, corresponding to the excitation energy of an electron outside the Fermi surface or of a hole inside the Fermi surface in the normal state. It takes the minimum value $|\Sigma_1|$ at the Fermi surface $\xi_k = 0$, so that the spectrum has a gap (see Sect. 4.9).

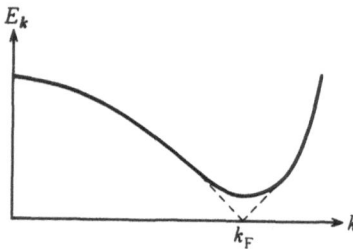

Fig. 6.25. One-electron excitation energy in the superconducting state

In the case of strong coupling, where $k_B T_C$ is not small compared with ω_0, we have to solve integral equations (6.8.15), (6.8.16) numerically. It is known that we can account for various properties of the superconductor by substituting the phonon spectrum observed through the neutron experiment into $\mathscr{D}$ and also by including the effect of Coulomb repulsion appropriately.

7. Interaction Between Elementary Excitations and Spectral Line Shapes

As was mentioned in Chap. 5, two kinds of elementary excitations with bilinear interaction can be brought, through a linear transformation, into two mixed modes which no longer interact with each other. Such an elimination is impossible for nonlinear interaction, which in fact gives rise to a variety of essentially new effects. In this chapter, we will study how the dynamics of nonlinear interactions within the matter are reflected in the spectral line shapes of absorption, emission, and scattering of light. Various nonlinear effects will be described in Sect. 7.1 with the use of the two-oscillator model, while later sections will be devoted to the theory of spectral line shapes in the electron (exciton)–phonon systems.

7.1 What Happens with Nonlinear Interactions?

Let us represent the fields x and y by simple classical oscillators with respective frequencies ω_1 and ω_2, and with interaction higher than bilinear, say, of the form: $(\gamma\omega_1^2/2)x^2y$. The equations of motion corresponding to (5.1.3) for the bilinear case are given by

$$\ddot{x} + \omega_1^2 x + \gamma\omega_1^2 xy = 0 , \tag{7.1.1a}$$

$$\ddot{y} + \omega_2^2 y + \frac{1}{2}\gamma\omega_1^2 x^2 = 0 . \tag{7.1.1b}$$

If the zeroth-order solution: $x = X \exp(\pm i\omega_1 t)$ and $y = Y \exp(\pm i\omega_2 t)$, obtained by neglecting nonlinear terms, is put into the third term in (7.1.1a), there appear in the x oscillation the satellite components with frequencies $|\omega_1 \pm \omega_2|$ in addition to the principal frequency ω_1. By successive approximation, one obtains a series of components with frequencies $|\omega_1 + n\omega_2|$ with integer n.

Under oscillating external force $F \exp(-i\omega t)$ acting upon (7.1.1a), the resonance absorption takes place not only at $\omega = \omega_1$ but also at $|\omega_1 + n\omega_2|$: the external force directly excites a quantum $\hbar\omega_1$ of the x oscillator, which in turn causes, through the nonlinear interaction, multiple production of quanta $\hbar\omega_2$ of the y oscillator. This part of the absorption spectra corresponding to simultaneous excitation of another oscillator not directly subject to the external force is sometimes called the *sideband*. The examples are phonon, magnon, and plasmon sidebands in the photoabsorption spectra of single-electron excitation.

The main band ($n = 0$) corresponding to direct excitation alone is called, for example, the *zero-phonon line*.

If the interaction is not small, it is not appropriate to decompose the absorption spectra into successive sidebands. An alternative way is to put together the second and third terms in (7.1.1a) and to regard $\Omega(t) \equiv \omega_1[1 + \gamma y(t)]^{1/2} = \omega_1 + \delta\omega_1(t)$ as the instantaneous frequency. Such an interpretation is meaningful if the motion of the frequency shift $\delta\omega_1(t)$ is slow enough. In this *adiabatic* or *slow-modulation* limit, the absorption spectrum shows the width around ω_1 which reflects the time-averaged distribution of $\delta\omega_1(t)$ (namely, the nth moment of the spectrum is given by $\langle\delta\omega_1(t)^n\rangle$).

How slow should the motion of $\delta\omega_1(t)$ be for the adiabatic picture to be valid? According to the uncertainty principle, $\delta\omega_1(t)$ should stay around that value for a time large compared to $(\delta\omega_1)^{-1}$ in order that $\delta\omega_1(t)$ by perceived as a frequency shift at that moment. Denoting the average frequency and the average amplitude of $\delta\omega_1(t)$ by $\omega_2'(\sim\omega_2)$ and $\delta(\delta^2 \equiv \langle\delta\omega_1(t)^2\rangle)$, respectively, one should have $\omega_2' \ll \delta$ for the adiabaticity. Otherwise, the distribution of $\delta\omega_1(t)$ cannot manifest itself to the full extent in the spectrum . In the other extreme: $\omega_2' \gg \delta$—*rapid-modulation* case, $\delta\omega_1(t)$ will stay around that value only for a time $(\omega_2')^{-1}$ which is much shorter than the minimum time δ^{-1} required for the perception of the instantaneous frequency shift $\delta\omega_1(t)$. Because of this rapid modulation, the observed spectral width is given by δ^2/ω_2', being smaller than the adiabatic width δ by the effectivity factor δ/ω_2' mentioned above. This is the *motional narrowing*, the phenomenon well-known in magnetic-resonance studies. Another example will be found in Sect. 7.3.

Let x be the electronic polarization with static polarizability α_0. Putting on the right hand side of (7.1.1a) the external electric force $\alpha_0\omega_1^2 E \exp(-i\omega t)$, and regarding $y(t)$ as an adiabatic parameter whose motion is slow enough ($\omega_2 \ll \omega, \omega_1$), one obtains the forced-oscillation solution:

$$x(t) = \alpha(\omega, y)E \exp(-i\omega t) , \tag{7.1.2}$$

$$\alpha(\omega, y) = \frac{\alpha_0}{(1 + \gamma y) - (\omega/\omega_1)^2} . \tag{7.1.3}$$

Since the polarizability $\alpha(\omega, y)$ varies with time through $y(t) \approx Y \exp(\pm i\omega_2 t)$, the solution (7.1.2) comprises the oscillatory components with frequencies $\omega - n\omega_2$ (n: integer), of which the first-order term ($n = \pm 1$) is given by

$$x^{(1)}(t) = \left[\frac{\partial}{\partial y} \alpha(\omega, y)\right]_{y=0} EY\exp[-i(\omega \mp \omega_2)t] . \tag{7.1.4}$$

These components as dipolar sources will emit the electromagnetic waves with frequencies $\omega' = \omega - n\omega_2$. This is the nth order *Raman scattering*: the incident photon with energy $\hbar\omega$ excites n quanta of the oscillator y through the modulation of electronic polarizability α, the remaining energy $\hbar\omega' = \hbar(\omega - n\omega_2)$ being emitted as a photon. Of the first-order Raman-scattered light, the component with $\omega - \omega_2$ is called the *Stokes line* and that with $\omega + \omega_2$ the *anti-Stokes line*. The $n = 0$ component (elastic scattering) is called the *Rayleigh line*. The oscillator y may be any elementary excitations with lower frequencies than those of electronic excitations, such as intramolecular vibrations in molecules or phonons and magnons in insulators. In macroscopic systems, y consists of a great number of oscillators with continuously distributed frequencies, and the Raman spectra consist of continuous bands although the Rayleigh component remains a line spectrum. The relation between the Rayleigh line and the Raman bands has some similarity to that between the zero-phonon line and the phonon sidebands in the absorption spectra, except that the zero-phonon line has a finite width because the electronic transition is involved.

Since the Raman scattering takes place through the modulation of polarizability α by y, and not through the direct excitation of y by light, the infrared inactive modes of lattice vibrations (or other oscillators) can be *Raman active*. In systems with inversion symmetry, the infrared active mode x (accompanied by electric polarization) is odd against the inversion, while the Raman active mode y is even in order that the interaction energy proportional to $x^2 y$ (anharmonicity) be invariant. Namely, the first-order infrared absorption spectra and the first-order Raman scattering spectra present information completely orthogonal to each other. Even though the Raman active mode y can also be infrared active in systems without inversion symmetry, its direct interaction with the electromagnetic field has a negligible effect as far as $\omega \gg \omega_2$. In magnetic insulators, the Raman scattering by magnons takes place because the electronic polarizability depends on the spin directions (y) of localized electrons through their exchange interactions. Direct interaction between the spin magnetic moment and the magnetic field of light is usually negligible.

Since the interaction between matter and a radiation field is very small, the Raman scattering which is the second-order optical process has usually much smaller cross section that that of the photoabsorption which is the first-order process. Intense laser light with frequency ω away from the absorbing region ($\sim\omega_1$) is favorable for the observation of a purely second-order process.

As ω approaches ω_1, however, the Raman cross section increases as $(\omega - \omega_1)^{-4}$ according to the semiclassical treatment: (7.1.2–4) (note that the cross-section is proportional to $(\partial\alpha/\partial y)^2$). In quantum mechanics, the corresponding factor comes from the square of the energy denominator $(\hbar\omega - \hbar\omega_1)(\hbar\omega - \hbar\omega_1 \mp \hbar\omega_2)$ of the third-order perturbation process—the incident photon ($\hbar\omega$) is absorbed to virtually excite the electron ($\hbar\omega_1$), then the low-energy quantum ($\hbar\omega_2$) is emitted or absorbed, and finally the electron is de-excited to emit the secondary photon ($\hbar\omega' = \hbar\omega \mp \hbar\omega_2$). (Note that ω_2 was neglected against ω_1 in the semiclassical treatment.) By choosing ω of the incident light nearly resonant to ω_1 of the electronic excitation, one can enhance the Raman cross section so significantly that the observation is easily possible with an ordinary light source.

Another important aspect of this resonance Raman scattering should be noted here. When $|\omega - \omega_1|$ is within the absorption bandwidth caused by the motion of the oscillator y, say the interatomic or lattice vibrations, the electronic excitation by photoabsorption can take place as a real process conserving energy. The excited electron interacts with the oscillator y, and lets it relax towards its new equilibrium position, the so-called *relaxed excited state*. During this relaxation process, the electron may return to the ground state by emitting a secondary photon $\hbar\omega'$. The resonance Raman scattering can be considered as successive real processes—absorption, relaxation, and emission—under certain conditions. If, in addition, the lifetime of the excited state is much longer than the relaxation time mentioned above, the correlation between absorption and emission disappears so that the scattering spectrum (function of ω and ω') factorizes into absorption (ω) and emission (ω') spectra. In other words the correlation between ω and ω' in the scattering spectrum represents the incompleteness of relaxation during the lifetime. The part of secondary light which is emitted before the completion of relaxation is sometimes called *hot luminescence*. For these reasons, the resonance Raman scattering provides a very powerful tool of investigating the relaxation in the excited state. In this connection, the *time-resolved spectroscopy* of secondary light after pulsed-light excitation enables one to pursue the evolution of relaxation process, within the limitation of the uncertainty principle between time and energy.

Among the second-order optical processes is the *two-photon absorption*, besides the Raman scattering mentioned above. The electromagnetic wave brings the electric polarization into forced oscillation, $x \propto E \exp(-i\omega t)$, which in turn brings y, through (7.1.1b), into forced oscillation with frequency 2ω (in addition to the static displacement). A strong resonance absorption by the y oscillator takes place when $2\omega = \omega_2$. This is the two-photon absorption—two photons with energy $\hbar\omega$ are simultaneously absorbed to create an elementary excitation with energy $\hbar\omega_2$. The probability for this second-order optical process to occur is evidently proportional to I^2 where $I(\propto E^2)$ is the intensity of the incident light. Although the proportionality constant is very small, this process is easily observed with the use of intense laser light.

The simultaneous incidence of two light beams (ω, I) and (ω', I') will give rise to two-photon absorption of three possible combinations: 2ω, $\omega + \omega'$ and $2\omega'$, with respective probabilities $\propto I^2, II'$, and I'^2. If $I \gg I'$, the absorption constant for the second beam should be proportional to $II'/I' = I$. The easiest way for two-photon spectroscopic study of elementary excitations is to prepare the laser light with fixed $\hbar\omega$ and the ordinary light with variable $\hbar\omega'$, each being in the nonabsorbing region, and to measure the absorption constant of the latter beam as a function of $\hbar(\omega + \omega')$.

Since the two-photon absorption obeys the same selection rule as the Raman scattering, its spectra provide complementary information to those of one photon absorption, especially in systems with inversion symmetry. In crystals, the dependence of two-photon absorption cross section upon the polarization directions of the two beams relative to crystallographic axis provides us with further information on the symmetries of initial and final states of the transition.

Returning to the case of one-beam excitation, we note that the indirect forced oscillation of y with frequency 2ω, even for off-resonance $2\omega \neq \omega_2$, will emit an electromagnetic wave with frequency 2ω provided y is accompanied by electric polarization. This oscillation of y, combined with the forced oscillation of x with frequency ω, will give rise, through the nonlinear term in (7.1.1a), to the component of x oscillation with frequency 3ω and to the electromagnetic wave of the same frequency. In this way, the electromagnetic wave with frequency ω incident upon a nonlinear system can, in general, generate electromagnetic waves with frequency $m\omega$ (m is integer). This phenomenon is called *higher-harmonics generation*.

In systems with inversion symmetry, the invariant interaction of the type $x^2 y$ is possible only when y is even against the inversion, namely, y is not accompanied by electric polarization. Only the harmonics of odd order ($m = 3, 5, \ldots$) can be generated in such systems.

Except for the selection rules which depend upon the symmetry of the system, one can think of any higher-order optical processes which satisfy the energy conservation rule (in addition to the wave-number conservation which is not mentioned here):

$$n_1 \omega_1 + n_2 \omega_2 + \cdots + m\omega + m'\omega' + \cdots = 0 \ . \tag{7.1.5}$$

Here, $\omega_1, \omega_2 \cdots$ represent the frequencies of elementary excitations, $\omega, \omega' \cdots$ the frequencies of electromagnetic waves, the integers $n, \cdots, m, \cdots$ the number of emitted or absorbed quanta according to whether the sign is positive or negative. As for the electromagnetic waves, ($m = \mp 1$, other m's vanishing) represent absorption and emission of one photon, ($m = -1, m' = +1$) the Raman scattering, ($m = -2$) or ($m = -1, m' = -1$) the two photon absorption, and ($m < -1, m' = +1, \omega' = |m|\omega$) higher-harmonics generation. As for the material system, ($n_1 = 1, n_2 = 0$) and ($n_1 = 1, n_2 \neq 0$) represent for example the zero-phonon line and its phonon sidebands in the optical spectra specified by m's.

While the electromagnetic waves in vacuum behave mutually independently according to the principle of superposition, they can combine with each other inside matter as described in (7.1.5), leading to a variety of nonlinear optical phonemena. All the optical nonlinearity originates from nonlinear interactions in the material system, as is evident from the arguments in this section and in Chap. 5.

7.2 The Absorption and Emission Spectra of a Localized Electron in the Phonon Field

As a simplest model system for the line-shape study of photoabsorption or emission, we consider an electron localized around a *lattice imperfection* in an insulator, such as an impurity atom, interstitial atom, or vacancy. Both the initial and final electronic states of the optical transition are assumed to be discrete states within the energy gap between the valence and conduction bands (see Fig. 7.1).

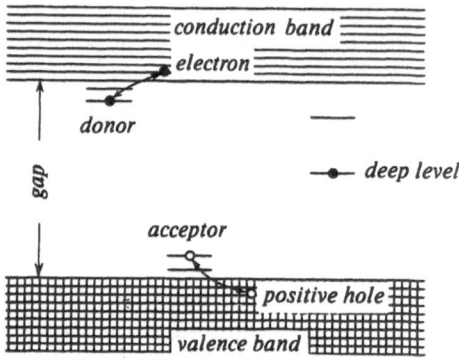

Fig. 7.1. Bands and impurity levels in a semiconductor

7.2.1 A Variety of Localized Electrons

The donor electron in a semiconductor is perhaps the best known among the various types of localized electron. In the IVth column semiconductors such as Ge and Si and the III–V compound semiconductors such as GaP and InSb, each atom has four covalent bonds with its four neighbors. If a host atom in a Si crystal is replaced by an impurity atom with one valence difference such as P (with five valence electrons) or B (with three valence electrons), a surplus electron not participating in the covalent bonds or a positive hole—a deficit electron resulting from the completion of the covalent bonds will be liberated into the conduction or valence band as a *charge carrier* contributing to the electric current. One can vary the electric conductivity over many orders of magnitude by controlling the impurity concentration (*extrinsic semiconductors*). The impurity atom with greater (smaller) valence than the host atom is called *donor* (*acceptor*) since it gives away (takes) an electron to (from) the conduction (valence) band (see Fig. 7.1).

To be more exact, however, the remaining impurity is charged with $\pm e$, exerting attractive Coulomb potential $- e^2/\epsilon r$ on the carrier with $\mp e$, where ϵ is the dielectric constant of the host crystal. At low temperature, most of the carriers will be bound by donor (acceptor) ions, forming neutral donors (neutral acceptors). If the effective mass m^* of the carrier is isotropic, one expects the hydrogen-like level scheme, with ground state (1s) binding energy $R = (m^*/m) \cdot \epsilon^{-2} \cdot R_H$ and orbital radius $a = (m/m^*) \cdot \epsilon \cdot a_H$. Here $R_H \equiv me^4/2\hbar^2 = 13.6$ eV and $a_H = \hbar^2/me^2 = 0.53$ Å are respectively the Rydberg constant and the orbital radius of the hydrogen atom. In most of IV and III–V semiconductors,

$\epsilon > 10$ and $m^* \lesssim m$ because of the small energy gap; as a result, the effective Rydberg R is as small as several or several tens of meV and the orbital radius a is as large as 10–100 Å. The fact that a is much larger than the lattice constant $d(\sim$ a few Å) justifies a posteriori the use of the dielectric continuum model (macroscopic ϵ for microscopic screening) and effective mass approximation.

An opposite extreme to the donor electron as regards the spatial extension is a transition metal ion with incomplete d shell or a rare earth ion with incomplete f shell, put as impurity into the ionic crystal. The orbital radius a of d or f electrons participating in the optical transition is much smaller than the lattice constant d of the host crystal. The electronic state of such an impurity ion can be treated, in good approximation, as an isolated ion subject to the crystalline field of lower symmetry originating from surrounding host ions (crystal-field theory). Systematic studies have been made on the optical spectra of intrashell (d $\leftrightarrow$ d, f $\leftrightarrow$ f) and intershell (d $\leftrightarrow$ f) transitions.

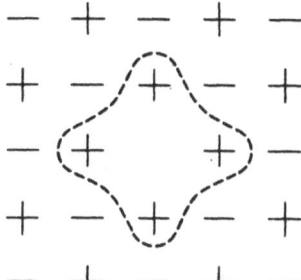

Fig. 7.2. The structure of the F center in alkali halide, of which the electron is extended as shown by the broken line

The intermediate situation $(a \sim d)$ between the two extremes mentioned above may be represented by the F *center* in alkali halide, which is a neutral center consisting of a conduction electron trapped around an anion vacancy (see Fig. 7.2). In fact, the analysis of hyperfine structures in the electron-spin-resonance spectra revealed that the electron is symmetrically extended to the six cations around the anion vacancy. While the stoichiometric alkali halide crystal is optically transparent since the fundamental absorption starts in the ultraviolet region, the crystal containing the F centers has a broad bell-shaped absorption band, the so-called F *band*, in the visible region, the peak energy depending on the host alkali halide. As a result, the crystal shows complementary color as seen in the light, which fact was the origin of the German name: Farbzentrum. The F band corresponds to the electronic transition from the totally symmetric s-like ground state to the triply degenerate p-like states.

In this way, the orbital radius a of a localized electron varies over a wide range from $a \gg d$ to $a \ll d$, depending on the type of imperfection as well as on the host crystal. As will be seen later, the characteristic features of phonon-induced broadening and structures of the absorption spectrum of the localized electron depend sensitively upon its spatial extension.

7.2.2 The Generating Function for the Optical Spectra and Their Moments

Let $H_L = K_L(p) + U_L(q)$ be the Hamiltonian for the lattice vibrations of the crystal containing a lattice imperfection, and $\mathcal{H}_e(p, r; q)$ be the Hamiltonian for the electron localized around the imperfection. r and p denote respectively the coordinate and the momentum of the electron, while multidimensional $q \equiv (q_1, q_2, \ldots)$ and $p \equiv (p_1, p_2, \ldots)$ denote those of the lattice vibrations. $\mathcal{H}_e$ depends upon q since the potential for the electron varies as the atoms around the imperfection displace. The total Hamiltonian of this electron–phonon system is given by $H_{\text{tot}} = \mathcal{H}_e + H_L$.

If an electronic state λ is separated from other electronic states by energy much larger than the phonon energies, one can resort to the adiabatic approximation. With q as adiabatic parameter, one solves the Schrödinger equation for the electron:

$$\mathcal{H}_e(p, r; q)\,\Phi_\lambda(r; q) = \varepsilon_\lambda(q)\,\Phi_\lambda(r; q) \ . \tag{7.2.1}$$

Then, with the use of the adiabatic potential $W_\lambda(q) \equiv \varepsilon_\lambda(q) + U_L(q)$, one solves the Schrödinger equation for the lattice vibrations:

$$H_\lambda \chi_{\lambda n}(q) \equiv [K_L(p) + W_\lambda(q)]\,\chi_{\lambda n}(q) = E_{\lambda n}\chi_{\lambda n}(q) \ , \tag{7.2.2}$$

where n denotes for the vibrational state. The energy and the wave function of the electron–phonon system are then given by $E_{\lambda n}$ and $\Psi_{\lambda n}(r, q) = \Phi_\lambda(r; q) \cdot \chi_{\lambda n}(q)$, respectively. The adiabatic potential $W_\lambda(q)$ and the vibronic levels $E_{\lambda n}$ are shown schematically in Fig. 7.3 for the one-dimensional coordinate q.

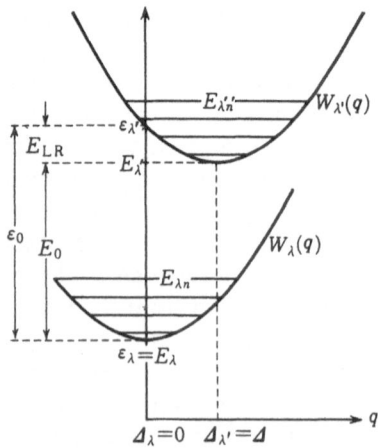

Fig. 7.3. The adiabatic potentials and the vibronic levels for a localized electron

In thermal equilibrium, the ensemble of the system is distributed over vibronic states (λn) with the probabilities $\omega_{\lambda n} \propto \exp(-\beta E_{\lambda n})$. As far as the thermal energy $k_B T = \beta^{-1}$ is much smaller than the energy separation between the ground (λ) and the excited (λ') electronic states, the system is certainly in the ground state, and the optical absorption spectrum corresponding to the electronic transition $\lambda \to \lambda'$ is given, apart from unimportant factors, by

$$F(E) = \sum_n \sum_{n'} w_{\lambda n} |P_{\lambda n, \lambda' n'}|^2 \delta(E - E_{\lambda' n'} + E_{\lambda n}) , \tag{7.2.3}$$

where E is the photon energy and P is the component of the electron dipole moment $- er$ along the direction of polarization of the light.

It is convenient to introduce the *generating function* $f(t)$ which is the Fourier transform of $F(E)$:

$$f(t) \equiv \int_{-\infty}^{+\infty} dE\, F(E) \exp\left(-\frac{i}{\hbar} Et\right) . \tag{7.2.4}$$

Inserting (7.2.3) and making use of the density matrix $\rho_\lambda(\beta) = \exp(-\beta H_\lambda)$ with H_λ defined in (7.2.2), one can rewrite (7.2.4) in terms of the trace operation tr_L over the lattice system alone:

$$f(t) = \mathrm{tr}_L \left[\rho_\lambda(\beta) \left\{ \exp\left(\frac{i}{\hbar} H_\lambda t\right) P_{\lambda \lambda'} \exp\left(-\frac{i}{\hbar} H_{\lambda'} t\right) \right\} P_{\lambda' \lambda} \right] / \mathrm{tr}_L[\rho_\lambda(\beta)]$$

$$= \mathrm{tr}_L \left[\rho_\lambda \left(\beta - \frac{i}{\hbar} t\right) P_{\lambda \lambda'} \rho_{\lambda'} \left(\frac{i}{\hbar} t\right) P_{\lambda' \lambda} \right] / \mathrm{tr}_L[\rho_\lambda(\beta)] . \tag{7.2.5}$$

The expression in the braces is the (λ, λ') element of the Heisenberg representation $P(t)$ of the operator P at time t, and the generating function turns out to be the correlation function of the transition dipole moment: a simple example of fluctuation-dissipation theorem. The last expression is nothing but the Schrödinger representation in which the density matrix, instead of dynamical variables, varies with time. $P_{\lambda \lambda'} \equiv \int dr \Phi_\lambda^*(r; q) P \Phi_{\lambda'}(r; q)$, which in general depends on q, is an operator of the lattice system.

If the generating $f(t)$ is known, the spectrum $F(E)$ is immediately obtained as its inverse Fourier transform. Moreover, the *m*th *moment* of the latter is directly related to the *m*th derivative of the former:

$$\langle E^m \rangle \equiv \int_{-\infty}^{+\infty} E^m F(E) dE \bigg/ \int_{-\infty}^{+\infty} F(E)\, dE \tag{7.2.6}$$

$$= \frac{(i\hbar)^m f^{(m)}(0)}{f(0)} ,$$

as is evident from (7.2.4)

7.2.3 A Model Calculation of the Generating Function

A few simplifications will be introduced to make the calculation of the generating function feasible. In the first place, we take (I) harmonic approximation for the lattice vibrations:

$$H_L = \sum_j \frac{1}{2}(p_j^2 + \omega_j^2 q_j^2) = \sum_j \left(b_j^\dagger b_j + \frac{1}{2}\right)\hbar\omega_j \ , \tag{7.2.7}$$

Here q_j and p_j ($j = 1, 2, \dots$) represent the normal coordinates and their conjugate momenta of the crystal lattice containing an imperfection, while b_j and $b_j^\dagger$ are the phonon annihilation–creation operators of the corresponding modes, defined as in (1.4.7). In the next place, we expand $\mathcal{H}_e(p, r; q)$ in a Taylor series around $q_j = 0$ and take into account up to (II) the linear term:

$$\mathcal{H}_e = H_e(p, r) + H'(r, q) \ , \tag{7.2.8}$$

$$H' = -\sum_j c_j(r)\, q_j = -\sum_j \gamma_j(r)(b_j + b_j^\dagger) \ . \tag{7.2.9}$$

H' is the linear electron–phonon interaction, of which the coefficients are related by $\gamma_j(r) = (\hbar/2\omega_j)^{1/2}\, c_j(r)$. Now we assume that the purely electronic eigenvalue problem has been solved:

$$H_e \phi_\lambda(r) = \varepsilon_\lambda \phi_\lambda(r) \ . \tag{7.2.10}$$

If one expands the electron-annihilation operator as $\Psi(r) = \sum_\lambda a_\lambda \phi_\lambda(r)$, one can also rewrite H_e of (7.2.10) and H' of (7.2.9) in the second quantized form:

$$H_e = \sum_\lambda \varepsilon_\lambda a_\lambda^\dagger a_\lambda \ , \tag{7.2.11}$$

$$H' = -\sum_{\lambda, \lambda'} \gamma_{j\lambda\lambda'} a_\lambda^\dagger a_{\lambda'}(b_j + b_j^\dagger) \ , \tag{7.2.9'}$$

where $\gamma_{j\lambda\lambda'} \equiv \int dr\, \phi_\lambda^*(r)\gamma_j(r)\phi_{\lambda'}(r)$. The interaction Hamiltonian (7.2.9') is of the x^2y form discussed in Sect. 7.1 if a's and b's are replaced by x and y, respectively. In fact, we shall find that the effects of nonlinear interaction as were studied in that section are realized in the electron–phonon system.

Calculating the eigenenergies $\varepsilon_\lambda(q)$ of (7.2.1) up to the second-order perturbation in H', and adding $U_L(q)$, one obtains the adiabatic potential:

$$W_\lambda(q) = \varepsilon_\lambda + \sum_j \left(\frac{1}{2}\omega_j^2 q_j^2 - c_{j\lambda\lambda} q_j\right)$$

$$+ \sum_j \sum_{j'} \left(\sum_{\lambda'(\neq\lambda)} \frac{c_{j\lambda\lambda'} c_{j'\lambda'\lambda}}{\varepsilon_\lambda - \varepsilon_{\lambda'}}\right) q_j q_{j'} \ . \tag{7.2.12}$$

If H' is small enough, one can (III) neglect the second-order perturbation in (7.2.12), to obtain

$$W_\lambda(q) = E_\lambda + \sum_j \frac{1}{2}\omega_j^2(q_j - \Delta_{j\lambda})^2 \tag{7.2.13}$$

where

$$\Delta_{j\lambda} \equiv \frac{c_{j\lambda\lambda}}{\omega_j^2}, \quad E_\lambda = \varepsilon_\lambda - \sum_j \frac{1}{2}\omega_j^2\Delta_{j\lambda}^2. \tag{7.2.14}$$

$(\Delta_{1\lambda}, \Delta_{2\lambda}, \dots)$ represents the equilibrium configuration of the lattice in the electronic state λ where the adiabatic potential $W_\lambda(q)$ takes the minimum value E_λ. In this first-order approximation, the optical spectrum for $(\lambda \leftrightarrow \lambda')$ transition is governed only by the relative displacement of the equilibrium configurations of the two states:

$$\Delta_j = \Delta_{j\lambda'} - \Delta_{j\lambda} = \left(\frac{2}{\hbar\omega_j^3}\right)^{1/2}\gamma_j, \quad \gamma_j \equiv \gamma_{j\lambda'\lambda'} - \gamma_{j\lambda\lambda}. \tag{7.2.15}$$

If one chooses the equilibrium configuration in the state λ as the origin of (q_j) so that $\Delta_{j\lambda} = 0$ in (7.2.13), one can write the adiabatic potential for the state λ' as

$$W_{\lambda'}(q) = E_{\lambda'} + \sum_j \frac{1}{2}\omega_j^2(q_j - \Delta_j)^2, \tag{7.2.16}$$

$$E_0 \equiv E_{\lambda'} - E_\lambda = (\varepsilon_{\lambda'} - \varepsilon_\lambda) - \sum_j \frac{1}{2}\omega_j^2\Delta_j^2 \equiv \varepsilon_0 - E_{LR}, \tag{7.2.17}$$

as shown schematically in Fig. 7.3. The vertical energy difference ε_0 at the origin represents the *optical-excitation energy* in the *Franck–Condon approximation*, while the energy difference E_0 between the minima of the two adiabatic potentials represents the *thermal-excitation energy*. Their difference E_{LR} represents the *lattice-relaxation energy* released after the optical excitation. In the present approximation with the same curvature of $W(q)$ for the two electronic states, E_{LR} is also equal to the lattice relaxation energy released after the optical de-excitation from the relaxed excited state $(\lambda'; q_j = \Delta_j)$ to the ground state λ, and hence, equal to half of the *Stokes shift*—the difference between the peak energies of the absorption and the emission bands.

If we (IV) neglected q dependence of the electronic wave function $\Phi_\lambda(r; q)$, we can take the constant $|P_{\lambda\lambda'}|^2$ out of the operation tr_L in (7.2.5). Since the vibrational Hamiltonians H_λ and $H_{\lambda'}$ of the two electronic states are diagonalized with the common normal coordinates p_j and q_j according to (7.2.13, 16) [the approximation (III) is essential for this], $\text{tr}_L [\dots]$ in (7.2.5) consists of the product over j of the same quantities for one-dimensional harmonic oscillators. Two different ways of calculating it will be shown below.

The density matrix $\rho(\beta) = \exp(-\beta h)$ for a one-dimensional harmonic oscillator with Hamiltonian $h = (p^2 + \omega^2 q^2)/2$ satisfies the *Bloch equation*: $\partial\rho/\partial\beta + h\rho = 0$, which can be written in q representation as

$$\left(\frac{\partial}{\partial\beta} - \frac{\hbar^2}{2}\frac{\partial^2}{\partial q^2} + \frac{\omega^2}{2}q^2\right)(q|\rho(\beta)|\bar{q}) = 0 . \tag{7.2.18}$$

The solution of this differential equation satisfying the initial condition: $\lim_{\beta\to+0}(q|\rho(\beta)|\bar{q}) = \delta(q - \bar{q})$ is given by

$$(q|\rho(\beta)|\bar{q}) = \left(\frac{2\pi\hbar}{\omega}\sinh(\beta\hbar\omega)\right)^{-1/2}$$

$$\exp\left[-\left(\frac{\omega}{4\hbar}\tanh\frac{\beta\hbar\omega}{2}\right)(q + \bar{q})^2 - \left(\frac{\omega}{4\hbar}\coth\frac{\beta\hbar\omega}{2}\right)(q - \bar{q})^2\right] \tag{7.2.19}$$

as is easily confirmed. Noting that the equilibrium position of the jth oscillator is displaced by Δ_j between λ' and λ states, and that $\text{tr}_L[\ldots]$ in (7.2.5) means $\int dq(q|\ldots|q)$, we finally obtain

$$f(t) = |P_{\lambda'\lambda}|^2\exp\left[-\frac{i}{\hbar}E_0t - S + S_+(t) + S_-(t)\right] \tag{7.2.20}$$

where

$$S_\pm(t) \equiv \int_0^\infty dE' s(E')\left[\frac{N(E') + 1}{N(E')}\right]\exp\left(\mp\frac{i}{\hbar}E't\right) , \tag{7.2.21}$$

$$S \equiv S_+(0) + S_-(0) = \int_0^\infty dE' s(E')[2N(E') + 1] , \tag{7.2.22}$$

$$s(E') \equiv \sum_j \frac{1}{2\hbar}\omega_j\Delta_j^2\delta(E' - \hbar\omega_j) = \sum_j \frac{\gamma_j^2}{\hbar^2\omega_j^2}\delta(E' - \hbar\omega_j) . \tag{7.2.23}$$

$N(E') = [\exp(\beta E') - 1]^{-1}$ denotes the average number of phonons per mode of energy E' in thermal equilibrium.

The second method resorts to the operator algebra for b and $b^\dagger$. With the use of the second quantized expressions for the vibrational Hamiltonians (with suitable choice of the origin of energy)

$$H_\lambda = \sum_j \hbar\omega_j b_j^\dagger b_j, \quad H_{\lambda'} = E_0 + \sum_j \hbar\omega_j\left(b_j^\dagger - \frac{\gamma_j}{\hbar\omega_j}\right)\left(b_j - \frac{\gamma_j}{\hbar\omega_j}\right) , \tag{7.2.24}$$

we must calculate

$$f(t) = |P_{\lambda'\lambda}|^2 \left\langle \exp\left(\frac{i}{\hbar} H_\lambda t\right) \exp\left(-\frac{i}{\hbar} H_{\lambda'} t\right) \right\rangle_L \tag{7.2.25}$$

where $\langle \ldots \rangle_L$ denotes the statistical average over the vibrational states in the ground electronic state. Similarly to (6.3.2), we introduce the unitary operator

$$U \equiv \exp\left[\sum_J \frac{\gamma_J}{\hbar\omega_J} (b_J^\dagger - b_J) \right], \tag{7.2.26}$$

with which b_J is transformed to $U^{-1}b_J U = b_J + \gamma_J/\hbar\omega_J$ and $H_{\lambda'}$ to $U^{-1}H_{\lambda'}U = E_0 + H_\lambda$. With the use of the Heisenberg representation $U(t) \equiv \exp(iH_\lambda t/\hbar)U \cdot \exp(-iH_\lambda t/\hbar)$, one obtains

$$f(t) = |P_{\lambda'\lambda}|^2 \exp\left(-\frac{i}{\hbar} E_0 t\right) \langle U(t)U^{-1}(0)\rangle_L \tag{7.2.27}$$

$\langle \ldots \rangle_L$ in (7.2.27) is of the same form as (1.5.3). Making use of (1.5.12) and (1.5.10), one immediately obtains (7.2.20).

In the Mössbauer effect described in Chap. 1, we were concerned with the way in which the recoil kinetic energy and momentum of the unstable nucleus as it emits the γ-rays are transferred to the crystal lattice. In the present problem, the optical transition of the localized electron results in the sudden change of the gradient of the adiabatic potential which is nothing but the generalized force (in the multidimensional q space) upon the crystal lattice. The optical spectrum with phonon sidebands describes how the crystal lattice disposes of this sudden force as vibrational energy.

7.2.4 Phonon Sidebands and Zero-Phonon Line

Let us normalize the total intensity of the spectrum to unity $[f(0) = |P_{\lambda\lambda'}|^2 = 1]$ and expand the generating function (7.2.20) in a power series of $S_\pm(t)$. Then, the term $S_+(t)^{n'}S_-(t)^n$ represents the $(n' + n + 1)$th perturbation process in which n' phonons are emitted and n phonons absorbed simultaneously with the optical transition of the electron. In fact, the Fourier transform of this term gives the phonon sideband consisting of the spectral lines at $E = E_0 + \sum^{n'} \hbar\omega_{J'} - \sum^n \hbar\omega_J$, each with intensity given by the product over the $(n' + n)$ participating modes of dimensionless coupling constant $(\gamma_J/\hbar\omega_J)^2$ times $[(N(\hbar\omega_J) + 1)$ or $N(\hbar\omega_J)]$. In particular, the term with $n = n' = 0$ gives the zero-phonon line at $E = E_0$, with intensity $\exp(-S)$ which is a decreasing function of temperature according to (7.2.22). The remaining intensity $1 - \exp(-S)$ is distributed over the phonon sidebands.

At absolute zero of temperature where $N(\hbar\omega) = 0$, only the phonon emission sidebands appear on the high energy side of the zero-phonon line. With the position of this line being chosen as the origin of energy, the one-phonon sideband is given by $\exp(-S) \cdot s(E')$, the two-phonon sideband by $\exp(-S)/2! \cdot \int s(E'')s(E' - E'')dE''$, and the n'-phonon sideband by $\exp(-S)/n'!$ times the n'th-order convolution of $s(E')$, as shown schematically in Fig. 7.4. The integrated intensities of these sidebands obey the Poisson distribution $\exp(-S) \cdot S^{n'}/n'!$, and hence, the average number of phonons emitted is given by S. It should be noted in passing that the overlap integral $(\chi_{\lambda 0}, \chi_{\lambda' 0})$ of the zero-point vibrational states of the two electronic states λ and λ' is equal to $\exp(-S/2)$ as is evident from (7.2.3). In this way, all information on the line shape of the absorption spectrum, within the approximations (I) through (IV) mentioned above, is contained in S and $s(E')$, which will be called the *coupling strength* and the *spectral coupling*, respectively.

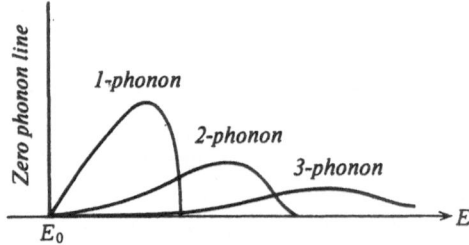

Fig. 7.4. The zero-phonon line and the phonon sidebands

The optical emission spectrum corresponding to the electronic de-excitation $\lambda' \to \lambda$ can be calculated by replacing E by $-E$ and interchanging $\lambda n \leftrightarrow \lambda' n'$ in (7.2.3), provided that the system is initially in thermal equilibrium as regards the distribution over the vibrational states n'. Equations (7.2.4, 20) remain unchanged if $\mp$ in the exponent of (7.2.21) is replaced by $\pm$. $s(E')$ is invariant against the interchange $\lambda \leftrightarrow \lambda'$ since it depends on the squares of the relative displacements $(\Delta_{j\lambda'} - \Delta_{j\lambda})$. Consequently, the emission and absorption spectra are the mirror reflection of each other against the common zero-phonon line at $E - E_0$. It should be noted that this mirror symmetry holds only under the assumptions (I) through (IV).

7.2.5 Strong-Coupling Limit and Configuration-Coordinate Model

If the coupling strength S is large compared to unity, the intensity of the zero-phonon line, $\exp(-S)$, is extremely small while a host of higher-order sidebands appear overlapping with each other. We then expect a broad and smooth line shape. With this in mind, we expand (7.2.21) in powers of t, put it in (7.2.20) and make use of (7.2.17, 22, 23), to obtain

$$f(t) = |P_{\lambda'\lambda}|^2 \exp\left[-\frac{i}{\hbar}\varepsilon_0 t - \frac{D^2 t^2}{2\hbar^2} + O(t^3)\right],$$

(7.2.28)

$$D^2 \equiv \int_0^\infty dE's(E')[2N(E') + 1]E'^2 \; . \tag{7.2.29}$$

Neglecting t^3 and higher-order terms in the exponent of (7.2.28) under the assumption that t is much smaller than the reciprocal of the average phonon frequency ω, we obtain, after Fourier transformation, the Gaussian absorption band:

$$F(E) = \frac{|P_{\lambda'\lambda}|^2}{\sqrt{2\pi}D} \exp\left[-\frac{(E - \varepsilon_0)^2}{2D^2}\right] \; . \tag{7.2.30}$$

Since (7.2.28) decays out at $t \gtrsim \hbar/D$, the neglect of higher powers of t is justified when D is much larger than the average phonon energy $\hbar\bar{\omega}$. The width D is proportional to $\sqrt{T}$ at high temperatures: $k_B T \gg \hbar\bar{\omega}$, as is easily seen from (7.2. 29).

The result (7.2.30) for the strong-coupling case can be derived much more easily with the use of the *configuration-coordinate model*. Let us rescale the configuration coordinates as $\omega_j q_j = q'_j (j = 1, 2, \ldots)$, and from this set we introduce a new set of coordinates $Q_l(l = 1, 2, \ldots)$ by an orthogonal transformation such that Q_1, in particular, is given by

$$cQ_1 = \sum_j \omega_j \Delta_j q'_j \; , \quad c^2 \equiv \sum_j \omega_j^2 \Delta_j^2 \; . \tag{7.2.31}$$

The adiabatic potentials for the excited and the ground electronic states are then written as

$$W_{\lambda'} = \varepsilon_\lambda + \sum_i \frac{1}{2} Q_i^2 - cQ_1, \quad W_\lambda = \varepsilon_\lambda + \sum_i \frac{1}{2} Q_i^2 \; .$$

According to the Franck–Condon approximation in which the lattice coordinates are considered to be unchanged during the optical transition of the electron, the absorption spectrum is given by

$$F(E) \propto \prod_j \int dq_j \exp[-\beta W_\lambda(q)]\delta[W_{\lambda'}(q) - W_\lambda(q) - E]$$

$$\propto \int dQ_1 \exp\left(-\frac{1}{2}\beta Q_1^2\right) \delta(\varepsilon_0 - cQ_1 - E) \tag{7.2.32}$$

which is nothing but the high-temperature limit of (7.2.30). The situation is shown schematically in Fig. 7.5.

The merit of introducing Q_1 is that we have only to deal with the one dimensional configuration coordinate Q_1 (instead of multidimensional $q_1, q_2 \ldots$) as far as the classical approximation for the absorption spectrum is concerned. Q_1 is called the *interaction mode*. It is a mode as localized as the electron is, but is not a normal mode. After the electronic excitation by a pulse of photons

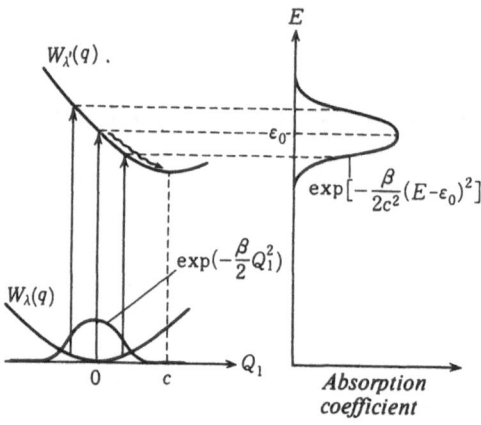

Fig. 7.5. The configuration-coordinate model with interaction mode Q_1, the Franck–Condon principle, and the absorption line shape

with energy ε_0 at $t = 0$ and $Q_1 = 0$, the classical-mechanical motion of Q_1 is given by

$$Q_1(t) = \sum_j \frac{1}{c}\,\omega_j^2 \Delta_j^2 (1 - \cos \omega_j t)$$

since each normal coordinate oscillates as $q_j(t) = \Delta_j(1 - \cos \omega_j t)$. If there are no localized modes among the normal coordinates q_j of the imperfect crystal, (ω_j) form a continuous spectrum, and hence, $Q_1(t)$ will finally approach the new equilibrium position c (relaxed excited state), since the sum of the cosine terms vanishes as $t \to \infty$ because of the dephasing. This means that the energy of the Q_1 mode is finally transferred to the remaining modes $Q_2, Q_3 \ldots$, namely, that the heat released by optical excitation is dispersed all over the crystal. The energy dissipation takes place only because we have projected the whole q space onto the configuration coordinate space of the interaction mode which is not a normal mode.

The one- or few-dimensional configuration-coordinate model, such as is depicted in Fig. 7.5, is very often used, not only for the spectral analysis but also for the dynamical study of molecular or lattice relaxation after the electronic transition. The abscissa in such a model should be interpreted as the interaction mode rather than a representative normal mode, especially when we are concerned with large molecules or macroscopic systems.

7.2.6 A Model Calculation of Coupling Strength

It is of interest to see how the coupling strength S characterizing the spectral line shape depends on the spatial extention a of the localized electron, for, a extends over a wide range depending on the imperfection as well as on the host crystal, as mentioned in Sect. 7.2.1.

The existence of lattice imperfection causes local perturbation of normal modes of lattice vibrations which otherwise would be the plane-wave modes specificed by the wave vector q. Neglecting this perturbation for simplicity, we take into account the imperfection only as a potential source [to be included in H_e of (7.2.8)] localizing the electron. The wave function of the localized electron (hole) can be expressed as a linear combination of the Bloch functions of the conduction (valence) band if its energy level is not far from the bottom (top) of the band. Then it suffices to consider the interaction between the conduction electron and the unperturbed lattice for H' of (7.2.8). It can be written in the form of (6.1.15), namely, (7.2.9) is to be rewritten as

$$ - \sum_q [\gamma_q(r) \, b_q + \gamma_q^*(r) \, b_q^\dagger] $$

with

$$ \gamma_q(r) = - V_q \exp(iq \cdot r) \tag{7.2.33} $$

as far as the complex running-wave modes (instead of the real standing-wave modes) are made use of.

For the polar optical modes of lattice vibrations in ionic crystal, the coupling coefficient V_q is given by (6.1.17). For the acoustic modes of long wavelengths which are nothing but the elastic waves, we consider an elastic continuum with displacement $\xi(r)$ at the position r, and hence, with strain tensor $\epsilon_{ij}(r) = (\partial \xi_i / \partial \gamma_j + \partial \xi_j / \partial \gamma_i)/2$ $(i, j = 1, 2, 3)$. The bottom energy of the conduction band is then subject to a change, which is a linear function

$$ \delta E_c(r) = \sum_{i,j} E_{ij} \epsilon_{ij}(r) \tag{7.2.34} $$

if the strain is small. Making use of (1.4.15, 16), ϵ_{ij} can be expressed as a linear function of b_q and $b_q^\dagger$. Equation (7.2.34) can be regarded as a potential to which the conduction electron is subject, namely, as the electron–phonon interaction H' (see Fig. 7.6). In order that this picture be meaningful, $\epsilon_{ij}(r)$ must vary with r so slowly that the "local band" is well definable, namely, (7.2.34) can be used only for long-wavelength components ($q \ll$ reciprocal lattice). Equation (7.2.34) or its coefficients E_{ij} are called the *deformation potential*.

In cubic crystals and with the nondegenerate conduction-band bottom situated at $k = 0$, we have $E_{11} = E_{22} = E_{33} = E_d$ and $E_{23} = E_{31} = E_{12} = 0$. The interaction can be written as

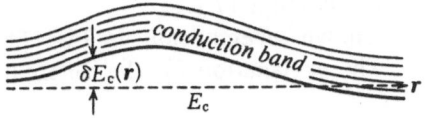

Fig. 7.6. Deformation potential

$$H' = E_d \Delta(r) \tag{7.2.35}$$

where $\Delta(r) = \text{div}\xi(r) = \epsilon_{11} + \epsilon_{22} + \epsilon_{33}$ is the local dilation of the lattice. If we assume furthermore that the acoustic modes consist of purely longitudinal and purely transverse waves, we immediately find that only the longitudinal modes have nonvanishing coupling, with coefficient:

$$V_q = -i\left(\frac{\hbar E_d^2}{2NMc_s}\right)^{1/2} q^{1/2} . \tag{7.2.36}$$

To calculate the coupling strength S for the optical transition of the localized electron, we assume a 1s-like wave function $\phi_\lambda = (\pi R_s^3)^{-1/2} \exp(-r/R_s)$ for the ground state and 2p-like wave function $\phi_{\lambda'} = (\pi R_p^5)^{-1/2} r \cos\theta \exp(-r/R_p)$ for the excited state. We consider the limiting cases of (I) small radii ($R_s < R_p \ll d$) and (II) large radii ($d \ll R_s < R_p$). In calculating $\gamma_q = \gamma_{q\lambda'\lambda'} - \gamma_{q\lambda\lambda}$ with the use of (7.2.15, 33), (6.1.17) or (7.2.36), we immediately find that the state with radius R closer to the lattice constant d makes the larger contribution. Keeping this larger contribution alone [from λ' in (I) and λ in (II)] for simplicity, we calculate the coupling strength S at $T = 0$ K for the acoustic and optical modes with the use of (7.2.22, 23). The results are given, in terms of the mean-square-root radii ($a_s = \sqrt{3}R_s$, $a_p = \sqrt{15/2}R_p$), as follows:

$$S_{ac} = \frac{E_d^2 d}{\hbar M c_s^3} \times \begin{cases} 0.46\left(\dfrac{a_p}{d}\right)^4 & \text{(I)} \\[2ex] 0.051\left(\dfrac{a_s}{d}\right)^{-2} & \text{(II)} \end{cases}$$

$$S_{op} = \frac{e^2}{\hbar\omega_l d}\left(\frac{1}{\epsilon_\infty} - \frac{1}{\epsilon_0}\right) \times \begin{cases} 1.8\left(\dfrac{a_p}{d}\right)^4 & \text{(I)} \\[2ex] 0.54\left(\dfrac{a_s}{d}\right)^{-1} & \text{(II)} \end{cases}$$

Putting typical values for the material constants, we obtain the result shown schematically in Fig. 7.7. The reason why S becomes greatest at $a \sim d$ is as follows: the localized electron interacts most effectively with those phonons whose wavelength $2\pi/q$ is comparable to the orbital radius a, whereas the phonon frequency spectrum has its peak near the zone boundary ($q \sim \pi/d$).

If the initial and final states have not much difference in a (as is the case for the intrashell transition mentioned later), S is expected to be smaller than the above estimation because of the significant cancellation in (7.2.15).

The calculated spectral coupling $s(E')$ is shown schematically in Fig. 7.8 for the cases (I) and (II) mentioned above. The explanation for the different behaviors in the two cases is left for the reader.

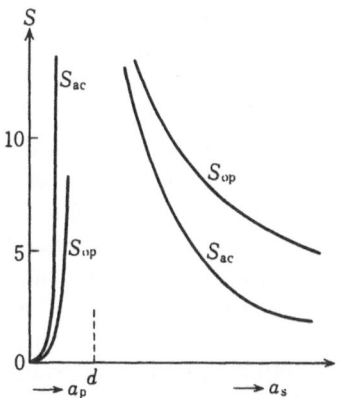

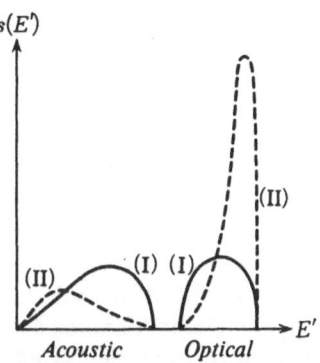

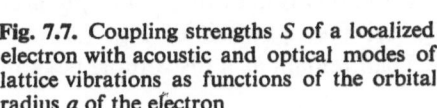

Fig. 7.7. Coupling strengths S of a localized electron with acoustic and optical modes of lattice vibrations as functions of the orbital radius a of the electron

Fig. 7.8. Spectral coupling of a localized electron in the phonon field, with solid and broken lines representing the cases of small and large radius ($a \lessgtr d$), respectrvely

We will now show typical examples of the observed spectra corresponding to (I), (II), and intermediate cases. The emission spectrum of MgO: V^{2+} is shown in Fig. 7.9 where the abscissa is taken towards decreasing photon energy. It originates from the d–d transition within the incomplete 3d shell of V^{2+}, and is representative of case (I) except that it is not an electric-dipole-allowed transition but for the electron–phonon interaction. On the low-energy side of the sharp zero-phonon line, there appears one phonon sideband consisting of two main peaks due to acoustic and optical modes. The higher-order sidebands are much weaker, namely, $S \ll 1$. This is partly due to the small orbital radius ($a \ll d$) of the d electron, and partly due to the intrashell transition accompanied by little change in charge distribution. In fact, in the case of (4f) $\rightleftarrows$ (5d) transitions in rare earth ions in alkali halides, multiple phonon sidebands appear with S as large as 5.

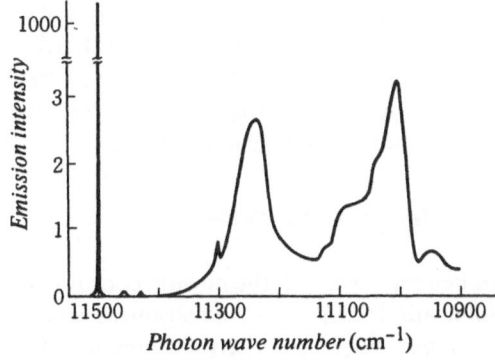

Fig. 7.9. Emission spectra of V^{2+} ion in MgO. [M.D. Sturge: Phys. Rev. **130**, 639 (1963)]

As an example of case (II), we show in Fig. 7.10 the absorption and emission spectra of AgBr: I⁻, which correspond to the creation and the annihilation of a *bound exciton* at isoelectronic impurity I⁻. In the lowest excited state with one electron–hole pair, the hole is bound with moderately large radius ($a > d$) around the I⁻ ion because of the electron affinity difference between I and Br, and the electron is bound by the Coulomb potential of this hole center, with larger radius because of the large dielectric constant of the host crystal. (In the one electron picture, λ and λ' correspond to the wave functions of the bound hole and the bound electron, respectively.) In Fig. 7.10, the one-phonon sideband of the emission spectra on the low-energy side of the zero-phonon line (denoted as "0") consists of a weak, broad acoustic phonon band and a strong, sharp optical phonon band (denoted as "1") while the multiple phonon sidebands have apparently the form of $\hbar\omega_{ac} + n\hbar\omega_{op}$ with repetition of the optical phonon only. This seems partly because $S_{op} > S_{ac}$ (see the right half of Fig. 7.7) and partly because only the sharp peak in $s(E')$ can remain as a discernible peak in the convolution.

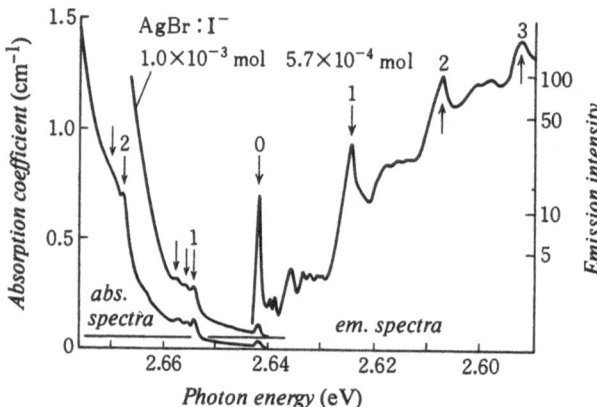

Fig. 7.10. The absorption and emission spectra of AgBr: I⁻ at $T = 2$ K, with the most prominent equidistant multiphonon structures being numbered starting with zero-phonon line common to both spectra. [H. Kanzaki, S. Sakuraki: J. Phys. Soc. Jpn. **27**, 109 (1969)].

The *F* center whose structure was shown in Fig. 7.2 belongs to the intermediate case: $a \sim d$, and is expected, according to Fig. 7.7, to have $S = S_{op} + S_{ac}$ as large as several tens—the strong-coupling situation described in Sect. 7.2.5. In fact, as shown in Fig. 7.11, the F absorption and emission spectra are broad Gaussian bands with $\sqrt{T}$-type widths at high temperature, separated by large Stokes shift, without any observable zero-phonon line between them.

7.2.7 The Effect of Curvature Difference in the Adiabatic Potentials

If the second-order term in (7.2.12) is taken into account, the normal coordinates (directions of principal axes of the quadratic form) as well as eigenfrequencies become dependent on the electronic states. Then, the zero-phonon line has

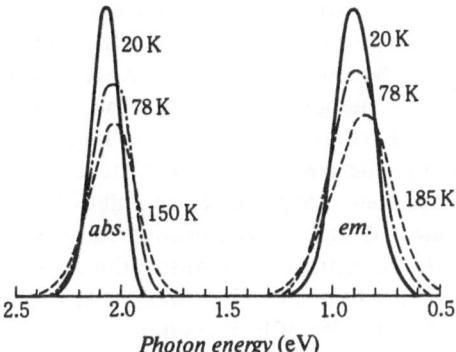

Fig. 7.11. Temperature-dependent absorption and emission spectra of the F center in KBr. [W. Gebhardt, H. Kühnert: Phys. Lett. **11**, 15 (1964)]

finite width at finite temperature, and the mirror-reflection relationships between absorption and emission spectra break down. A brief remark will be made here on the latter effect as an underplot of final-state interaction described in Sect. 7.4.

Let us consider a one-dimensional configurate coordinate q, with two adiabatic potentials with different equilibrium positions and different curvatures as shown schematically in Fig. 7.12a. At $T = 0$ K, the multiphonon lines in the absorption spectra are spaced with frequency ω' of the excited electronic state while those in the emission spectra are spaced with ω of the ground state, as shown in Fig. 7.12b—the break down of the mirror reflection rule. In Fig. 7.10, the energy interval $\hbar\omega'$ of the optical phonon sidebands (numbered as 1, 2, ... towards the left) in the absorption spectra is about 20% smaller than that of the corresponding structures (1, 2, ... towards the right) in the emission spectra which is approximately equal to $\hbar\omega_l$ of the longitudinal optical phonon of the host crystal.

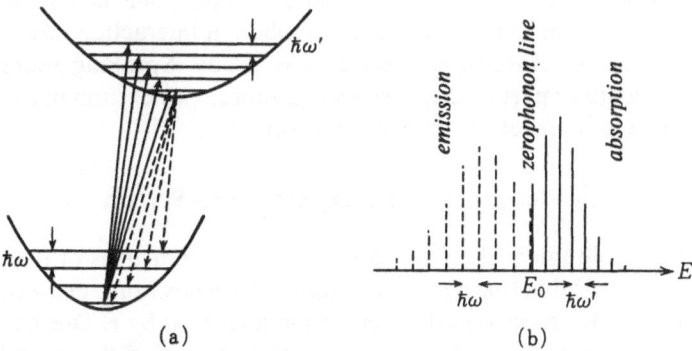

Fig. 7.12. a,b. The absorption and emission spectra at $T = 0$ K of a localized electron, with different curvatures of the adiabatic potentials between the ground and the excited states, and hence, with broken symmetry of mirror reflection

The lowest excited state is usually followed, within a range of energy smaller than its distance from the ground state, by a host of higher excited (and/or ionized) states whose contribution to the frequency tensor in (7.2.12) is predominantly negative because of small negative energy denominators. In contrast, the second-order term is much smaller for the ground state because of its greater distance from the excited states. Therefore, the frequency of the main (most strongly coupled) mode in the lowest excited state is expected to be usually lower than the corresponding frequency in the ground state. This is in conformity with a number of observations. For instance, the structures in the absorption spectra of Fig. 7.10 with interval $\hbar\omega' \approx 0.8\ \hbar\omega_l$ seem to be due to a localized mode which is split off from the continuum of the optical modes of host crystal through the second-order term of (7.2.12) as local perturbation. A similar effect is found in the intrinsic exciton absorption, too, as will be mentioned in Sect. 7.4.1.

7.3 Exciton-Phonon Interaction and Fundamental Absorption Spectra

We have studied the effect of electron–phonon interaction on the optical absorption spectra of a localized electron in an imperfect crystal. How about the fundamental absorption spectra, namely, when a host electron is excited? An important difference is that equivalent localized electron can be found at any host atom and the excitation energy is transferred from atom to atom through the interatomic interaction—the exciton described in Sect. 2.3. The exciton also interacts with the surrounding lattice as the localized excitation (Sect. 7.2) does; however, this instantaneous interaction will more or less be averaged out because of the translational motion of the exciton, resulting in the motional narrowing of the overall width of the absorption spectra and in the motional reduction of the phonon sideband intensities.

Because of its electrical neutrality and short lifetime, the exciton is more difficult than the electron or the hole to observe by transport phenomena. On the other hand, one can observe the dynamics of exciton–phonon interaction rather directly through the exciton absorption spectra as a probe. Speaking more mathematically, the imaginary part of the Green's function for an exciton in the phonon field gives the line shape of the absorption spectra.

7.3.1 The Generating Function for the Fundamental Absorption Spectra

Let H_{tot} be the total Hamiltonian of an insulating crystal. The degrees of freedom consist of all the atomic displacements from equilibrium positions, denoted en bloc by q, and all the electronic coordinates, denoted en bloc by r. One can use the adiabatic approximation for the ground electronic state of the crystal on the assumption that its energy separation from the lowest excited state, namely, the energy of the exciton, is much larger than the phonon energies. Denoting the kinetic energy of the atoms by K_L, one solves the eigenvalue problem of $(H_{tot} - K_L)$ for the electrons with q being fixed as adiabatic parameters; the lowest eigenvalue and the corresponding eigenfunction are denoted as $U_L(q)$ and $\Psi_g(r; q)$, respectively. $U_L(q)$ is the adiabatic potential for the ground electronic state of the crystal, and hence, $H_L = K_L + U_L$ is the Hamiltonian for the lattice vibrations. The remainder $\mathcal{H}_e(r, q)$, defined by

$$H_{tot} = \mathcal{H}_e + H_L ,\tag{7.3.1}$$

satisfies the equation: $\mathcal{H}_e \Psi_g = [(H_{tot} - K_L) - U_L(q)] \Psi_g = 0$.

As is evident from the definition (7.3.1), the Hamiltonian $\mathcal{H}_e(r, q)$ represents the electronic excitation energy for fixed q. For the rigid lattice: $q = 0$, the low-lying excited states are the one-electron excitations denoted by $|\lambda K\rangle$ in Sect. 2.3. The excitation energy $E_{\lambda K}$, namely, the nonvanishing eigenvalue of $\mathcal{H}_e(r, 0)$, will be denoted by $\varepsilon_{\lambda K}$ in the present section. λ is the quantum number for the electron–hole relative motion, and consist of discrete spectra of the bound states (exciton) and continuous spectra of the ionized states. K is the wave vector of the translational motion of the pair. $\varepsilon_{\lambda K}$ is a continuous function of K, forming in particular the exciton band for a discrete λ. The level scheme of $\varepsilon_{\lambda K}$ is shown schematically in Fig. 7.13.

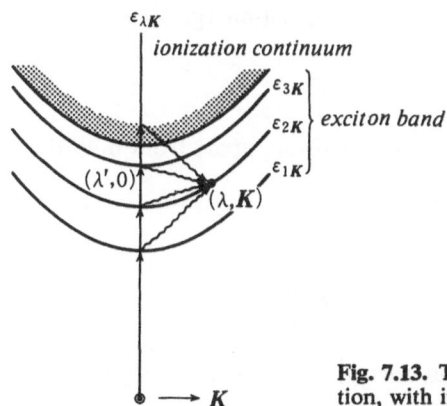

Fig. 7.13. The energy-level scheme for one-electron excitation, with indirect transition channels being shown by a set of successive solid and wavy arrows

Writing $\mathcal{H}_e(r, q)$ as the sum of $\mathcal{H}_e(r, 0) \equiv H_e$ and the remainder H', and projecting them on the subspace of one electron excitation, one can put

$$H_e = \sum_{\lambda K} |\lambda K\rangle \varepsilon_{\lambda K} \langle \lambda K| ,\tag{7.3.2}$$

$$H' = \sum_{\lambda K} \sum_{\lambda' K'} |\lambda K\rangle H'_{\lambda K, \lambda' K'} \langle \lambda' K'| .\tag{7.3.3}$$

To sum up, the Hamiltonian for the subspace of one-electron excitation can be written in the form of the interacting electron (exciton)–phonon system

$$H = H_e + H_L + H' \equiv H_0 + H' , \qquad (7.3.4)$$

while that for the ground state is simply given by H_L.

Let E_{Ln} and $\langle q | n \rangle \equiv \chi_n(q)$ be respectively the eigenvalues and eigenfunctions of H_L, then the total wave function for the ground electronic state is given by $\Psi_g(r; q) \cdot \chi_n(q)$ in the adiabatic approximation. Because of the q dependence of $\Psi_g(r; q)$, H_{tot} has nonadiabatic matrix elements, originating from K_L, between the ground and excited electronic states. However, they are usually much smaller than the energy difference ($\approx \varepsilon_{\lambda K}$) and have no significant effect. Neglecting the q dependence of Ψ_g for this reason, one can denote Ψ_g by the electronic eigenvector $|g\rangle$, and the eigenvectors of the total system in the ground electronic state by direct products: $|g\rangle |n\rangle$.

With the same notation, the subspace of one electron excitation is spanned by direct product vectors: $|\lambda K n \rangle \equiv |\lambda K \rangle |n\rangle$. Let E_j and $|j\rangle$ be the eigenvalues and the eigenvectors, respectively, of H in (7.3.4). $|j\rangle$ can be written as linear combination of $|\lambda K n\rangle$.

Under the neglect of the photon wave vector, the optical transition from $|g\rangle$ is allowed only to those excited states with $K = 0$, namely, to $|\lambda 0\rangle$, due to the wave vector conservation. Denoting this electronic matrix element by $P_{g\lambda}$, one can write the total matrix element for the transition $|gn\rangle \rightarrow |j\rangle$ as

$$P_{gn,j} = \sum_\lambda P_{g\lambda} \langle \lambda 0 n | j \rangle .$$

The fundamental absorption spectra can then be written, apart from unimportant factors, as

$$\begin{aligned} F(E) &= \sum_{n,j} w_n | P_{gn,j} |^2 \delta(E - E_j + E_{Ln}) \\ &= \sum_{nj\lambda\lambda'} w_n P_{g\lambda} \langle \lambda 0 n | j \rangle \delta(E - E_j + E_{Ln}) \langle j | \lambda' 0 n \rangle P_{\lambda' g} , \end{aligned} \qquad (7.3.5)$$

which is of the same form as (7.2.3); the corresponding generating function [see (7.2.4) for definition] is given by

$$\begin{aligned} f(t) &= \sum_{n\lambda\lambda'} w_n P_{g\lambda} \langle \lambda 0 n | \exp\left(\frac{i}{\hbar} H_L t\right) \exp\left(-\frac{i}{\hbar} H t\right) | \lambda' 0 n \rangle P_{\lambda' g} \\ &= \sum_{\lambda\lambda'} P_{g\lambda} \langle\!\langle \lambda 0 | \exp\left(\frac{i}{\hbar} H_L t\right) \exp\left(-\frac{i}{\hbar} H t\right) | \lambda' 0 \rangle\!\rangle_L P_{\lambda' g} \end{aligned} \qquad (7.3.6)$$

which is again of the same form as (7.2.5 or 25). Here, $\langle\!\langle \lambda 0 | \ldots | \lambda' 0 \rangle\!\rangle_L$ means the statistical average over vibrational states (E_{Ln}) of electronic matrix element $\langle \lambda 0 | \ldots | \lambda' 0 \rangle$.

Expanding the matrix element $H'_{\lambda K ,\lambda' K'}$ of the electron–phonon interaction (7.3.3) in powers of the lattice displacements and keeping only the linear term, one can put

$$H'_{\lambda K, \lambda' K'} = V_{\lambda K \ \lambda' K'}(b^{\dagger}_{-K+K'} + b_{K-K'}) \tag{7.3.7}$$

with the use of phonon annihilation–creation operators b_q and $b^{\dagger}_q$. The relations: $V_{\lambda K,\lambda' K'} = V_{\lambda' K',\lambda K}^*$ must hold for H' to be Hermitian. One can put $H_L = \Sigma_q \hbar\omega_q b^{\dagger}_q b_q$ in the harmonic approximation.

7.3.2 Spectral Narrowing due to the Translational Motion of the Exciton

The host excitation is different from the localized excitation in that the final state of the former—exciton—has the translational degree of freedom, K. To study this effect upon the spectral shape, let us consider a system with a single exciton band, omitting the suffix λ. Inserting $H_L = - H_e + H_0$ into (7.3.6) and making use of (7.3.2), one obtains

$$f(t) = |P|^2 \exp\left(-\frac{i}{\hbar} \varepsilon_0 t\right)\langle\!\langle 0| U(t) |0\rangle\!\rangle_L \tag{7.3.8}$$

where $U(t)$ is a unitary operator defined by

$$U(t) \equiv \exp\left(\frac{i}{\hbar} H_0 t\right) \exp\left(-\frac{i}{\hbar}(H_0 + H')t\right) . \tag{7.3.9}$$

Consider the system to be photo-excited from $|gn\rangle$ to $|0n\rangle$ at $t = 0$ (note that the phonon state n is not changed since the photon is assumed to interact directly with electrons only). The system will thereafter evolve with the Hamiltonian (7.3.4), namely, the wave function at time t will be given by $\exp[- i(H_0 + H')t/\hbar] |0n\rangle$. But for the interaction H', the system would stay in the same state except for the change in the phase factor: $\exp(- iH_0 t/\hbar)|0n\rangle = \exp[-i(\varepsilon_0 + E_{Ln})t/\hbar]|0n\rangle$. Therefore, $\langle 0n| U(t)|0n\rangle$ represents the probability amplitude that the exciton–phonon system, having started from $|0n\rangle$ at $t = 0$, is found in the same state at t. Similarly, $\langle Kn' | U(t)|0n\rangle$ represents the probability amplitude that the same system is found in the state $| Kn' \rangle$. In this way, the unitary operator $U(t)$ describes the scattering of the exciton by phonons.

From (7.3.9) follows the differential equation

$$\frac{dU(t)}{dt} = -\frac{i}{\hbar} H'(t)U(t) \tag{7.3.10}$$

where $H'(t)$ is the Heisenberg motion of the perturbation H' driven by the nonperturbed Hamiltonian $H_0 = H_e + H_L$:

$$H'(t) \equiv \exp\left(\frac{i}{\hbar} H_0 t\right) H' \exp\left(-\frac{i}{\hbar} H_0 t\right) . \tag{7.3.11}$$

Integrating (7.3.10) over a very short time interval (t_i, t_{i+1}) and neglecting the terms of $O[(t_{i+1} - t_i)^2]$, one obtains

$$U(t_{i+1}) \approx U(t_i) + \left(-\frac{i}{\hbar}\right)(t_{i+1} - t_i) H'(t_i)U(t)$$

$$\approx \exp\left[-\frac{i}{\hbar} \int_{t_i}^{t_{i+1}} dt' H'(t')\right] U(t_i) . \tag{7.3.12}$$

Iterating this infinitesimal integration from 0[when $U(0) = 1$] to t with due care being paid to the noncommutativity of the operators, one obtains symbolically

$$U(t) = T \exp\left[-\frac{i}{\hbar} \int_0^t dt' H'(t')\right] \tag{7.3.13}$$

with the use of the Wick's T product introduced in Sect. 1.8. By definition, (7.3.13) is a product of the exponentials of the type given in (7.3.12) over the whole time interval $(0, t)$, ordered in such a way that time increases towards the left.

Equation (7.3.10) can also be solved by repeated integration and substitution, as follows:

$$U(t) = 1 + \left(-\frac{i}{\hbar}\right) \int_0^t dt' H'(t')U(t')$$

$$= 1 + \left(-\frac{i}{\hbar}\right) \int_0^t dt_1 H'(t_1) + \left(-\frac{i}{\hbar}\right)^2 \int_0^t dt_1 \int_0^{t_1} dt_2 H'(t_1)H'(t_2) + \dots .$$

$$\tag{7.3.14}$$

In the nth-order term, the product of n $H'(t_j)$'s is ordered in such a way that time is increasing towards the left: $H'(t_1)H'(t_2) \dots H'(t_n)$ with $t_1 > t_2 \dots > t_n$. To extend the region of integration to the n-dimensional cube: $0 < t_j < t$ $(j = 1, 2, \dots, n)$ after putting Wick's T on this product is equivalent to $n!$ times repetititon of the original integration. The nth-order term in (7.3.14) can thus be written as

$$\frac{(-i/\hbar)^n}{n!} \int_0^t dt_1 \int_0^t dt_2 \dots \int_0^t dt_n T[H'(t_1)H'(t_2) \dots H'(t_n)] ,$$

the sum of which over n is nothing but the power-series expansion of (7.3.13).

According to (7.3.8, 13), the energy of optical excitation at each instant is given by $\varepsilon_0 + H'(t)$, with the instantaneous shift $H'(t)$ varying as (7.3.11) [note that this interpretation is justified only because (7.3.13) is a T product]. Can one then expect that the distribution of $H'(t)$ over a long time interval—which should be equal to the ensemble average according to the ergodicity—manifests itself directly as the spectral width of the optical absorption?

To answer this question, let us calculate the expectation value in (7.3.8) with the use of the expansion (7.3.14). From (7.3.3, 7, 11), one obtains

$$H'(t) = \sum_{KK'} |K\rangle \exp\left[\frac{i}{\hbar} (\varepsilon_K - \varepsilon_{K'})t\right] \langle K'|$$

$$\cdot V_{KK'}[b^\dagger_{-K+K'} \exp(i\omega_{-K+K'}t) + b_{K-K'} \exp(-i\omega_{K-K'}t)] \ . \tag{7.3.15}$$

Since it is linear in the phonon operators b and $b^\dagger$, the expectation value of the nth-order term in (7.3.14) vanishes when n is an odd number. In the second-order term, we change the integration variables (t_1, t_2) into $(t_1, \tau \equiv t_1 - t_2)$. We have to deal with the autocorrelation of $H'(t)$, which, according to (7.3.15), is given by

$$\langle\langle 0| H'(t_1)H'(t_1 - \tau)|0\rangle\rangle_L = \langle\langle 0|(H')^2|0\rangle\rangle_L g(\tau)$$

$$= \sum_{K\pm}| V_{K0}|^2 \left(N(\hbar\omega_{\mp K}) + \frac{1}{2} \pm \frac{1}{2}\right)\exp\left[\frac{i}{\hbar} (\varepsilon_0 - \varepsilon_K \mp \hbar\omega_{\mp K})\tau\right] \ . \tag{7.3.16}$$

The expectation value of the second-order term can then be written as

$$-\frac{1}{\hbar^2} \langle\langle 0|(H')^2|0\rangle\rangle_L \int_0^t (t - \tau)g(\tau)\, d\tau \equiv - L(t) \ . \tag{7.3.17}$$

Rearranging the expectation value of (7.3.14) as the expansion of the exponent, one can write (7.3.8) as

$$f(t) = |P|^2 \exp\left[-\frac{i}{\hbar} \varepsilon_0 t - L(t) + O(H'^4)\right] \ . \tag{7.3.18}$$

Let us neglect for the moment the fourth- and higher-order terms in H', as will in fact be justified later in the limiting cases of weak and strong coupling. Assuming that the correlation (7.3.16) vanishes for $\tau \to \infty$, we put

$$\int_0^\infty g(\tau)\, d\tau = \tau_c + i\tau'_c \ . \tag{7.3.19}$$

Since $g(0) = 1$, τ_c and τ'_c represent the duration times of the correlation (the correlation times) of H' with itself. Considering this in the integration of (7.3.17), one can put $g(\tau) \approx g(0) = 1$ when $t \ll \tau_c$, while in the opposite case: $t \gg \tau_c$, one can replace the upper bound t of the integration by ∞ after taking out the factor $(t - \tau) \approx t$ as a constant. As the result, one obtains

$$L(t) = \begin{cases} \dfrac{t^2}{2\hbar^2}\, D^2 & (t \ll \tau_c) \qquad\qquad (7.3.20) \\[3mm] \dfrac{i}{\hbar}\, t(\varDelta_0 - i\varGamma_0) & (t \gg \tau_c) \qquad (7.3.21) \end{cases}$$

with

$$D^2 \equiv \langle\!\langle 0|(H')^2|0\rangle\!\rangle_L = \sum_{\boldsymbol{K}} |V_{\boldsymbol{K}0}|^2 [2N(\hbar\omega_{\boldsymbol{K}}) + 1] \ , \qquad (7.3.22)$$

$$\varDelta_0 - i\varGamma_0 \equiv \frac{1}{i\hbar} \int_0^{\infty} \langle\!\langle 0|H'(\tau)H'|0\rangle\!\rangle_L \, d\tau = \frac{D^2}{\hbar}(\tau_c' - i\tau_c)$$

$$= \sum_{\boldsymbol{K}\pm} |V_{\boldsymbol{K}0}|^2 \left(N(\hbar\omega_{\mp\boldsymbol{K}}) + \frac{1}{2} \pm \frac{1}{2}\right)\left[\frac{P}{\varepsilon_0 - \varepsilon_{\boldsymbol{K}} \mp \hbar\omega_{\mp\boldsymbol{K}}}\right.$$

$$\left. - i\pi\delta(\varepsilon_0 - \varepsilon_{\boldsymbol{K}} \mp \hbar\omega_{\mp\boldsymbol{K}})\right], \qquad (7.3.23)$$

where use has been made of (7.3.16) and the identity

$$\lim_{\eta\to+0} \int_0^{\infty} \exp(i\omega\tau - \eta\tau)\, d\tau = iP\frac{1}{\omega} + \pi\delta(\omega)$$

on the principal part P and the δ function.

According to (7.3.20, 21), the real part of $L(t)$ rises as t^2 and then increases as $\varGamma_0 t/\hbar$. The correlation function (7.3.16) will decay out beyond the relaxation time τ_R defined by $\mathrm{Re}\{L(\tau_R)\} = 1$. In calculating the line shape of the absorption spectra given by the Fourier transform of (7.3.18), one has to make use of (7.3.20 or 21) accordingly as (a) $\tau_c \gg \tau_R$ or (b) $\tau_c \ll \tau_R$ (see Fig. 7.14a, b. With the use of (7.3.20, 21, 23), this criterion can be written in various forms:

$$\tau_c \gtrless \tau_R \leftrightarrow \frac{\tau_c D}{\hbar} \gtrless 1 \leftrightarrow \frac{\tau_c \varGamma_0}{\hbar} \gtrless 1 \leftrightarrow \varGamma_0 \gtrless D \ . \qquad (7.3.24)$$

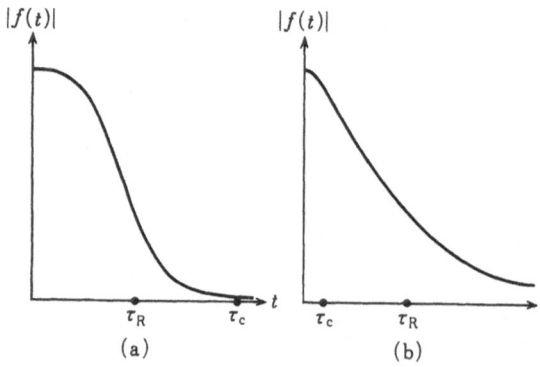

Fig. 7.14. a,b. The relaxation of the transition dipole moment in the two limiting situations: (a) $\tau_c \gg \tau_R$ (b) $\tau_c \ll \tau_R$

Making use of the relation $f(-t) = f^*(t)$ which follows from (7.2.4) and the reality of $F(E)$, one can calculate $F(E)$, by the inverse transform of (7.2.4), as

(a) $\quad F(E) = \dfrac{|P|^2}{\sqrt{2\pi}D} \exp\left[-\dfrac{(E - \varepsilon_0)^2}{2D^2} \right]$ \hfill (7.3.25)

(b) $\quad F(E) = \dfrac{|P|^2}{\pi} \dfrac{\Gamma_0}{(E - \varepsilon_0 - \Delta_0)^2 + \Gamma_0^2}$ \hfill (7.3.26)

in the two limiting cases.

The physical meaning of these results is as follows. The instantaneous shift $H'(t)$ of the optical-excitation energy is fluctuating in time with average amplitude D according to (7.3.22). In order that $H'(t)$ is perceived as an instaneous shift of the spectral line and hence that the amplitude D manifests itself directly as the spectral width, $H'(t)$ must remain constant at least for a time interval $\Delta t \approx \hbar/D$, according to the uncertainty principle.

According to (7.3.16), the correlation time τ_c of the fluctuation $H'(t)$ is given by

$$\frac{\hbar}{\tau_c} \approx \text{Max.} \left\{ \left| \varepsilon_0 - \varepsilon_K \right|, \hbar\bar{\omega}_K \right\} . \hfill (7.3.27)$$

Namely, $H'(t)$ varies significantly as the exciton is transferred to the neighboring site or as the lattice vibrates.

If τ_c is much smaller than Δt mentioned above, one is justified in treating $H'(t)$ in a quasistatic way; namely, the statistical distribution of $H'(t)$ is directly reflected in the Gaussian line shape (7.3.25). The result agrees exactly with the absorption line shape (7.2.30) of the strongly coupled localized electron–phonon system, as it should since in the present case (a) the effective width $\left| \varepsilon_0 - \varepsilon_K \right|$ of the exciton band and the average phonon energy $\hbar\bar{\omega}_K$ are both much smaller than the fluctuation D of the local excitation energy [D defined in (7.3.22) is the same as D defined in (7.2.29) as is evident from (7.2.23, 33)]. This agreement justifies a posteriori the neglect of $O(H'^4)$ in (7.3.18) in the present limit (a) of strong coupling (slow modulation).

In the opposite limit (b) of weak coupling (rapid modulation), $H'(t)$ will lose its memory in time τ_c which is much smaller than Δt, the minimum time needed for the perception of $H'(t)$ as instantaneous spectral shift. As a result, the spectral width is given by $\Gamma_0 = (D^2/\hbar)\tau_c$ [see (7.3.23, 26)], which is equal to the amplitude D times the effectivity factor $\tau_c/\Delta t (\ll 1)$ for the perception mentioned above. This is an example of the *motional narrowing* mentioned in Sect. 7.1.

In typical inorganic insulators, the width of the exciton band, like that of the valence band, is of the order of eV. Being much larger than the phonon energies, it determines $\hbar/\tau_c$. On the other hand, D is of the order of a tenth of eV, being much smaller than $\hbar/\tau_c$. Hence, the absorption spectrum is usually of the motionally narrowed Lorentzian form given by (7.3.26).

According to (7.3.23), the peak energy shift Δ_0 and the full width at half-maximum $2\Gamma_0$ of the absorption spectrum represent respectively the energy shift and the lifetime broadening $\hbar/\tau_0$ of the exciton with $K = 0$ in the phonon field [see also (6.2.2, 11)]. At moderate and high temperatures T where the phonon number $N(\hbar\omega)$ is proportional to T, D is proportional to $\sqrt{T}$ and hence $\Gamma_0 \propto D^2$ to T.

7.3.3 Direct and Indirect Transitions with Their Interference

Whereas we have studied the line-shape problem from the viewpoint of lifetime broadening of the absorption line in (b), we shall now study the same problem from a different side, namely as the intermediate-state resonance of the indirect transitions. This will provide us with a heuristic derivation of a more general and rigorous formula for the line shape, which in turn will be put on a more sound basis in (d).

If $H' = 0$ in (7.3.4), the light can excite the electronic system only—the first-order process or the so-called *direct transition*—with the absorption spectra

$$F^{(0)}(E) = \sum_{\lambda} |P_{\lambda g}|^2 \delta(E - \varepsilon_{\lambda 0}) . \tag{7.3.28}$$

The next process induced by H' is the second-order process or the so-called *indirect transition* in which the electronic system is excited from $|g\rangle$ to $|\lambda'0\rangle$ by absorbing light and then the exciton is scattered from $|\lambda'0\rangle$ to $|\lambda K\rangle$. Considering the fact that the electronic system can pass through various intermediate states $|\lambda'0\rangle$ to reach the same final state $|\lambda K\rangle$ as shown by straight and wavy arrows in Fig. 7.13, one obtains the normalized [in conformity with (7.3.28)] spectra for the indirect absorption

$$\begin{aligned}
F^{(2)}(E) = \sum_{\lambda K \pm} \left| \sum_{\lambda'} \frac{V_{\lambda K \lambda'0} P_{\lambda'g}}{E - \varepsilon_{\lambda'0}} \right|^2 &\left[N(\hbar\omega_{\mp K}) + \frac{1}{2} \pm \frac{1}{2} \right] \\
\cdot \delta(E - \varepsilon_{\lambda k} \mp \hbar\omega_{\mp k}) &.
\end{aligned} \tag{7.3.29}$$

It is easy to see that (7.3.28, 29) (in the off-resonance region) correspond to the zeroth- and the second-order terms, respectively, of (7.3.14).

Equation (7.3.29) can be written in a compact form by introducing the shift Δ_K and the broadening Γ_K which were originally defined in (7.3.23) but are now generalized to energy-dependent matrices with λ for their indices

$$\begin{aligned}
\Delta_{\lambda'\lambda''K'}(E) - i\Gamma_{\lambda'\lambda''K'}(E) \equiv \sum_{\lambda K \pm} V_{\lambda'K',\lambda K} V_{\lambda K,\lambda''K'} &\left[N(\hbar\omega_{\mp(K-K')}) + \frac{1}{2} \pm \frac{1}{2} \right] \\
\cdot \left[\frac{P}{E - \varepsilon_{\lambda K} \mp \hbar\omega_{\mp(K-K')}} - i\pi\delta(E - \varepsilon_{\lambda K} \mp \hbar\omega_{\mp(K-K')}) \right] &.
\end{aligned} \tag{7.3.30}$$

By putting $E = \varepsilon_{\lambda'K'}$ in the diagonal element $\lambda' = \lambda''$, one obtains the shift and broadening of the exciton $|\lambda' K'\rangle$. Denote by $\Delta_{K'}(E)$ and $\Gamma_{K'}(E)$ the matrices whose (λ', λ'') elements are given by real and imaginary parts, respectively, of (7.3.30). They are easily found to be Hermitian: $\Delta_K = \Delta_K^\dagger$, $\Gamma_K = \Gamma_K^\dagger$. Correspondingly, we introduce a diagonal matrix H_K° whose (λ, λ) element is given by $\varepsilon_{\lambda K}$, and a single-column matrix P whose λth element is given by $P_{\lambda g}$, and its Hermitian conjugate $P^\dagger$ which is a single-row matrix. (7.3.29) can then be written as

$$F^{(2)}(E) = \frac{1}{\pi} \sum_{\lambda'\lambda''} P_{g\lambda'} \frac{1}{E - \varepsilon_{\lambda'0}} \Gamma_{\lambda'\lambda''0}(E) \frac{1}{E - \varepsilon_{\lambda''0}} P_{\lambda''g} \tag{7.3.31}$$

$$= \frac{1}{\pi} P^\dagger \frac{1}{E - H_0^\circ} \Gamma_0(E) \frac{1}{E - H_0^\circ} P \ . \tag{7.3.31'}$$

Decomposing the cross terms $(\lambda' \neq \lambda'')$ in (7.3.31) as

$$\frac{1}{E - \varepsilon_{\lambda'0}} \frac{1}{E - \varepsilon_{\lambda''0}} = \frac{1}{\varepsilon_{\lambda'0} - \varepsilon_{\lambda''0}} \left(\frac{1}{E - \varepsilon_{\lambda'0}} - \frac{1}{E - \varepsilon_{\lambda''0}} \right) ,$$

one can rearrange the double summation in it into a single summation

$$F^{(2)}(E) = \frac{1}{\pi} \sum_\lambda f_\lambda \frac{\Gamma_{\lambda\lambda0}(E) + A_\lambda(E)(E - \varepsilon_{\lambda0})}{(E - \varepsilon_{\lambda0})^2} \tag{7.3.32}$$

where

$$f_\lambda \equiv |P_{\lambda g}|^2 \ , \tag{7.3.33}$$

$$f_\lambda A_\lambda(E) \equiv \sum_{\lambda'(\neq\lambda)} \frac{2\mathrm{Re}\{P_{g\lambda}\Gamma_{\lambda\lambda'0}(E)P_{\lambda'g}\}}{\varepsilon_{\lambda0} - \varepsilon_{\lambda'0}} \ . \tag{7.3.34}$$

Neglecting the asymmetry terms with $A_\lambda(E)$ in (7.3.32) for simplicity, we note that $F^{(2)}(E)$ diverges as $(E - \varepsilon_{\lambda0})^{-2}$ when the photon energy E approaches an intermediate state energy $\varepsilon_{\lambda0}$. This type of divergent spectra, inherent in the second-order perturbation process, should in fact connect smoothly to the broadened spectral lines of the first-order process since the virtual transition gradually changes to the real transition in the resonance region. In other words, the spectral line λ in (7.3.28) should in fact be lifetime broadened into a Lorentzian (7.3.26) with width $\Gamma_{\lambda\lambda0}$, whose tail part should be identified with (7.3.32). The correct formula for $F(E)$ is then supposed to be obtained by replacing the denominators in (7.3.32) by

$$[E - \varepsilon_{\lambda0} - \Delta_{\lambda\lambda0}(E)]^2 + [\Gamma_{\lambda\lambda0}(E)]^2$$

with energy shift Δ and broadening Γ being considered, whereby the unphysical divergence is automatically removed.

On the other hand, the term with A_λ in (7.3.32) is a new one which was not found in (7.3.26). As is seen from (7.3.34), this term originates from the phonon-induced influence from other exciton bands $\lambda'(\neq \lambda)$ represented by the non-diagonal elements $\Gamma_{\lambda\lambda'0}(E)(\lambda' \neq \lambda)$ of the matrix $\Gamma(E)$. From the formal consistency, one may well suppose that the resonance divergence should be removed by supplementing Δ and Γ in the matrix form, rather than by preferring their diagonal elements as was prescribed above. This can in fact be done in the matrix expression (7.3.31'), almost in a unique way, as follows:

$$F(E) = \frac{1}{\pi} P^\dagger \frac{1}{E - H_0^s - \Delta_0(E) + i\Gamma_0(E)} \Gamma_0(E) \frac{1}{E - H_0^s - \Delta_0(E) - i\Gamma_0(E)} P$$

$$= \frac{i}{2\pi} P^\dagger \left[\frac{1}{E - H_0^s - \Delta_0(E) + i\Gamma_0(E)} - \frac{1}{E - H_0^s - \Delta_0(E) - i\Gamma_0(E)} \right] P .$$

$$(7.3.35)$$

As will be shown later, (7.3.35) is in fact the exact expression (exact up to the infinite order of perturbation theory) for the absorption spectra provided that exactly defined $\Delta_0(E)$ and $\Gamma_0(E)$ [instead of their lowest approximation (7.3.23)] are used in it.

Leaving the proof of (7.3.35) to Sect. 7.3.4, let us rewrite this matrix expression into ordinary form. Let us first diagonalize the renormalized energy matrix $H_{\boldsymbol{K}}^s + \Delta_{\boldsymbol{K}}(E) - i\Gamma_{\boldsymbol{K}}(E)$ by transforming it through a matrix $T_{\boldsymbol{K}}(E)$ (not unitary)

$$T_{\boldsymbol{K}}(E)[H_{\boldsymbol{K}}^s + \Delta_{\boldsymbol{K}}(E) - i\Gamma_{\boldsymbol{K}}(E)]T_{\boldsymbol{K}}(E)^{-1} = \tilde{H}_{\boldsymbol{K}}(E) - i\tilde{\Gamma}_{\boldsymbol{K}}(E) . \qquad (7.3.36)$$

$\tilde{H}_{\boldsymbol{K}}(E)$ and $\tilde{\Gamma}_{\boldsymbol{K}}(E)$ are real, diagonal matrices whose λth elements will be denoted by $\tilde{\varepsilon}_{\lambda\boldsymbol{K}}(E)$ and $\tilde{\Gamma}_{\lambda\boldsymbol{K}}(E)$, respectively [so numbered that $\tilde{\varepsilon}_{\lambda\boldsymbol{K}}(E) \to \varepsilon_{\lambda\boldsymbol{K}}$ as $H' \to 0$]. With the use of (7.3.36), one can rewrite (7.3.35) as

$$F(E) = \frac{i}{2\pi} \sum_\lambda \left[\frac{(P^\dagger T_0^{-1})_\lambda (T_0 P)_\lambda}{E - \tilde{\varepsilon}_{\lambda 0} + i\tilde{\Gamma}_{\lambda 0}} - (\text{c.c.}) \right]$$

$$= \sum_\lambda \tilde{f} \frac{1}{\pi} \frac{\tilde{\Gamma}_{\lambda 0} + \tilde{A}_\lambda(E - \tilde{\varepsilon}_{\lambda 0})}{(E - \tilde{\varepsilon}_{\lambda 0})^2 + \tilde{\Gamma}_{\lambda 0}^2}$$

$$(7.3.37)$$

where the renormalized intensity $\tilde{f}_\lambda$ and the renormalized degree of asymmetry $\tilde{A}_\lambda$ are defined by

$$(P^\dagger T_0^{-1})_\lambda (T_0 P)_\lambda = \tilde{f}_\lambda - i\tilde{f}_\lambda \tilde{A}_\lambda . \qquad (7.3.38)$$

The symbol $\sim$ is put on the renormalized quantities in which the effect of H' is taken into account up to infinite order: they are functions of E (photon energy in the present problem).

The final expression (7.3.37) carries a variety of implications. In the first place, the direct and the indirect absorption spectra are unified into a single expression, as is evident from its derivation. In the second place, the fine structures as well as the gross contour are contained in the same expression. The former reside in the energy dependence of renormalized quantities, and will be described in Sect. 7.3.5. Under the neglect of this E dependence, the gross contour of each component (λ) of (7.3.37) is an asymmetric Lorentzian with a dip on one side of the peak. Of course, the total absorption is positive since this dip is superposed onto a smooth background of indirect absorption from other ($\lambda' \neq \lambda$) intermediate states, namely, onto the sum of other λ' components in (7.3.37). A typical example is found in Fig. 2. 10: each of the absorption bands (2p, 3p, . . .) is strongly asymmetric with a dip on the high energy side ($\tilde{A}_\lambda < 0$).

As is seen from the approximate equation (7.3.34) for $\tilde{A}_\lambda$, the dip originates from the cross terms ($\lambda \neq \lambda'$) in (7.3.31). Namely, it results from the interference of the two waves coming through different intermediate states: $|\lambda 0 \rangle$ and $|\lambda' 0 \rangle$. It is a special case of the *Fano effect*, well-known in atomic physics, which takes place as the result of interference between line and continuous spectra. In the present case, the line spectrum corresponds to the direct absorption peak λ and the continuum to the indirect absorption band coming through all other intermediate states $\lambda'(\neq \lambda)$.

7.3.4 Renormalization of Exciton–Phonon Interaction

We shall now give an exact derivation of the line-shape expression (7.3.35) starting from the basic formula (7.3.5).

For simplicity, we neglect the internal degree of freedom (λ) of the exciton for the moment, as we did in Sect. 7.3.2. The Hamiltonian is then given by

$$H = H_0 + H' = H_{\rm e} + H_{\rm L} + H' , \tag{7.3.39}$$

$$H_{\rm e} = \sum_{\boldsymbol{K}} |\boldsymbol{K}\rangle \varepsilon_{\boldsymbol{K}} \langle \boldsymbol{K}| , \quad H_{\rm L} = \sum_{\boldsymbol{K}} \hbar \omega_{\boldsymbol{K}} b_{\boldsymbol{K}}^\dagger b_{\boldsymbol{K}} , \tag{7.3.40}$$

$$H' = \sum_{\boldsymbol{K}\boldsymbol{K}'} |\boldsymbol{K}\rangle V_{\boldsymbol{K}\boldsymbol{K}'}(b_{-\boldsymbol{K}+\boldsymbol{K}'}^\dagger + b_{\boldsymbol{K}-\boldsymbol{K}'})\langle \boldsymbol{K}'| . \tag{7.3.41}$$

Let us introduce the *resolvent* operator $R(z)$ and its diagonal part $D(z) = [R(z)]_{\rm d}$ defined respectively by

$$R(z) \equiv \frac{1}{z - H} , \tag{7.3.42}$$

$$\langle \boldsymbol{K}n| D(z) | \boldsymbol{K}'n'\rangle \equiv \delta_{\boldsymbol{K}\boldsymbol{K}'}\delta_{nn'}\langle \boldsymbol{K}n| R(z) | \boldsymbol{K}n\rangle , \tag{7.3.43}$$

with complex variable z. The zeroth approximation for them is given by replacing H by H_0 in (7.3.42) and (7.3.43):

$$R^{(0)}(z) \equiv \frac{1}{z - H_0} = D^{(0)}(z) \ . \tag{7.3.44}$$

Expanding (7.3.42) in power series in H': $R = R^{(0)} + R^{(0)} H' R^{(0)}$ $+ R^{(0)} H' R^{(0)} H' R^{(0)} + \dots$, and taking the diagonal part, one obtains

$$D(z) = D^{(0)}(z) + D^{(0)}(z) S(z) D^{(0)}(z) \ , \tag{7.3.45}$$

$$S(z) \equiv [H' D^{(0)} H' + H' D^{(0)} H' D^{(0)} H' D^{(0)} H' + \dots]_d \ , \tag{7.3.46}$$

since the terms of odd order in H' have no diagonal part. The lowest-order term in (7.3.46) reads

$$\langle \boldsymbol{K}n | H' D^{(0)} H' | \boldsymbol{K}n \rangle$$

$$= \sum_{\boldsymbol{K}'n'} V_{\boldsymbol{K}\boldsymbol{K}'} \langle n | b^{\dagger}_{-\boldsymbol{K}+\boldsymbol{K}'} + b_{\boldsymbol{K}-\boldsymbol{K}'} | n' \rangle \frac{1}{z - E_{\boldsymbol{K}'n'}^{(0)}} \langle n' | b^{\dagger}_{-\boldsymbol{K}'+\boldsymbol{K}} + b_{\boldsymbol{K}'-\boldsymbol{K}} | n \rangle V_{\boldsymbol{K}'\boldsymbol{K}}$$

$$= \sum_{\boldsymbol{K}'\pm} \frac{|V_{\boldsymbol{K}\boldsymbol{K}'}|^2 (n_{\mp(\boldsymbol{K}'-\boldsymbol{K})} + 1/2 \pm 1/2)}{z - E_{\boldsymbol{K}'n'}^{(0)}} \ . \tag{7.3.47}$$

Expressing the exciton by a solid line, the phonons by a broken line and the vertex by a solid dot, one can represent (7.3.47) by the graph of Fig. 7.15a, indicating thereby that the exciton–phonon system is scattered from $|\boldsymbol{K}n\rangle$ to an intermediate state $|\boldsymbol{K}'n'\rangle$ through emission or absorption of a phonon with wave vector $\mp(\boldsymbol{K}' - \boldsymbol{K})$ and is then scattered back to $|\boldsymbol{K}n\rangle$ through the reverse process. Assign the matrix element V to a vertex, $(n_{\boldsymbol{K}-\boldsymbol{K}'} + 1)$ or $n_{-\boldsymbol{K}+\boldsymbol{K}'}$ to the phonon line, and the free propagator $D^{(0)}(z)_{\boldsymbol{K}'n'} = (z - E_{\boldsymbol{K}'n'}^{(0)})^{-1}$ to the intermediate state $|\boldsymbol{K}'n'\rangle$; take their product and sum over all possible intermediate states; then one obtains (7.3.47).

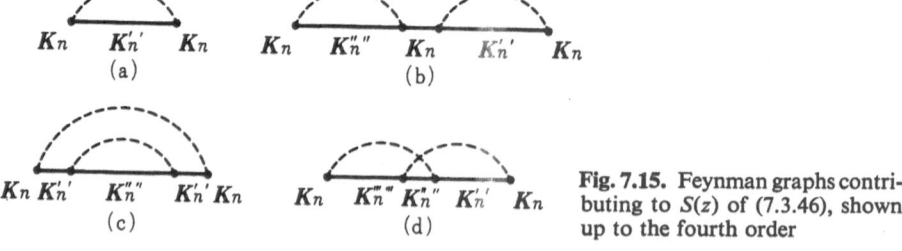

Fig. 7.15. Feynman graphs contributing to $S(z)$ of (7.3.46), shown up to the fourth order

According to this representation, the fourth-order term of (7.3.46) is contributed by three types of graphs shown in Fig. 7.15b-d. Generally speaking, the graphs of any order can be classified into two groups according to whether they can (as in Fig. 7.15b) or cannot (as in Fig. 7.15a, c, d) be decomposed into lower-order graphs by breaking an exciton line, namely, according to whether the system does or does not return to the initial state as an intermediate step. The grphas of the latter type will be denoted collectively by a white circle and their contribution to (7.3.46) by $\Sigma (z)$. Since any graph of the former type can be constructed by connecting graphs of the latter type by solid lines of the initial state $| \boldsymbol{K} n \rangle$, one obtains

$$S(z) = \Sigma + \Sigma D^{(0)} \Sigma + \Sigma D^{(0)} \Sigma D^{(0)} \Sigma + \cdots = \frac{1}{1 - \Sigma D^{(0)}} \Sigma$$

by exhausting all posible ways of construction. If one denotes the whole graphs symbolically by a black circle, the above equation can be expressed diagrammatically as in Fig. 7.16a. Putting this equation into (7.3.45) and making use of (7.3.44), one obtains

$$D(z) = \frac{1}{z - H_0 - \Sigma(z)} \; . \tag{7.3.48}$$

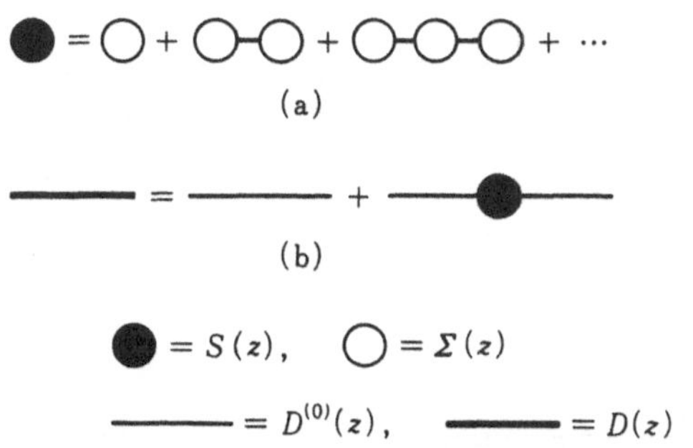

(a)

(b)

$$\bullet = S(z), \quad \bigcirc = \Sigma(z)$$
$$\text{———} = D^{(0)}(z), \quad \text{———} = D(z)$$

Fig. 7.16. a,b. Diagrammatic representations of $S(z)$ and $D(z)$

$\Sigma(z)$ as defined above still includes the "reducible" graphs, such as Fig. 7.15c, in which (at least) a set of phonon lines are interposed between the same intermediate state ($| \boldsymbol{K}'n' \rangle$ in the figure) without interlocking with other phonon lines. Consider an *irreducible* graph, such as Fig. 7.15a, d, in which no phonon line is separated from others. To interpose all possible graphs $S(z)$ in one of the intermediate states of the irreducible diagram is equivalent to replacing the free propagator $D^{(0)}(z)$ of this intermediate state by $D^{(0)}(z)S(z)D^{(0)}(z)$. Adding this contribution of the *reducible* graphs to that of the irreducible graph and considering (7.3.45), we find that the replacement of the free propagators $D^{(0)}(z)$ in all the intermediate states of the irreducible graph by the exact propagator $D(z)$ is equivalent to the inclusion of contributions from all possible reducible graphs constructed from that irreducible graph. This is shown diagrammatically in Fig. 7.16b, where the thick solid line denotes the exact propagator. Since any reducible graph can be constructed from one and only one irredicible graph in the way mentioned above, one can write the *self-energy* $\Sigma(z)$ as the sum over all irreducible graphs

$$\Sigma(z) = [H'D(z)H' + H'D(z)H'D(z)H'D(z)H' + \dots]_{id} \qquad (7.3.49)$$

where the suffix id is the abbreviation of "irreducible" and "diagonal." The irreducible graph can also be defined as the graph in which all the intermediate states are different from each other as well as from the initial state. For clarity, all the irreducible graphs of the second, fourth and sixth orders contributing to $\Sigma(z)$ are shown in Fig. 7.17.

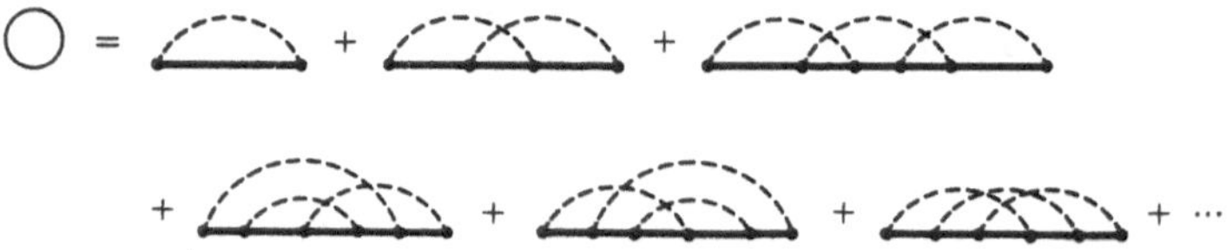

Fig. 7.17. Irreducible graphs contributing to the self-energy $\Sigma(z)$

Equations (7.3.48, 49) are the self-consistent equations by which $D(z)$ and $\Sigma(z)$ are to be determined for given $H = H_0 + H'$. Taking their Kn element and putting

$$z = E + E_{Ln} + i\eta , \qquad (7.3.50)$$

one obtains:

$$\langle Kn | D(z) | Kn \rangle = \frac{1}{E + i\eta - \varepsilon_K - \langle Kn | \Sigma(z) | Kn \rangle} , \qquad (7.3.51)$$

$$\langle Kn | \Sigma(z) | Kn \rangle = \sum_{K'\pm} \frac{|V_{KK'}|^2 (n_{\mp(K'-K)} + 1/2 \pm 1/2)}{E + i\eta - \varepsilon_{K'} \mp \hbar\omega_{\mp(K'-K)} - \langle K'n' | \Sigma(z) K'n' \rangle}$$
$$+ \dots . \qquad (7.3.52)$$

The intermediate phonon state n' in the first term of (7.3.52) is different from the initial phonon state n in the phonon number (by ± 1) of a mode $\mp(K' - K)$ among the $3N$ modes where N is the number of atoms in the crystal. Therefore, $\langle K'n' | \Sigma(z) | K'n' \rangle$ can be replaced by $\langle K'n | \Sigma(z \mp \hbar\omega_{\mp(K'-K)} | K'n \rangle$ within an error of $O(N^{-1})$, since $|V_{KK'}|^2$ in (7.3.52) is $O(N^{-1})$.

Taking statistical averages of (7.3.51, 52) and taking their limiting values at $\eta \to +0$, one obtains

$$D_K(E) \equiv \lim_{\eta \to +0} \langle\!\langle Kn | D(z) | Kn \rangle\!\rangle_L = \frac{1}{E - \varepsilon_K - \Sigma_K(E)} , \qquad (7.3.51')$$

$$\Sigma_K(E) \equiv \lim_{\eta \to +0} \langle\!\langle Kn \,|\, \Sigma(z) \,|\, Kn \rangle\!\rangle$$

$$= \sum_{K'\pm} \frac{|V_{KK'}|^2 [N(\hbar\omega_{\mp(K'-K)}) + 1/2 \pm 1/2]}{E - \varepsilon_{K'} \mp \hbar\omega_{\mp(K'-K)} - \Sigma_{K'}(E \mp \hbar\omega_{\mp(K'-K)})} + \dots . \quad (7.3.52')$$

The replacement of $\langle Kn | \Sigma | Kn \rangle$ by its average in the denominators of (7.3.51) and (7.3.52) is justified since Σ, being the sum of N terms of $O(N^{-1})$, has fluctuation of $O(N^{-1/2})$. Write explicity the real and imaginary parts of the self-energy

$$\Sigma_K(E) = \Delta_K(E) - i\Gamma_K(E) . \quad (7.3.53)$$

$\Gamma_K(E)$ is easily shown to be positive. If one put $\eta \to -0$, one would obtain $\Delta_K + i\Gamma_K$. The discontinuity of $\Sigma_K(E)$ and $D_K(E)$ across the real axis takes place because of the continuous distribution of the eigenvalues of H on that axis.

The absorption line shape (7.3.5), within the single-band model, is then given, with the use of (7.3.42, 43, 50, 51'), by

$$-\frac{1}{\pi} \lim_{\eta \to +0} \sum_{n,j} w_n \langle 0n | j \rangle \operatorname{Im} \left[\frac{1}{E + E_{Ln} + i\eta - E_j} \right] \langle j | 0n \rangle$$

$$= -\frac{1}{\pi} \lim_{\eta \to +0} \operatorname{Im} \{ \langle\!\langle 0n | R(z) | 0n \rangle\!\rangle_L \}$$

$$= -\frac{1}{\pi} \operatorname{Im} \left[\frac{1}{E - \varepsilon_0 - \Delta_0(E) + i\Gamma_0(E)} \right]$$

which is the single-band version of (7.3.35). In order to take into account the internal degree of freedom with quantum number λ, one has to generalize $D(z)$ and $\Sigma(z)$ into matrices which are not diagonal in the index λ [the lowest-order approximation for Σ is given by (7.3.30)], as well as to generalize ε_K into diagonal matrix H_K^e. It can be confirmed, with due care to the noncommutativity of these matrices, that the above argument on renormalization can be generalized so as to lead to the matrix expression (7.3.35) for the multiband exciton spectra.

In calculating $\Sigma_K(E) = \Delta_K(E) - i\Gamma_K(E)$, one has to take all the irreducible graphs with renormalized energy $H_K^e + \Sigma_K(E')$ for the propagator where E' is equal to the photon energy E minus the energy of phonons emitted or absorbed $(-)$ in the intermediate state concerned. The phonon energy (in constrast to the exciton energy) is not renormalized in our problem since we are concerned with one exciton in a large crystal.

7.3.5 Phonon Strucutures in the Absorption Spectra

Considering again the single-band case in (7.3.37) ($\tilde{A}_\lambda = 0$), let us study how the Lorentzian expression

$$F(E) = \frac{1}{\pi} \frac{\Gamma_0(E)}{[E - \varepsilon_0 - \Delta_0(E)]^2 + [\Gamma_0(E)]^2} \quad (7.3.54)$$

contains the substructures originating from E dependence of $\Gamma_0(E)$ and $\Delta_0(E)$.

If $\Gamma_K(E)$ is small enough, the pole of (7.3.51′) is given by the solution $E = E_K$ of the equation

$$\varepsilon_K + \Delta_K(E) = .E \ , \tag{7.3.55}$$

which is the "renormalized energy" of the exciton. If one expands E_K as

$$E_K = E_0 + \frac{\hbar^2 K^2}{2\bar{M}} + \cdots , \tag{7.3.56}$$

the coefficient $\bar{M}$ represents the "renormalized mass."

Again assuming $\Gamma_K(E')$ to be small enough on the right-hand side of (7.3.52′) one obtains the spectral broadening function $\Gamma_0(E)$ at $T = 0$ K, in the lowest approximation, as

$$\Gamma_0(E) = \pi \sum_K |V_{0K}|^2 \delta[\varepsilon_K + \Delta_K(E - \hbar\omega_{-K}) + \hbar\omega_{-K} - E] \tag{7.3.57}$$

With the use of (7.3.55), the argument of the δ function in (7.3.57) can be expanded, around its zero point, as

$$(E_K + \hbar\omega_{-K} - E)\left[1 - \frac{d\Delta_K(E_K)}{dE_K}\right] + \cdots .$$

Let us neglect the effect of $d\Delta_K/dE_K$ in (7.3.57) which itself is already a small quantity. Denoting the photon energy measured from E_0 by

$$E - E_0 \equiv E' \ , \tag{7.3.58}$$

we can write

$$\Gamma_0(E) = \pi \sum_K |V_{0K}|^2 \delta\left[\hbar\omega_{-K} + \frac{\hbar^2 K^2}{2\bar{M}} - E'\right] . \tag{7.3.59}$$

As will be shown later, $\Gamma(E) = 0$ when $E' < 0$, while $\Gamma(E) \propto E'^3$ as $E' \to +0$, approaching zero much faster than $E - \varepsilon_0 - \Delta_0(E) \approx [1 - d\Delta_0(E_0)/dE_0]E'$. Hence, we neglect the second term in the denominator of (7.3.54) in the whole range of E' under the assumption of weak exciton–phonon coupling.

At finite temperature, $\Gamma_0(E)$ is finite at $E' = 0$ because of phonon absorption, which tends to $+0$ as $T \to 0$ K. Therefore, (7.3.54) gives a line spectrum at $E' = 0$. The above results can be summarized, with the neglect of higher-order terms, as

$$F(E) = \left[1 + \frac{d\Delta_0(E_0)}{dE_0}\right]\delta(E') + \mathfrak{s}(E') \ , \tag{7.3.60}$$

$$\hat{s}(E') = \frac{1}{\pi} \frac{\Gamma_0(E)}{E'^2} \tag{7.3.61}$$

$$= \sum_{K} \frac{|V_{0K}|^2}{[\hbar\omega_{-K} + (\hbar^2 K^2 / 2\bar{M})]^2} \delta\left(\hbar\omega_{-K} + \frac{\hbar^2 K^2}{2\bar{M}} - E'\right) . \tag{7.3.61'}$$

The first term of (7.3.60), representing the zero phonon line for the creation of exciton with $K = 0$, is subject to the energy shift $\Delta_0(E_0) = E_0 - \varepsilon_0$ and the intensity reduction $-d\Delta_0(E_0)/dE_0 (> 0)$, because the created exciton is dressed by phonons. The lost part of the intensity appears as the second term of (7.3.60) representing the phonon sideband. In fact, the integration of (7.3.61) gives $-d\Delta_0(E_0)/dE_0$ if the dispersion relation (2.2.12) between $\Delta_K(E)$ and $\Gamma_K(E)$ is considered.

The sideband corresponds to the indirect transition in which an exciton and a phonon are created simultaneously, and which needs additional (to E_0) energy

$$\hbar\omega_K + \frac{\hbar^2 K^2}{2\bar{M}} = \hbar\Omega_K , \tag{7.3.62}$$

that is, the phonon energy plus the recoil kinetic energy of exciton. Equation (7.3.61') is also obtained by replacing $\hbar\omega_K$ with $\hbar\Omega_K$ in (7.2.23) for the sideband of localized excitation ($j \rightarrow K$, $\gamma_j \rightarrow V_{0K}$), namely, by taking into account the recoil effect. Solve (7.3.62) for $\hbar\omega_K$ as a function of $\hbar\Omega_K$, considering the dispersion of $\hbar\omega_K$, as

$$\hbar\omega_K = \hat{E}(\hbar\Omega_K) . \tag{7.3.63}$$

Then, (7.3.61) can be related to the phonon sideband $s(E)$ of localized excitation (the limiting case $\bar{M} \rightarrow 0$ of exciton) by

$$\hat{s}(E') = s[\hat{E}(E')] \frac{d\hat{E}(E')}{dE'} \left[\frac{\hat{E}(E')}{E'}\right]^2 . \tag{7.3.64}$$

One can read, in Fig. 7.18, the functional form of (7.3.63) in the cases of acoustic and optical phonons. Then one finds from (7.3.64) how $s(E')$ changes to $\hat{s}(E')$ due to the recoil effect; the result is shown schematically in Fig. 7.19. The effective coupling strength, given by the sideband area: $\hat{S} = \int \hat{s}(E') dE'$, is much reduced from the value S for the localized excitation, because of the factor $(\hat{E}/E')^2$ in (7.3.64). The reduction is more remarkable for the acoustic-phonon sideband, which, moreover, appears so close to the zero-phonon line (separation being $\bar{M}c_s^2 \sim 10^{-4}$·eV) that they usually coalesce into a single absorption band.

In contrast, the optical-phonon sideband appears beyond the optical-phonon energy, being well separable. As an example, the absorption spectrum of NaI is shown in Fig. 7.20. At low temperatures, the spectrum consists of the strong zero-phonon line of the 1s exciton and the weak one-LO (longitudinal optical)-

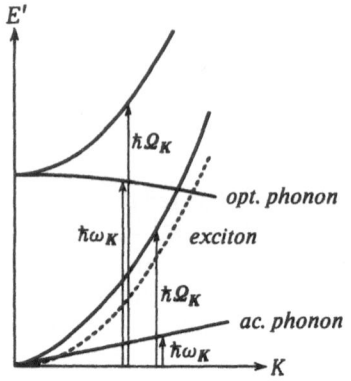

Fig. 7.18. Acoustic or optical phonon energy $\hbar\omega_k$, kinetic energy of recoiled exciton $\hbar^2K^2/2\bar{M}$, and their sum $\hbar\Omega_k$

phonon sideband, their separation being slightly larger than the LO-phonon energy $\hbar\omega_l$ because of the recoil effect. At high temperatures where $\Gamma_0(E)$ is larger than $\hbar\omega_l$, these substructures are overshadowed under the entire broadening of the asymmetric Lorentzian.

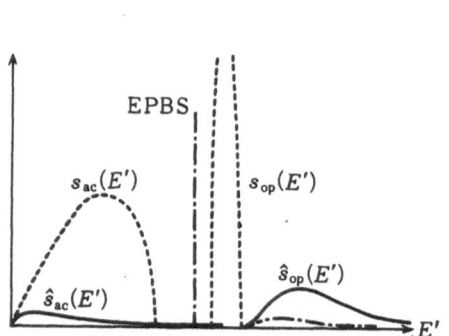

Fig. 7.19. The motional effect on the spectral coupling $s(E')$ shown with dotted line is for a localized electron, $\hat{S}(E')$ with solid line for an exciton. The latter may be modified into the chain line with exciton–phonon bound state if the final state interaction between exciton and optical phonon is taken into account (see Sect. 7.4.1)

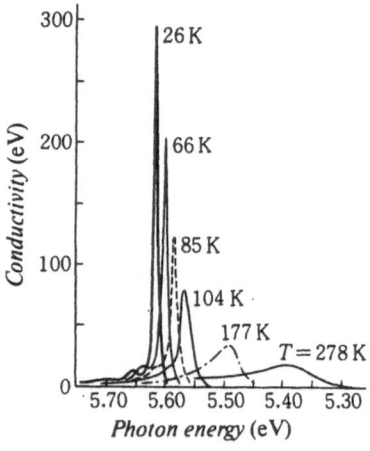

Fig. 7.20. The fundamental absorption spectra of NaI at various temperatures. [T. Miyata: J. Phys. Soc. Jpn. 31, 529 (1971)]

7.4 Final-State Interaction

Consider the optical process (absorption or scattering) in which two quasiparticles are simultaneously created. They should have opposite momenta p and $-p$ if the photon momenta are negligible, and the excitation energy is given by the sum of their respective energies

$$E(p) = \varepsilon_1(p) + \varepsilon_2(-p) \ . \tag{7.4.1}$$

The optical spectra will consist of a continuum of (7.4.1) corresponding to the continuous range of p.

The elementary excitation—the quasiparticle—is the concept which was introduced on the basis of their independent nature implied by (7.4.1). In general, however, this independence is only approximate; their interaction is not negligible when the two quasiparticles created simultaneously stay in the neighborhood of each other. This effect of *final-state interaction* plays an important role in the optical spectra of simultaneous excitation.

A typical example of final-state interaction is the exciton described in Sect. 2.3. According to the one-electron picture for the interband transition, the fundamental absorption spectra of the insulator are given [see (2.4.6)] by

$$F(E) = \sum_k |P_{cv}(k)|^2 \delta[\varepsilon_c(k) - \varepsilon_v(k) - E] \ , \tag{7.4.2}$$

where $\varepsilon_{c,v}(k)$ is the energy of the conduction or valence band and $P_{cv}(k)$ the transition dipole moment. Equation (7.4.2) is an example of the independent-particle picture, (7.4.1), since this process can be considered as the simultaneous creation of a conduction electron with momentum $\hbar k$ and energy $\varepsilon_c(k)$ and a positive hole with momentum $-\hbar k$ and energy $-\varepsilon_v[-(-k)]$. In fact, however, electron and the hole are subject to the attractive Coulomb force, as the result of which not only the continuum (7.4.2) is enhanced towards its low-energy edge but also a series of discrete lines due to bound states—exciton—appear below it.

In this section, we will present a few different examples of the final state interaction in the optical spectra. Sect. 7.4.1 is devoted to an example in which the internal degree of freedom of a composite particle acts as an attractor for the third quasiparticle to form a higher-order composite particle. Section. 7.4.2 is concerned with the spectral anomaly of the core electron excitation in metal caused by the simultaneous redistribution of conduction electrons.

7.4.1 Exciton–Phonon Bound State

The phonon sideband of the exciton described in Sect. 7.3.5 represents the simultaneous excitation of an exciton with energy $E_K = E_0 + (\hbar^2 K^2/2\bar{M})$ and a phonon with energy $\hbar\omega_{-K}$[see (7.3.61′)]. Such LO-phonon sidebands have in fact been observed in a variety of ionic crystals.

According to (7.3.62), the peak-to-peak separation $\hbar\omega'$ of the LO-phonon sideband from the main band (the zero phonon line merged by its acoustic phonon sideband) is expected to be larger than the LO phonon energy $\hbar\omega_1$ (for $K = 0$) because of the recoil energy of exciton (see the full line $\hat{s}_{op}$ in Fig. 7.19). While this is the case in alkali halides, there has also been observed the separation $\hbar\omega'$ smaller than $\hbar\omega_l$ (by the order of ten percents), especially in those crystals with exciton binding energy R as small as $\hbar\omega_l$. As an example, the absorption spectrum of MgO is shown in Fig. 7.21. This fact implies that the sideband corresponds not to the free pair of exciton and LO phonon but to a bound state of them—*exciton-phonon bound state*—with binding energy $\hbar\omega_l - \hbar\omega'$.

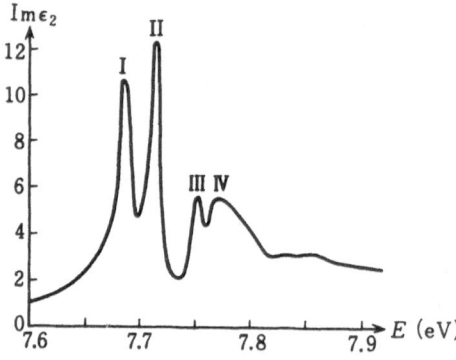

Fig. 7.21. The fundamental absorption spectra of MgO at 20 K. The peaks I and II correspond to 1s excitons originating from spin–orbit split valence bands, while III and IV are assigned to the exciton–phonon bound states originating from I and II, respectively. [R. C. Whited, W.C. Walker: Phys. Rev. Lett. **22**, 1478 (1969)]

What is the origin of the force binding together the exciton and the LO phonon? The longitudinal polarization wave of LO-mode lattice vibration gives rise to an electric field which strongly couples with the electron and hole. As for the exciton, the electron–hole relative motion as well as the translational motion will be coupled to this electric field. Confining ourselves to the former, we expect the exciton to be polarized by the electric field, with energy shift by $-\alpha E^2/2$ if the electric field E can be taken as uniform over the exciton (valid for the wavelengths longer than a). The polarizability α is proportional to a^2/R where a and R are the radius and the binding energy, respectively, of the exciton. This energy stabilization acts as attractive force between the exciton and the LO phonon. It resembles the interatomic (intermolecular) Van der Waals force although only the exciton has an internal degree of freedom in our case. The fact that the exciton–phonon bound state is observed only in the shallow—large radius—exciton is consistent with the above mentioned dependence of α on R and a.

The distortion of an internal state λ of the exciton induced by lattice vibration is to be described as the virtual excitation to other states λ' through the matrix elements $V_{\lambda K, \lambda' K'}$ of (7.3.7). Since the energy denominator $(\varepsilon_{\lambda K} - \varepsilon_{\lambda' K'})$ of the second-order perturbation is of the order of R, the shallower exciton is subject to the greater change in energy and wave function.

It would be instructive to recapitulate the similar problem in localized excitation before considering the exciton. As was stated in Sect. 7.2.7, the adiabatic potential of the lowest excited state which is closely followed by a host of higher states has smaller curvature (and hence, smaller phonon energy) due to the second-order term in (7.2.12). In the case of AgBr: I (Fig. 7.10) mentioned as an example, the coefficient $c_{j\lambda\lambda'}(\lambda \neq \lambda')$ of the second-order term is the matrix element of electron–phonon interaction H' between different internal states of the bound exciton, and corresponds to $V_{\lambda K, \lambda' K'}$ of (7.3.7). The phonon structure with $\hbar\omega' \sim 0.8\ \hbar\omega_l$ observed in AgBr: I was ascribed to the localized mode of LO phonon resulting from the second-order term.

What would happen if the impurity ion I⁻ were suddenly replaced by a host ion Br⁻? The bound exciton would then be delocalized. The same would be the case for the higher state consisting of the bound exciton plus one localized LO phonon: Namely, these two quasiparticles, being kept together, would be translationally delocalized. In fact, the phonon was bound to the bound exciton (through the second-order effect mentioned above) rather than to the impurity itself, as is evident from the absence of LO-phonon energy shift in the emission spectrum in Fig. 7.10.

The adiabatic approximation is not appropriate in describing the exciton–phonon bound state, first because we are concerned with a moving entity instead of localized excitation and secondly because the energy separations of internal states of the exciton are as small as the LO-phonon energy. Therefore, we consider the resonance of zero- and one-phonon states as follows. Noting that the total wave vector K_{tot} is a constant of motion of the exciton-LO-phonon system: (7.3.4), we consider the excited states with $K_{tot} = 0$ which can be reached optically from the ground state $|g, 0\rangle$. The lowest state is (a) $1s\,(\lambda = 1)$ exciton with $K = 0$ and no phonon: $|10, 0\rangle$; the next lowest states are (b) $1s$ exciton and one phonon: $b^{\dagger}_{-K}|1K, 0\rangle$, and (c) higher exciton $(\lambda > 1)$ and no phonon: $|\lambda 0, 0\rangle$. When $R \sim \hbar\omega_l$, it would be a tolerable approximation to solve the eigenvalue problem for the second excited state within the subspace spanned by (b) and (c). Put the wave function

$$\Psi = \sum_K u_K b^{\dagger}_{-K}|1K, 0\rangle + \sum_{\lambda > 1} v_\lambda|\lambda 0, 0\rangle \tag{7.4.3}$$

into the equation $(H - E)\Psi = 0$, where H is given in (7.3.2–4, 7). Multiplying $\langle 1K; 0|b_{-K}$ and $\langle 0\lambda; 0|$ from the left of this equation, one obtains the equations for u_K and v_K. Eliminating v_K, one obtains the equation

$$(\varepsilon_{1K} + \hbar\omega_{-K})u_K + \sum_{K'}\left(\sum_{\lambda > 1} \frac{V_{1K\,\lambda 0}V_{\lambda 0\,1K'}}{E - \varepsilon_{\lambda 0}}\right)u_{K'} = Eu_K . \tag{7.4.4}$$

The first term represents the energy sum of independent quasiparticles with $K_{tot} = 0$, and the second term their interaction originating in the internal degree of freedom of the exciton as mentioned before. u_K is the K representation of the wave function for the exciton–phonon relative motion.

If $E < \varepsilon_{\lambda 0}$ for $\lambda > 1$, this interaction is certainly attractive. If the attractive potential is deep enough, there will appear an exciton–phonon bound state below the continuum of the free states: $\varepsilon_{1K} + \hbar\omega_{-K}$. It can be easily seen that the interaction term in (7.4.4) corresponds to the second-order term in (7.2.12).

The interaction term in (7.4.4) turns out to be small for the exciton with large binding energy: $R \gg \hbar\omega_l$ because of the large denominator. For exciton with binding energy as small as $\sim \hbar\omega_l$, there appear a host of internal excited states $\varepsilon_{\lambda 0}(\lambda > 1)$ in near reasonance with the one phonon states: $\varepsilon_K + \hbar\omega_{-K}$, making big contributions to the interaction term. Because of the distribution of $\varepsilon_{\lambda 0}$ towards the high-energy side, the interaction is expected to be attractive in effect. In fact, the exciton–phonon bound state has been observed only in those crystals of which the exciton binding energy R is of the same order as $\hbar\omega_l$.

When R is significantly smaller than $\hbar\omega_l$, the energy of the exciton–phonon bound state, $\varepsilon_{10} + \hbar\omega'$, may be larger than that of the band gap, $\varepsilon_g = \varepsilon_{10} + R$. The bound state is then a resonance state embedded in the ionization continuum—a quasibound state. The corresponding absorption spectrum is expected to be given by a Lorentzian form, as given by (7.3.54) for the exciton line, with lifetime broadening $2\Gamma(E)$ because of the disintegration of this composite particle (through absorption of its internal phonon) into an unbound pair of electron and hole. Being proportional to the spectral density of the final states, $\Gamma(E)$ behaves as a step function at $E \approx \varepsilon_g$ in the same way as in Fig. 2.8a. In the absorption spectrum of TlBr shown in Fig. 7.22, the exciton–phonon quasibound state manifests itself as a "truncated Lorentzian" as described above. It is noted in passing that the step itself is followed by its phonon sidebands up to the third order, with interval being equal to the LO-phonon energy $\hbar\omega_l$ of the bulk crystal.

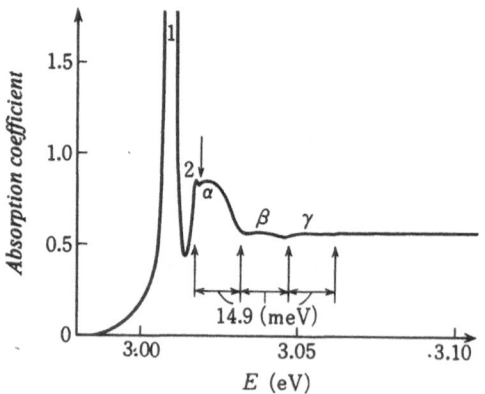

Fig. 7.22. The fundamental absorption spectra of TlBr. 1 and 2 represent 1s and 2s excitons, respectively, the latter being practically the edge of the ionization continuum. The broad band α is assigned to the quasibound state of the 1s exciton and LO phonon. Its low energy tail is sharply truncated below the peak 2. [S. Kurita, K. Kobayashi: J. Phys. Soc. Jpn. **28**, 1097 (1970)]

Finally, another example of a composite particle, in which a phonon is replaced by a magnon, will be mentioned. The continuous spectrum shown in Fig. 7.23 represents the *magnon sideband* of exciton (simultaneous excitation of an exciton and a magnon) of antiferromagnetic compound MnF_2. A sharp line slightly below the lower bound of the continuum is ascribed by Hopfield to the *exciton—magnon bound state*.

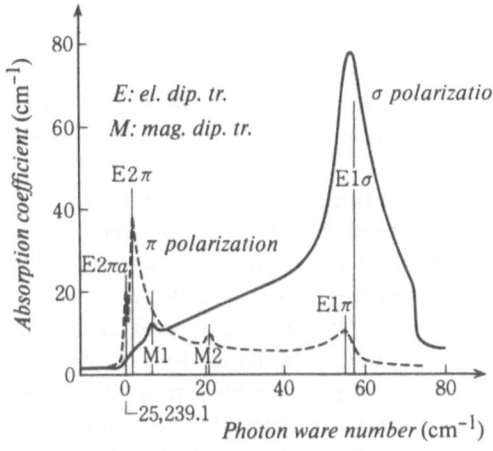

Fig. 7.23. The absorption spectra of MnF₂ around the magnetic dipole transition $^6A_{1g} \rightarrow {}^4A_{1g}$ ^4Eg (I). The continuous spec tra represent the magnon sideband of the exciton, while the very small peak at 25, 239.1 cm is assigned to the excitonmagnon bound state. [R.S. Meltzer, M.Y. Chen, D.S. McClure, M. Lowe-Pariseau: Phys. Rev. Lett. **21**, 913 (1968)]

7.4.2 The Edge Anomalies in the Soft X-Ray-Absorption Spectra of Metals

The energy necessary to excite an inner-core electron to the conduction band is usually in the soft-X-ray region. The energy width of the band formed by inner core states (whose energy will be denoted by ε_i) is usually so small that one can assume the positive hole to be well localized on a particular atom in the final state of the optical transition. In metals, the conduction band states are occupied up to the Fermi level ε_F as shown in Fig. 7.24a, and the final states available for the optically excited electron are confined to above ε_F. Therefore, one would expect the absorption spectrum to rise stepwise at $E_0 = \varepsilon_F - \varepsilon_i$ towards higher energy, as shown schematically in Fig. 7.24b. Similary, the emission spectrum due to the recombination of a conduction electron with an inner-core hole (which has been produced in some way beforehand) would rise stepwise at E_0 towards lower energy as shown in Fig. 7.24c. We will confine ourselves to absolute zero of temperature throughout this subsection.

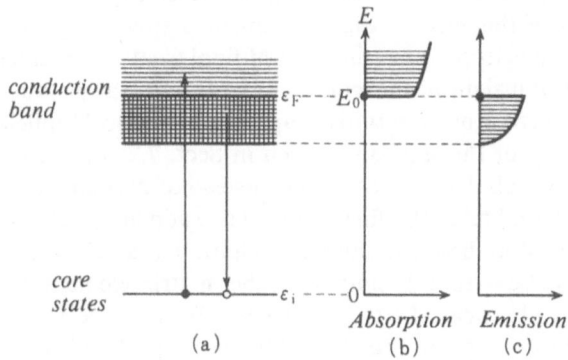

Fig. 7.24. a-c. The conduction band and the core states in metal **(a)**, and the soft X-ray absorption **(b)** and emission **(c)** spectra

In the above argument, we have neglected the electron–hole interaction, which, however, plays a role in metals as important as, and much more complicated than, that in insulators. We shall describe only the final-state interaction in the absorption spectra, since the initial-state interaction in the emission spectra can be treated in a similar way.

In the first place, we note that the Coulomb potential $-e^2/r$ due to the positive hole is screened into a short ranged potential $V(r)$ because of screening by conduction electrons [see (2.2.33)]. In the second place, the N (very large number, being proportional to the volume of the crystal) conduction electrons, which have been moving under periodic potential, as well as the electron excited from the inner core, are subject to the common scattering potential $V(r)$ in the final state. The wave function of each conduction electron is subject to modification of the order of N^{-1} because of the local potential $V(r)$, provided no bound state appears. It is the charge redistribution of N electrons, each with such modification, that screens the Coulomb potential of the core hole into the short-range potential $V(r)$. Therefore, $V(r)$ is to be determined self-consistently from the first principle.

For this reason, the optical transition of our concern is a $(N + 1)$-electron problem rather than a one-electron problem (the core hole being put aside simply as a potential source). The transition matrix element of electric dipole moment of the $(N + 1)$-electron system should be different from that of the one-electron excitation from the core state, since each of the N conduction electrons is subject to a change of N^{-1} accumulating to a finite charge redistribution. More important is the fact, inherent only in metals, that the Fermi degenerate electron system has high density of low-energy excitations due to the finite density of one-electron states around ε_F. One may well suppose that the sideband by simultaneous excitation of conduction electrons becomes very large towards vanishing energy, giving rise to an anomaly in the absorption edge of core electron excitation at $\varepsilon_F - \varepsilon_i$.

For simplicity, we neglect Coulomb interaction between electrons, except that it has been taken into account in the self-consistent determination of $V(r)$ which, however, is assumed to be given in our problem. While the dynamics is thereby reduced to a one-body problem of scattering, our difficulty is how to calculate the matrix elements of the electric dipole moment between the Slater determinants for the $(N + 1)$ electrons in the initial and final (scattered) states.

In order not to lose physical insight in the complicacy of such mathematical manipulations, let us present here a more intuitive approach, due to Hopfield, which is based upon the theory of the sideband given in Sect. 7.2. Let us first consider a fictitious system in which there are $(N + 1)$ (instead of N) conduction electrons in the initial state as well as in the final state. The sudden appearance of the potential $V(r)$ due to the core hole will induce excitation of an electron in state k_1 below ε_F into state k_2 above ε_F. For simplicity, the matrix element $V_{kk'}$ of the scattering potential $V(r)$ between the two states will be assumed to be a constant V_0 because of its short-range nature. The differential probability that an electron is excited by energy E' is then given, in the lowest-order perturbation theory, by

$$s(E') = \sum_{k_1}^{\varepsilon(k_1)<\varepsilon_F} \sum_{k_1}^{\varepsilon(k_2)>\varepsilon_F} \left| \frac{V_0}{\varepsilon(k_2) - \varepsilon(k_1)} \right|^2 \delta[E' - \varepsilon(k_2) + \varepsilon(k_1)]$$

$$= \frac{[N(\varepsilon_F) V_0]^2}{E'} = \frac{A}{E'} \quad (E' > 0) \tag{7.4.5}$$

if the probability for no excitation is normalized to unity. In deriving the second line of (7.4.5), E' has been assumed to be small enough so that $\varepsilon(k_1)$ and $\varepsilon(k_2)$ are confined to the vicinity of the Fermi level where the state density is given by $N(\varepsilon_F)$. δ given by

$$\delta \approx \tan \delta = - \pi N(\varepsilon_F) V_0 = \pi\sqrt{A} \tag{7.4.6}$$

represents the phase shift of the S-wave scattering by $V(r)$.

In the optical transition of the real system, the core electron should first be excited to (just above) ε_F in order to reach the initial state of the fictitious system. Equation (7.4.5) could therefore be taken as the spectral coupling for the sideband of conduction electron excitations [as was defined in (7.3.23) for the phonon sideband] starting from the zero line at $E_0 = \varepsilon_F - \varepsilon_i$. It is easy to see that the excitation energy $\varepsilon(k_2) - \varepsilon(k_1)$ of the conduction electron corresponds to the phonon energy $\hbar\omega_j$ and the scattering matrix element $V_{kk'} = V_0$ to the coefficient γ_j of electron–phonon coupling. If the low-energy excitations in the Fermi degenerate conduction electrons could be treated in the same way as those in harmonic oscillators, the total absorption spectra would be obtained as the Fourier transform of the generating function of the form (7.2.20).

An important difference of the present system from the electron–phonon system is that the coupling strength $S = \int_0^\infty s(E')dE'$ [see (7.2.22)] diverges logarithmically at the lower bound $E' = 0$ because of the E' dependence of $s(E')$ given by (7.4.5) (valid asymptotically as $E' \to 0$ as mentioned above). Considering the width of the conduction band and the (k, k') dependence of $V_{kk'}$, one can practically put an upper bound $E' = E_c$ for the integral while using (7.4.5) for the whole range, without causing any essential change in the argument which follows.

As an analogue to the phonon sideband, S represents the average number of conduction electrons excited simultaneously with core excitation. The above mentioned *infrared divergence* means that this number is infinitely large, predominantly because of low energy excitations. The fractional intensity $\exp(-S)$ of the zero line vanishes, indicating that the ground states of the N conduction electron system with and without the local potential $V(r)$ are orthogonal to each other—the effect being sometimes called the *orthogonality catastrophe*.

According to the convolution theorem for the higher-order sidebands as mentioned in Sect. 7.2.4, the more conduction electrons will be excited as the photon energy approaches the threshold energy from above, namely, the more terms in the perturbation theoretical expansion would be needed for convergence—the situation which might seem almost intractable. Fortunately, however, we have a closed form (7.2.20) for the generating function under the assumption of harmonic approximation, from which one can obtain the spectral shape without difficulty of divergence since the exponent

$$-S + S_+(t) = - \int_0^{E_c} dE' s(E') \left[1 - \exp\left(-\frac{i}{\hbar} E' t\right) \right]$$

as a whole does not show any divergence. We are interested in the asymptotic behavior of this integral at $|t| \to \infty$. The contribution of the exponential term can then be neglected, because of its rapid oscillation, except in a small region $|E'| \lesssim \hbar/t$ where the integrand tends to a finite value [instead of increasing as $(E')^{-1}$] due to the existence of this term. The latter situation can be taken into account by replacing the lower bound of the integrand by $\hbar/|t|$. Then the integral turns out to be $-A \ln|E_c t/\hbar|$ asymptotically as $|t| \to \infty$. Taking the Fourier transform of the generating function (7.2.20), one finds the asymtotic form of the absorption spectrum at the Fermi edge $E \to E_0 + 0$:

$$F(E) \propto (E - E_0)^{-(1-A)} . \tag{7.4.7}$$

(It can easily be shown that $F(E)$ vanishes at $E < E_0$ as it should.) According to the *Friedel sum rule*, $\sqrt{A} = \delta/\pi$ represents the S wave component of the extra number of conduction electrons brought over into the vicinity of the core hole through $V(r)$.

In the "fictitious" initial state considered above, we had $(N + 1)$ conduction electrons without perturbation $V(r)$. In the "real" initial state, however, one of the electrons was in the core state. This can be taken into account by considering a potential $U(r)$ just enough to bind one electron on that atom, òr equivalently (according to the Friedel sum rule mentioned above) by considering the phase shift $\delta = \pi$, in the initial state. It is the phase "difference" between the final and the initial states that governs the exponent of the edge singularity (7.4.7): in the expression $A = (\delta/\pi)^2$, one has to replace δ by the difference $\delta - \pi$. Hence, the real form of the edge singularity is given by

$$F(E) \propto (E - E_0)^{-2(\delta/\pi)+(\delta/\pi)^2} \quad (E > E_0) . \tag{7.4.8}$$

According to the more exact theory by Nozières and De Dominicis, the asymtotic form of the edge singularity is given by

$$F(E) = \sum_{l,m} |W_{lm}|^2 \left(\frac{E - E_0}{E_c}\right)^{-\alpha_l} , \tag{7.4.9}$$

$$\alpha_l = 2\frac{\delta_l}{\pi} - 2\sum_{l'}(2l'+1)\left(\frac{\delta_{l'}}{\pi}\right)^2 , \tag{7.4.10}$$

where δ_l is the scattering phase shift, due to $V(r)$, of the partial wave with orbital angular momentum $l(=0, 1, 2, \ldots)$ and W_{lm} the matrix element of electric dipole moment connecting the core state and the (l, m) component of the conduction band states at ε_F. The factor 2 in the second term of (7.4.10) comes from the spin degeneracy. Since δ_l/π represents the number of electrons with angular momentum (l, m) brought over into the vicinity of the core hole, their sum over orbital (l, m) and spin states should be equal to unity so as to fulfill the electrical neutrality (*Friedel sum rule*):

$$\sum_l 2(2l+1)\frac{\delta_l}{\pi} = 1 . \tag{7.4.11}$$

If the potential $V(r)$ is of very short range, significant scattering will take place only in the S wave which has nonvanishing amplitude at $r = 0$; namely, one expects that δ_0 is large compared to $\delta_l(l \neq 0)$, and hence, from (7.4.10, 11), that $\alpha_0 > 0$ and $\alpha_l < 0$ $(l \neq 0)$. The absorption-edge singularities in the two cases are shown schematically in Fig. 7.25a, b, together with the corresponding singularity of the emission edge which should be the mirror reflection of the former. According to the optical selection rule on the orbital angular momentum: $\Delta l = \pm 1$, the transition to the s components of the conduction band states is allowed [$W_{00} \neq 0$ in (7.4.9)] only from the p-like core states. In fact, the edge singularity of type (a) has been observed in the $L_{\mathrm{II, III}}$ spectra (from the 2p core) of Na, and that of type (b) in the K spectra (from the 1s core) of Li.

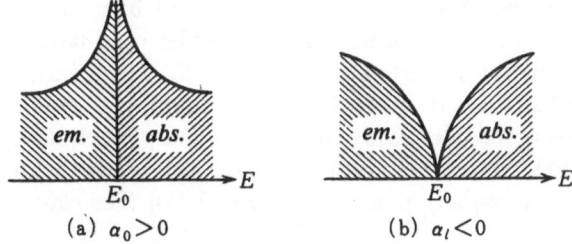

(a) $\alpha_0 > 0$ (b) $\alpha_l < 0$

Fig. 7.25. a,b. The edge anomalies in the soft X-ray absorption and emission spectra of metal

7.5 Self-Trapping

7.5.1 Polaron vs Self-Trapped Electron

As has been stated in the preceding and the present chapters, electrons, holes, and excitons moving in the crystal lattice are dressed with phonons induced by themselves, resulting in an effective mass change—the so-called renormalization. Especially strong is the electrostatic coupling between an electron (hole) and the optical phonons in ionic crystals, necessitating intermediate- or strong-coupling theory. In this case, however, the effective mass of the phonon-dressed electron or the "polaron" increases continuously with increasing coupling constant, without any abrupt change. In contrast, in the deformation-potential-type interaction [see (7.2.35, 36)] between electron and acoustic phonons, the renormalized effective mass increases abruptly by several orders of magnitude when the coupling constant exceeds a certain critical value, although the renormalization effect below this critical value is usually much smaller than in the case of optical phonons. The electron with such an enormous effective mass should be considered to be practically immobile, and is called the *self-trapped electron*. Even a small local perturbation would readily destroy the coherent motion of this phonon-dressed electron since the bandwidth of this quasiparticle (inversely proportional to the effective mass) is extremely small.

The concept of self-trapping was first introduced by Landau, who noticed the importance of the electric polarization of the ionic lattice. However, it turned out later, through the development of polaron theory, that such an extreme situation as occurs with self-trapping is realized only for unrealistically large coupling constants of electron-optical-phonon interaction. Only the deformation potential by acoustic phonons can be a trigger for the catastrophic change into the self-trapped state, a state well distinguishable from the polaron state.

This difference between the two types of interactions originates from their difference in the *force range*: the electrostatic interaction between polarization (optical phonons) and charge (electron) is *long range* ($\propto |r - r'|^{-1}$), whereas the deformation potential (acoustic phonons) for the electron is *short range* [$\propto \delta(r - r')$]—nonvanishing only where the lattice deforms [see (7.2.35)]. Let us consider an extra electron in the otherwise empty conduction band in insulating crystal. The effective mass of the conduction band (under rigid lattice) is denoted by m, and the deformation potential coefficient by E_d. The acoustic mode of lattice vibrations will be replaced by an elastic continuum, with cutoff for the crystal lattice being considered later. Only the dilation component, $\Delta(r)$, of the strain tensor is considered, since other components do not interact with the electron in the standard case [see the statement before (7.2.35)]. The elastic energy is then given by

$$\frac{1}{2} C \int \Delta(r)^2 dr \tag{7.5.1}$$

with appropriate elastic constant C. Let the medium deform in such a way that $\Delta(r) = \Delta \neq 0$ only within a sphere of radius R. The electron is then subject to the square well potential

$$V(r) = \begin{cases} E_d \Delta & (r < R) , \\ 0 & (r > R) . \end{cases} \tag{7.5.2}$$

if the sign of Δ is so chosen that $E_d\Delta < 0$. The electron can be bound by this potential well only when

$$\frac{2m}{\hbar^2} R^2(-E_d\Delta) > \left(\frac{\pi}{2}\right)^2 , \tag{7.5.3}$$

according to elementary quantum mechanics. In the extreme situation where the inequality in (7.5.3) is replaced by $\gg$, the binding energy is given by

$$E(\Delta, R) = |E_d\Delta| - \frac{\hbar^2}{2m}\left(\frac{\pi}{R}\right)^2 , \tag{7.5.4}$$

namely, the depth of the well minus the kinetic energy due to the uncertainty principle. Whereas, $E(\Delta, R) = 0$ when the inequality sign in (7.5.3) is reversed.

Measuring the electronic energy from the bottom of the conduction band as origin, one can write the adiabatic potential for the lowest electronic state of our system as a function of Δ and R, as follows:

$$W(\Delta, R) = \frac{1}{2} C \cdot \frac{4\pi}{3} R^3\Delta^2 - E(\Delta, R) . \tag{7.5.5}$$

The equipotential map of (7.5.5) is shown in Fig. 7.26, where the arrow indicates the direction of increasing potential. As we leave from the R or Δ axis where $W = 0$, we have only the increasing elastic energy until we reach the broken line, beyond which the bound state appears according to (7.5.3), contributing the negative term to (7.5.5). In particular, W takes a negative value in the shaded region, becoming infinitely deep towards the lower right.

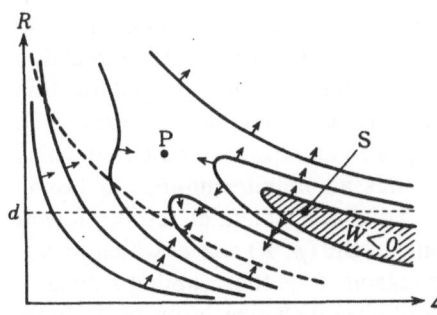

Fig. 7.26. The adiabatic potential for the lowest state of an electron embedded in the elastic continuum

In this way, our system of an electron in an elastic continuum always collapses into the self-trapped state with infinitely strong ($|\Delta| \to \infty$) and localized ($R \to 0$) distortion of the elastic medium, whereas the free state without distortion ($\Delta = 0$) is only metastable! In order to take into account the atomic structure of the lattice, however, we have to cut off the region $R < d$ (lattice constant) in Fig. 7.26, since we cannot conceive of elastic distortion with $R < d$.

If the shaded region ($W < 0$) is below the dotted line with $R = d$, the free state $F(\Delta = 0)$ is the stable state of the electron. If the shaded region extends above the dotted line as shown in the figure, the self-trapped state $S(R = d, \Delta \neq 0)$ is more stable than the free state. Making use of (7.5.4) as a rough estimation, we find that (7.5.5) with $R = d$ takes a minimum at

$$\Delta = -\frac{E_d}{C \cdot \frac{4\pi}{3} d^3} \, , \tag{7.5.6}$$

with the value

$$W_m = -\frac{E_d^2}{2C \cdot \frac{4\pi}{3} d^3} + \frac{\hbar^2}{2m} \left(\frac{\pi}{d}\right)^2 . \tag{7.5.7}$$

The self-trapped state S is more stable if the first term—the lattice relaxation energy due to the trapped electron [sum of (7.5.1, 2)] overcomes the second term—the kinetic energy of the electron due to localization. Otherwise, the free state F is more stable. These two types of states are rather distant from each other in the (Δ, R) space, being separated by a potential barrier (the saddle point denoted by P in the figure), and hence with lattice wave functions being almost orthogonal to each other. If one defines the electron–phonon coupling constant g by the ratio of the first and the second terms in (7.5.7), one finds that the stable state changes abruptly from F to S as g exceeds the critical value $g_c = 1$, although the other state can be a metastable state.

The existence of the potential barrier between the F and S states originates from the short-range nature of the interaction (7.2.35) which imposes the condition (7.5.3) for the electron to be bound. In the case of electrostatic interaction with the polarization field, one may consider $P(r)$ such that the polarization charge $-\text{div}P = \rho$ takes a constant positive value within the region: $r < R$, and zero otherwise. The electron is then subject to an attractive potential which is Coulombic at $r > R$ and which always has an infinite number of bound states. A simple argument shows that the adiabatic potential for the lowest electronic state has only one minimum point in the (ρ, R) space, which moves continuously towards increasing ρ and decreasing R as the coupling increases. The absence of potential barrier, and hence, of abrupt change, is due to the long-range nature of the electron–optical-phonon interaction—the Coulombic potential has always bound states.

In the adiabatic approximation described so far, we have ignored the kinetic energy of the lattice as well as the translational symmetry of the lattice. According to the polaron theory presented in Sects. 6.2, 3 in which they were properly considered from the beginning, the situation changes continuously as the coupling constant increases, in agreement with the above consideration although the adiabatic picture is valid only in the strong coupling limit—the static regime where the phonon energy plays the secondary role in the self-energy and the polaron radius. In the weak-coupling limit—the dynamical regime where the finiteness of the phonon energy plays an essential role, the self-energy can be calculated in the lowest-order perturbation by considering one phonon scattering of the free electron.

The transition from the dynamical regime to the static regime takes place abruptly at the critical coupling constant $(g = g_c)$ in the case of the short-range interaction, as mentioned above. The situation in the dynamical regime $(g < g_c)$ is similar to that of the polaron problem, except that the mass renormalization effect is much smaller in the present case. In the static regime $(g > g_c)$, the adiabatic approximation would give a self-trapped state at the nth lattice site, with wave function $\Psi_n = \phi_n(r)\chi_n(q)$. ϕ_n represents the electronic wave function localized around the site n, and χ_n the zero-point vibrational state of the lattice distorted around that site, both being to be determined self-consistently so as to minimize the total energy. Because of the translational symmetry, we have equivalent states at all other sites. The true eigenstate should be of a plane-wave-like linear combination of them: $N^{-1/2} \sum_n \exp(i\boldsymbol{K} \cdot \boldsymbol{R}_n)\Psi_n$ as in the tight-binding approximation for the Bloch states. The difference is that the bandwidth is reduced by the overlap integral of the lattice: (χ_n, χ_m) compared to the purely electronic problem. If, for instance, $\phi_n(r)$ is well localized within the nth atom, it is only the electronic Hamiltonian H which has matrix elements between Ψ_n and Ψ_m $(n \neq m)$:

$$(\Psi_n, H\Psi_m) = (\phi_n, H\phi_m)\,(\chi_n, \chi_m) \; ,$$

namely, the electronic transfer energy times the lattice-overlap integral.

χ_m is reached from χ_n through the following two steps. First remove the lattice distortion around the nth site so as to obtain $\chi^{(0)}$, the zero-point vibrational state of the undistorted lattice; secondly, set up the lattice distortion of the same pattern around the mth site so as to obtain χ_m. If the spatial overlap of the distortion patterns around the nth and mth sites is neglected (which is a rather crude approximation when m and n are nearest neighbors), one can put $|(\chi_n, \chi_m)| \approx |(\chi_n, \chi^{(0)})(\chi^{(0)}, \chi_m)| = |(\chi_n, \chi^{(0)})|^2$. According to Sect. 7.2, the lattice overlap integral between the distorted (χ_n) and undistorted $(\chi^{(0)})$ zero-point states (see Fig. 7.3) is given by $\exp(-S/2)$ where S is equal to the lattice relaxation energy E_{LR} divided by the average energy of the phonons [see (7.2.17, 22, 23)]. In the present case, E_{LR} is given by the first term of (7.5.7). In this way, we have $|(\chi_n, \chi_m)| \approx \exp(-S)$ as a crude estimation.

The wave function $\phi_n(r)$ of the self-trapped electron has spatial extention of the order of the lattice constant d according to Fig. 7.26, and hence, S is of the order of several tens according to Fig. 7.7. The reduction factor $\exp(-S)$ for the bandwidth due to the lattice distortion around the electron is so small, and the magnification factor $\exp(+S)$ for the effective mass is so large, that the electron can practically be considered to be immobile.

In this way, the consideration of translational symmetry with quantum-mechanical treatment of the lattice turns out to cause no essential change in the conclusion from the adiabatic picture that the stable state of the electron with short-range interaction with the lattice changes abruptly, at a critical value of the coupling constant, from F to S which are completely different in nature. They can be taken as essentially different quantum-mechanical states since their overlap integral $\exp(-S/2)$ is exceedingly small.

Under the coexistence of optical and acoustic modes as is the case in ionic crystals the F state is the polaron state dressed mainly with optical phonons, while both of the modes make comparable contributions to the S state (note that $S_{ac} \approx S_{op}$ at $a \sim d$ in Fig. 7.7) although the trigger for self-trapping is the acoustic mode. According to various kinds of observations reported so far including electrical conductivity, optical spectra, and spin resonance, electrons, holes, and excitons in ionic crystals and semiconductors can be rather distinctly classified into F and S types (molecular crystals with transfer energy as small as phonon energies are to be excluded since the quantum effect readily destroys the discontinuity). In almost all alkali halides, the electrons in the conduction bands belong to the F type while the holes in the valence bands to the S type. This difference seems to originate mainly from the difference in the band effective mass m of the rigid lattice [note that the coupling constant g is proportional to m according to (7.5.7)] between the two bands: the effective mass m of the valence band is several times the true electron mass m_0 in view of the band calculation available and the band width observed by soft-X-ray emission, while that of the conduction band is somewhat smaller than m_0 according to the cyclotron resonance experiments. It is then quite reasonable to have $g < 1$ for electrons and $g > 1$ for holes. It should be mentioned in passing that the self-trapped hole in the alkali halides is not of the one-center type as considered in our simplified model but of the two-center or molecular type, sometimes called V_K *center*: an electron is missing in the highest antibonding molecular state of a pair of nearest halide ions which are closer to each other than their regular distance, forming so to speak a molecular ion: Cl_2^- in the lattice. The situation is critical for the holes in silver halides although electrons are always mobile: the hole belongs to the S type in AgCl but to the F type in AgBr. This big difference in spite of general similarities between the two substances can only be understood by the discontinuity mentioned above: g is slightly above and below g_c in AgCl and AgBr, respectively. The self-trapped hole in AgCl is of the one center type, being located on Ag^+ (forming Ag^{++}) and not on Cl^-, with tetragonal distortion of the surrounding lattice. This is due to the 4d orbitals of Ag which strongly mix with 3p orbitals of Cl to form the valence band. According to the observation reported so far electrons and holes in II–VI and III–V compounds belong to the F type, presumably because of wide conduction and valence bands.

7.5.2 Free Exciton vs Self-Trapped Excitons

Excitons can also be classified distinctly into the F and S types in view of available observation. According to the Frenkel model, the exciton is not subject to long-range interaction with phonons because of its electrical neutrality, but this has no relevance to the existence of the F–S discontinuity which is caused by the short-range interaction. According to the Wannier–Mott model, one may well suppose that the self-trapping of the exciton is governed by that of its heavier (band) mass constituent—usually a positive hole. If the positive hole is self-trapped, the electron will also be trapped around it through the Coulomb potential unless the Coulomb binding is so strong as to liberate the hole from self-trapping. In fact, the excitons belong to the S type in alkali halides and AgCl, and to the F type in AgBr and II–VI, III–V compounds, in conformity with the behavior of the positive holes.

Since the exciton immediately after its optical creation is not yet subject to enough lattice relaxation, the effect of self-trapping will not manifest itself so significantly in the absorption spectra. On the other hand, it will be reflected much more directly in the emission spectra of the exciton annihilation which takes place after the lattice relaxation. The emission spectra of excitons belonging to the F type consist of sharp resonance line (about in the same position as the absorption line) and a few optical phonon sidebands, as shown in Fig.2.13 for CdS. In contrast, the emission spectra of the S type excitons behave similarly to those of the localized electron strongly coupled ($S \gg 1$) to the lattice as described in Sect. 7.2, namely, a broad Gaussian band with large Stokes shift from the absorption line, as is also seen from Fig.7.27. In this figure, the abscissa represents the configuration coordinate Q for a local lattice distortion (similar to the interaction mode introduce in Sect. 7.2.5, and the ordinate the adiabatic potential of the exciton-lattice system. As Q increases, a localized state of the exciton will split off from the continuum of the exciton band. This adiabatic potential in a

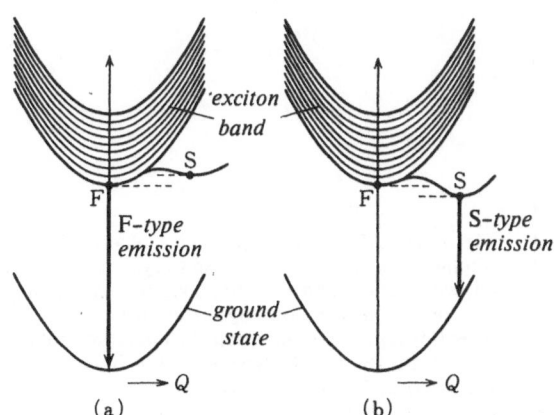

Fig. 7.27. a,b. The adiabatic potential for an exciton in a deformable lattice with local distortion Q, and the two different types of luminescence

one-dimensional configuration coordinate may be interpreted as a cross section of the two-dimensional adiabatic potential of Fig.7.26 along the line $R \sim d$. It will be evident from this figure how the emission spectra from the relaxed exciton behave differently according to whether the F state or the S state is more stable.

It is instructive to present here the observation of exciton emission spectra in the mixed crystals $AgBr_{1-x}Cl_x$ in which the transition from situation (a) to (b) of Fig.7.27 has been pursued by varying x. In Fig.7.28a,b are shown the emission spectra (solid lines) together with the indirect absorption edge (broken lines) for various concentrations. Only the F-type emission spectra are observed at $x \lesssim 0.4$, and only the S-type at $x \gtrsim 0.5$; their intensity ratio varies rapidly within a small range of x as shown in Fig.7.28c.

The rather abrupt transition of the emission spectra from the F type to the S type at $x \sim 0.45$ can be explained in terms of Fig.7.27 if the material constants E_d, C, m and d which are relevant to the coupling constant g [see (7.4.7)] are assumed to vary continuously with x. The small but finite range ($0.4 \lesssim x \lesssim 0.5$)

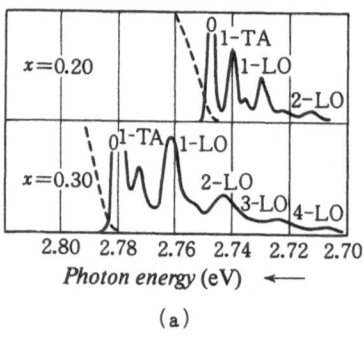

(a)

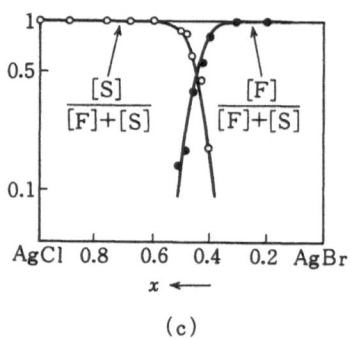

(c)

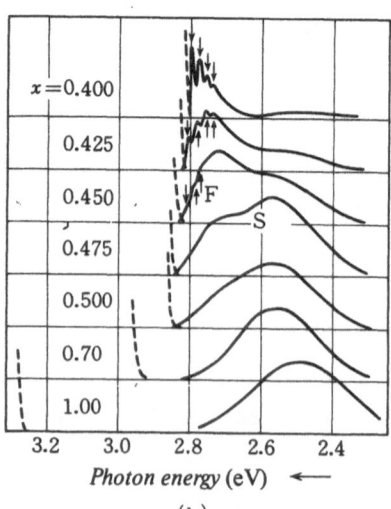

(b)

Fig. 7.28a-c. The emission spectra of exciton annihilation at $T = 2$ K in mixed crystals $AgBr_{1-x}Cl_x$ with various concentrations x. For comparison, the indirect absorption edge is shown by broken lines. (c) represents the quota of integrated intenties of the S and F type emission bands. [H. Kanzaki, S. Sakuragi, K. Sakamoto: Solid State Commun. **9**, 999 (1971)]

of F–S transition may be partly due to the local fluctuation of x and partly due to the incomplete thermalization between F and S excitons which are separated by a potential barrier.

7.5.3. The Electron Bubble and Exciton Bubble in Liquid Helium

The electrons, holes, and excitons in liquid helium are also self-trapped, the situation being even more drastic because of the fluidity. The bottom of the conduction band is known to be higher than the vacuum level by $V \sim 1.0$ eV, presumably because the electron feels a repulsive potential by the hard core of He atom.

This means that kinetic energy V must be supplied to an electron in order to inject it into liquid helium. Once injected, the electron will be stabilized by forming a cavity around itself since the potential energy is thereby lowered from V to 0. The optimum radius R of this *electron bubble* will be determined by minimizing the sum of the surface energy $4\pi\sigma R^2$ associated with surface tension σ of helium and the kinetic energy $(\hbar^2/2m)(\pi/R)^2$ of the electron confined within the cavity, the contribution from vapor pressure being negligibly small. The estimation gives $R \sim 20$ Å. The potential well of this size and depth (V) has a p-like excited state to which the electron can make optical transition, although such has not been observed yet.

The repulsive force between electron and helium atom corresponds to the negative sign of the deformation potential E_d since the local dilation ($\Delta > 0$) leads to the lowering of potential V. In contrast, the deformation potential for the valence-band hole seems to be positive, since the hole is supposed to form a *snow ball* in which the atomic density is higher than the normal value.

Even more interesting is the behavior of excitons. Liquid helium has a strong absorption band around 21 eV which presumably corresponds to the Frenkel exciton with (1s) (2p) configuration of the excited atom in view of its intensity and position. The emission band of the exciton is the S type broad band with Stokes shift as large as 5 eV; moreover, it resembles very much the emission band which is observed in helium gas and is well known to correspond to the de-excitation of He_2^* to its nonbonding ground state [an excited helium atom He* attracts another helium atom (unexcited) to form an excited molecule He_2^*]. This fact implies that an exciton in liquid helium relaxes into He_2^*, forming around it a cavity large enough to assure its independence from the surrounding medium. It should be noted that the interatomic distance of He_2^* molecule in gas is larger than the nearest-neighbor distance in liquid helium. On the other hand, it has been observed in superfluid helium that electrically neutral energy carriers, produced by α particle bombardment, at the bottom, can move almost unscattered to the upper surface where they seem to dissociate into He_2^+ ions and electrons, as the latter can be detected outside. It is not certain whether this neutral carrier is the electronic entity such as the metastable F type Frenkel exciton or the molecular entity such as the long-lived triplet state of He_2^*.

Bibliography

The following bibliography includes a number of books which may be useful for readers proceeding to more advanced studies of any subject discussed in the text. It also includes those original papers and review articles in which readers can find details of proofs and calculations omitted in the text to save space.

We should warn readers, however, that the list, being chosen rather arbitrarily, is by no means exhaustive.

General

(1) Kittel, C.: *Quantum Theory of Solids* (Wiley, New York 1963)
is perhaps one of the most standard books related to the whole volume of the present text.
(2) Landau, L. D., Lifshitz, E. M.: *Statistical Physics,* English transl. by J. B. Sykes, M. J. Kersley (Pergamon, Oxford, New York 1968)
Feynman, R. P.: *Statistical Mechanics,* ed. by J. Shaham (Benjamin, New York 1972)
contain some fine descriptions of elementary excitations.
(3) Thouless, D. J.: *The Quantum Mechanics of Many-Body Systems* (Academic Press, New York 1961)
is a brief description of mathematical methods and accessible to nonexperts. More advanced books on mathematical methods will be given below (see entries for Chap. 3).
(4) Pines, D.: *Elementary Excitations in Solids* (Benjamin, New York 1963)
was written, perhaps for the first time as a book, to present a solid as a system of elementary excitations.
(5) Anderson, P. W.: *Concepts in Solids* (Benjamin, New York 1963)
(6) Peierls, R. E.: *Quantum Theory of Solids* (Clarendon Press, Oxford 1964)
contain critical descriptions of fundamental concepts.

Chapter 1

A neat description of crystal symmetries is given in (2), Landau and Lifshitz. As for phonons, see (1), (4), and (6). The last includes physical effects caused by anharmonic terms. Somewhat more recent are

(7) Ludwig, W.: *Recent Developments in Lattice Theory,* Springer Tracts in Modern Physics, Vol. 43 (Springer, Berlin, Heidelberg, New York 1967)

(8) Cochran, W.: *Lattice Vibrations,* Reprint from Reports on Prog. in Phys. (Benjamin, New York 1969).

The latter volume also contains articles on effects of point defects by A. A. Maradudin and by I. M. Lifshitz and A. M. Kosevitch.

(9) Back, T. A. (ed.): *Phonons and Phonon Interactions* (Benjamin, New York 1964)

collects a series of lectures on the same subjects as in the present text.

(10) Guyer, J. P.: *Solid State Physics,* Vol. 23, ed. by F. Seitz, D. Turnbull (Academic Press, New York 1969)

(11) Varma, C. M., Werthamer, N. R.: In *Physics of Liquid and Solid Helium, Part I,* ed. by K. H. Bennemann, J. B. Ketterson (Wiley, New York 1976)

are review articles on solid helium as a quantum solid.

(12) Andreev, A. F.: LT 15, J. de Phys. *39,* C6–1257 (1978)

is a review on quantum defects.

Chapter 2

In connection with the optical mode of lattice vibrations, the following monograph is to be recommended because of its enlightening argument on the effective electric field within dielectrics:

(13) Fröhlich, H.: *Theory of Dielectrics – Dielectric Constant and Dielectric Loss* (Oxford Univ. Press, Oxford 1949).

The dielectric dispersion theory presented in Sects. 2.1.2 and 2.1.3 is mostly based upon the article

(14) Kurosawa, T.: J. Phys. Soc. Jpn. *16,* 1299 (1961).

A review article

(15) Cochran, W., Cowley, R. A.: „Phonons in Perfect Crystals", in *Light and Matter Ia.* Encyclopedia of Physics, Vol. XXV/2 a, ed. by S. Flügge (Springer, Berlin, Heidelberg, New York 1967) p. 59

provides a more systematic description of lattice vibrations and their dispersions and a brief account of ferroelectric transition.

As for the exciton,

(16) Knox, R. S.: *Theory of Excitons* (Solid State Physics, suppl. 5) (Academic Press, New York 1963)

is a standard textbook covering various aspects, while

(17) Davydov, A. S.: *Theory of Molecular Excitons,* English transl. by S. B. Dresner (Plenum Press, New York, London 1971)

gives details on molecular crystals. A series of lecture notes on excitons is found in

(18) Kuper, C. G., Whitfield, G. D. (eds.): *Polarons and Excitons* (Oliver and Boyd, Edinburgh 1963).

More recent developments are reviewed in

(19) Cho, K. (ed.): *Excitons,* Topics in Current Physics, Vol. 14 (Springer, Berlin, Heidelberg, New York 1979)

A review on excitonic molecules is presented by

(20) Hanamura, E.: In *Optical Properties of Solids,* ed. by B. Seraphin (North-Holland, Amsterdam 1976) p. 81,

while the dynamics of triplet excitons in molecular crystals is detailed in

(21) Avakian, P., Merrifield, R. E.: Mol. Cryst. *5,* 37 (1968).

Chapter 3

As a general reference for this chapter, (1) by Kittel will be most suitable. In particular, the detailed discussion of electron gas is provided in Chaps. 5 and 6 of this book.

(22) Pines, D. (ed.): *The Many-Body Problem* (Benjamin, New York 1962) is recommended to a reader who wants to study the many-body problem not restricted to the fermion system. This is a kind of collection of papers, and Landau's work on Fermi liquid theory is included in its original form.

(23) Pines, D., Nozières, P.: *The Theory of Quantum Liquids,* Vol. 1 (Normal Fermi Liquids) (Benjamin, New York 1966)

may be readable for beginners in this field.

(24) Nozières, P.: *Theory of Interacting Fermi Systems,* transl. by D. Hone (Benjamin, New York 1964)

presents the detailed description of general properties of Fermi liquid which was briefly referred to in Sect. 3.5.

In addition to the references mentioned above, the following books are suggested:

(25) Fetter, A. L., Walecka, J. D.: *Quantum Theory of Many-Particle Systems* (McGraw-Hill, New York 1971)

(26) Haken, H.: *Quantum Field Theory of Solids, An Introduction* (North-Holland, Amsterdam 1976).

As for the dilute solution ^3He in liquid ^4He, we presented only a short explanation in this book. The review article

(27) Ebner, C., Edwards, D. O.: Phys. Rep. (Sect. C of Phys. Lett.) *2,* 77 (1971)

is recommended to a reader interested in this problem.

Chapter 4

Chapters on symmetries and phase transitions in (2), Landau-Lifshitz, are illuminating. More details may be found in

(28) Forster, D.: *Hydrodynamic Fluctuations, Broken Symmetries, and Correlation Functions* (Benjamin, New York 1975).

As for the charge density wave state in one-dimensional conductors, see

(29) Barišić, S., Bjeliš, A., Cooper, J. R., Leontić, B. (eds.): *Quasi One-Dimensional Conductors*. Lecture Notes in Physics, Vols. 95, 96 (Springer, Berlin, Heidelberg, New York 1979).

For the spin ordering,

(30) Nagamiya, T.: *Solid State Physics,* Vol. 20, ed. by F. Seitz, D. Turnbull (Academic Press, New York 1967).

The original paper of the Holstein-Primakoff method is

(31) Holstein, T., Primakoff, H.: Phys. Rev. *58,* 1094 (1940); see also Dyson, F. J.: Phys. Rev. *102,* 1217, 1230 (1956).

The antiferromagnetic case is discussed in

(32) Anderson, P. W.: Phys. Rev. *86,* 694 (1952).

The concept of quasiaverage is given in

(33) Bogoliubov, N. N.: Physica *26,* 1 (1960).

An introduction to the mathematics of infinite quantum system is

(34) Haag, R.: "The Algebraic Approach to Quantum Statistical Mechanics. Equilibrium States and Hierarchy of Stability", in *Critical Phenomena,* ed. by J. Brey, R. B. Jones, Lecture Notes in Physics, Vol. 54 (Springer, Berlin, Heidelberg, New York 1976) p. 156

Quantum theory of optical coherence is reviewed, for instance, by

(35) Glauber, R. J.: In *Quantum Optics,* ed. by S. M. Kay, A. Maitland (Academic Press, New York 1970).

For the idea of regarding superfluid ^4He as a coherent de Broglie wave, see

(36) Anderson, P. W.: Rev. Mod. Phys. *38,* 298 (1966).

For details of superconductivity, see articles in

(37) Parks, R. D. (ed.): *Superconductivity,* 2 volumes (Dekker, New York, Basel 1969).

As for superfluid $^3He,$

(38) Leggett, A. J.: Rev. Mod. Phys. *47,* 331 (1975);
Anderson, P. W., Brinkman, W. F.: In *The Helium Liquids,* ed. by J. G. Armitage, I. E. Farquhar (Academic Press, New York 1975) p. 315;
Mermin, N. D.: In *Quantum Liquids,* ed. by J. Ruvalds, T. Regge (North-Holland, Amsterdam 1978) p. 196.

The free energy of the one-dimensional Ising model is found, for instance, in

(39) Thompson, C. J.: In *Phase Transitions and Critical Phenomena,* Vol. I, ed. By C. Domb, M. S. Green (Academic Press, New York 1972) p. 177.

The XY spin model of superfluid ^4He was discussed first by

(40) Matsubara, T., Matsuda, H.: Prog. Theor. Phys. *16,* 569 (1956); see also Nakajima, S.: "Superfluid and Excitonic States", in *Physics of Highly Excited States in Solids,* ed. by M. Ueta, Y. Nishina, Lecture Notes in Physics, Vol. 57 (Springer, Berlin, Heidelberg, New York 1976) p. 130.

Original papers concerning Goldstone modes are

(41) Goldstone, J.: Nuovo Cimento *19*, 154 (1961);
Anderson, P. W.: Phys. Rev. *112*, 1900 (1958) – superconductors;
Hugenholtz, N., Pines, D.: Phys. Rev. *116*, 489 (1959) – Bose superfluid;
Lange, R. V.: Phys. Rev. *146*, 301 (1966) – Heisenberg ferromagnets.
The idea of quantum hydrodynamics was given by
(42) Landau, L. D.: J. Phys. USSR *5*, 71 (1941).
Review articles on soft modes are
(43) Blinc, R., Zeks, B. L.: Adv. Phys. *21*, 693 (1972);
Scott, J. F.: Rev. Mod. Phys. *46*, 83 (1974);
Shirane, G.: Rev. Mod. Phys. *46*, 437 (1974).
The discussion on KH_2PO_4, given in the present text, is extremely simplified; for instance see
(44) Elliott, R. J., Young, A. P., Smith, S. R. P.: J. Phys. C4, L317 (1970)
Lattice dynamical approaches to central peaks are found in
(45) Krumhansl, J. A., Schrieffer, J. E.: Phys. Rev. B*11*, 3535 (1975);
Schneider, T., Stoll, E.: Phys. Rev. B*13*, 1216 (1976).
The random phase approximation of magnetic metals is discussed in
(46) Doniach, S., Sondheimer, E. H.: *Green's Functions for Solid State Physicists* (Benjamin, New York 1974).
As for the mean field theory of excitonic phase and electron-hole metal drop, see
(47) Halperin, B. I., Rice, T. M.: In *Solid State Physics,* Vol. 21, ed. by F. Seitz, D. Turnbull (Academic Press, New York 1968) p. 116;
Rice, T. M.: "Semiconductor-Metal Transitions", in *Physics of Highly Excited States in Solids,* ed. by M. Ueta, Y. Nishina, Lecture Notes in Physics, Vol. 57 (Springer, Berlin, Heidelberg, New York 1976) p. 114.
For Mott transitions,
(48) Mott, N. F.: *Metal Insulator Transitions* (Taylor and Francis, London 1974).
Original papers on Bogoliubov's inequality applied to low-dimensional systems are
(49) Wagner, H.: Z. Phys. *195*, 273 (1966) – crystals;
Mermin, N. D., Wagner, H.: Phys. Rev. Lett. *17*, 1133 (1966) – spins;
Hohenberg, P. C.: Phys. Rev. *158*, 383 (1967) – superfluid;
Mermin, N. D.: Phys. Rev. *176*, 250 (1968) – crystals.
Concepts of topological long-range order and phase transition as its destruction by vortices in two-dimensional systems were introduced by
(50) Kosterlitz, J. M., Thouless, D. J.: J. Phys. C6, 1181 (1973).
As for experimental checks, see
(51) Rudnick, I.: Phys. Rev. Lett. *40*, 1454 (1978);
Bishop, D. J., Reppy, J. D.: Phys. Rev. Lett. *40*, 1727 (1978);
Grimes, C. C., Adams, G.: Phys. Rev. Lett. *19*, 795 (1979).
The method of the renormalization group applied to the Kondo effect is described by

(52) Wilson, K. G.: Rev. Mod. Phys. *47*, 773 (1975).
The Kondo problem can also be attacked by using Anderson's Hamiltonian to describe the impurity in a metal and applying straightforward perturbational expansion; see
(53) Yosida, K., Yamada, K.: Prog. Theor. Phys. Suppl. *46*, 244 (1970); *53*, 1286 (1975);
      Yamada, K.: Prog. Theor. Phys. *53*, 970 (1975); *54*, 316 (1975).
For Moriya's new spin fluctuation theory of metallic magnetism, the following review article is most useful:
(54) Moriya, T.: Recent Progress in the Theory of Itinerant Electron Magnetism. J. Magn. Magn. Mater. *14*, 1 (1979).

Chapter 5

A brief review on the polariton in general is presented by
(55) Burstein, E., Mills, D. L.: Comments on Solid State Phys. *1*, 202 (1969); *2*, 93, 111 (1969); *3*, 12 (1970),
while the spatial dispersion of the exciton-polariton is detailed in
(56) Agranovich, V. M., Ginzburg, V. L.: *Spatial Dispersion in Crystal Optics and the Theory of Excitons* (Interscience, New York 1966).
Recent studies on resonant light scattering in the exciton-polariton region are reviewed in
(57) Permogorov, S.: Phys. Status Solidi (b) *68*, 9 (1975)
(58) Ulbrich, R. G., Weisbuch, C.: *Festkörperprobleme XVIII* (Vieweg, Braunschweig 1978) p. 217.

Chapter 6

A general indroduction to electron – phonon interaction is found in Chap. 7 of (1).
The references on the polaron problem besides (18) are
(59) Appel, J.: In *Solid State Physics*, Vol. 21, ed. by F. Seitz, D. Turnbull (Academic Press, New York 1968) p. 193
(60) Devreese, J. T. (ed.): *Polarons in Ionic Crystals and Polar Semiconductors* (North-Holland, Amsterdam 1972).
The recent developments in the method of path integral will be found in
(61) Papadopoulos, G. J., Devreese, J. T. (eds.): *Path Integrals and Their Applications in Quantum, Statistical, and Solid State Physics* (Plenum Press, New York, London 1978).
As for the electron-phonon interaction in metals,
(62) Schrieffer, J. R.: *Theory of Superconductivity* (Benjamin, New York 1964)

presents an instructive indroduction to this field. This book deals with the superconductivity but the article on the electron-phonon interaction may be understandable to a beginner.

(63) Hedin, L., Lundqvist, S.: In *Solid State Physics,* Vol. 23, ed. by F. Seitz, D. Turnbull (Academic Press, New York 1969) p. 1

describes how the electron-phonon or electron-electron interaction affects the properties of metals. A reader interested in the examples of these effects should refer to this monograph.

To a reader who wants to study further the Green's function method which was discussed after Sect. 6.4, besides (46) the following references are recommended:

(64) Abrikosov, A. A., Gor'kov, L. P., Dzyaloshinskii, I. Y.: *Quantum Field Theoretical Methods in Statistical Physics* (Pergamon, Oxford, New York 1965)

(65) Abe, R.: *Statistical Mechanics,* transl. Y. Takahashi (University of Tokyo Press, Tokyo 1975).

Furthermore,

(66) Kadanoff, L. P., Baym, G.: *Quantum Statistical Mechanics* (Benjamin, New York 1962)

is a good book for a reader interested in the Green's function method.

Chapter 7

(67) Bloembergen, N.: *Nonlinear Optics* (Benjamin, New York 1965)

is a standard textbook covering a variety of nonlinear optical effects. As for light scattering , a rapidly developing area, we refer to the topical reviews in

(68) Cardona, M. (ed.): *Light Scattering in Solids,* Topics in Applied Physics, Vol. 8 (Springer, Berlin, Heidelberg, New York 1975).

The optical properties of solids associated with electronic transitions are reviewed in

(69) Fan, H. Y.: "Photon-Electron Interaction, Crystals Without Fields", in *Light and Matter Ia,* Encyclopedia of Physics, Vol. XXV/2 a, ed. by S. Flügge (Springer, Berlin, Heidelberg, New York 1967) p. 157.

As for the radiative and nonradiative processes of localized electrons, we recommend the following books, each with its own merit:

(70) Fowler, W. B. (ed.): *Physics of Color Centers* (Academic Press, New York 1968)

(71) Stoneham, A. M.: *Theory of Defects in Solids* (Clarendon Press, Oxford 1975)

(72) Englman, R.: *Non-Radiative Decay of Ions and Molecules in Solids* (North-Holland, Amsterdam 1979).

The exciton-phonon dynamics described in Sect. 7.3 is based on

(73) Toyozawa, Y.: Prog. Theor. Phys. *20*, 53 (1958); J. Phys. Chem. Solids *25*, 59 (1964)

Rebane, K. K., Fedoseyev, V. G., Hizhnyakov, V. V.: In *Proc. 9th Intern. Conf. Semiconductors* (Nauka, Moscow 1968) p. 430,

while the renormalization theory used there is due to

(74) Van Hove, L.: Physica *21*, 901 (1955).

The general theory on the spectral line shape from the statistical mechanical viewpoint is presented by

(75) Kubo, R.: In *Fluctuation, Relaxation and Resonance in Magnetic Systems*, ed. by D. ter Haar (Oliver and Boyd, Edinburgh 1962).

A brief review on the final state interaction in general is found in

(76) Hopfield, J. J.: Comments Solid State Phys. *1*, 198 (1968),

while the description on the exciton-phonon bound state in Sect. 7.4.1 is based on

(77) Toyozawa, Y., Hermanson, J.: Phys. Rev. Lett. *21*, 1637 (1968).

The edge anomaly on the soft x-ray absorption spectra of metals is reviewed in

(78) Mahan, G. D.: In *Solid State Physics*, Vol. 29, ed. by H. Ehrenreich, F. Seitz, D. Turnbull (Academic Press, New York 1974) p. 75.

The discontinuous transitions from the nearly free state to the self-trapped state are described in

(79) Toyozawa, Y.: In *Vacuum Ultraviolet Radiation Physics*, ed. by E. Koch, R. Haensel, C. Kunz (Pergamon-Vieweg, Braunschweig 1974) p. 317

(80) Emin, D.: Adv. Phys. *22*, 57 (1973).

Theoretical arguments on electron bubbles in liquid helium are presented in

(81) Jortner, J., Kestner, N. R., Rice, S. A., Cohen, M. H.: J. Chem. Phys. *43*, 2614 (1965)

Fowler, W. B., Dexter, D. L.: Phys. Rev. *176*, 337 (1968),

while the experimental studies on the motion of excitons in liquid helium are reported in

(82) Surko, C. M., Reif, F.: Phys. Rev. Lett. *20*, 582 (1968)

Surko, C. M., Dick, G. J., Reif, F.: Phys. Rev. Lett. *28*, 842 (1969)

Stockton, M., Keto, J. W., Fitzsimmons, W. A.: Phys. Rev. A*5*, 372 (1972).

In this context, the behavior of excitons in other rare gases in the solid state is reviewed in

(83) Fugol' I. Ya.: Adv. Phys. *27*, 1 (1978).

Subject Index

Light Scattering in Solids

Editor: M. Cardona
1975. 111 figures, 3 tables. XIII, 339 pages
(Topics in Applied Physics, Volume 8)
ISBN 3-540-07354-X

Contents:
M. Cardona: Introduction. – *A. Pinczuk, E. Burstein:* Fundamentals of Inelastic Light Scattering in Semiconductors and Insulators. – *R. M. Martin, L. M. Falicov:* Resonant Raman Scattering. – *M. V. Klein:* Electronic Raman Scattering. – *N. H. Brodsky:* Raman Scattering in Amorphous Semiconductors. – *A. S. Pine:* Brillouin Scattering in Semiconductors. – *Y.-R. Shen:* Stimulated Raman Scattering.

Photoemission in Solids I

General Principles
Editors: M. Cardona, L. Ley
1978. 90 figures, 17 tables. XI, 290 pages
(Topics in Applied Physics, Volume 26)
ISBN 3-540-08685-4

Contents:
M. Cardona, L. Ley: Introduction. – *W. L. Schaich:* Theory of Photoemission. Independent Particle Model. – S. T. Manson: The Calculation of Photoionization Cross Sections: An Atomic View. – *D. A. Shirley:* Many-Electron and Final-State Effects: Beyond the One-Electron Picture. – *G. K. Wertheim, P. H. Citrin:* Fermi Surface Excitations in X-Ray Photoemission Line Shapes from Metals. – *N. V. Smith:* Angular Dependent Photoemission. – Appendix.

Photoemission in Solids II

Case Studies
Editors: L. Ley, M. Cardona
1979. 214 figures, 26 tables. XVIII, 401 pages
(Topics in Applied Physics, Volume 27)
ISBN 3-540-09202-1

Contents:
L. Ley, M. Cardona: Introduction. – *L. Ley, M. Cardona, R. A. Pollak:* Photoemission in Semiconductors. – *S. Hüfner:* Unfilled Inner Shells: Transition Metals and Compounds. – *M. Campagna, G. K. Wertheim, Y. Baer:* Unfilled Inner Shells: Rare Earths and Their Compounds. – *W. D. Grobman, E. E. Koch:* Photoemission from Organic Molecular Crystals. – *C. Kunz:* Synchrotron Radiation: Overview. – *P. Steiner, H. Höchst, S. Hüfner:* Simple Metals. – Appendix: Table of Core-Level Binding Energies. – Additional References with Titles. – Subject Index.

Solid-State Physics

1976. 152 figures, 35 tables. VIII, 279 pages
(Springer Tracts in Modern Physics, Volume 78)
ISBN 3-540-07774-X

Contents:
R. Dornhaus, G. Nimtz: The Properties and Applications of the $Hg_{1-x}Cd_xTe$ Alloy System: The Crystal. Band Structure. Transport Properties. Optical Properties. Infrared Devices. List of Important Symbols. Numerical Values of Important Quantities. References. – *W. Richter:* Resonant Raman Scattering in Semiconductors: Electric Susceptibility. Light Scattering. Experimental Methods. One-Phonon Deformation Potential Scattering. Infrared Active LO Phonons. Multiphonon Scattering. Conclusions. References. List of Symbols. – Subject Index.

Springer-Verlag
Berlin
Heidelberg
New York

Dynamics of Solids and Liquids by Neutron Scattering

Editors: S. W. Lovesey, T. Springer
1977. 156 figures, 15 tables. XI, 379 pages
(Topics in Current Physics, Volume 3)
ISBN 3-540-08156-9

Contents:
S. W. Lovesey: Introduction. – *H. G. Smith, N. Wakabayashi:* Phonons. – *B. Dorner, R. Comès:* Phonons and Structural Phase Transformations. – *J. W. White:* Dynamics of Molecular Crystals, Polymers, and Adsorbed Species. – *T. Springer:* Molecular Rotations and Diffusion in Solids, in Particular Hydrogen in Metals. – *R. D. Mountain:* Collective Modes in Classical Monoatomic Liquids. – *S. W. Lovesey, J. M. Loveluck:* Magnetic Scattering.

Excitons

Editor: K. Cho
1979. 118 figures, 8 tables. XI, 274 pages
(Topics in Current Physics, Volume 14)
ISBN 3-540-09567-5

Contents:
K. Cho: Introduction. – *K. Cho:* Internal Structure of Excitons. – *P. J. Dean, D. C. Herbert:* Bound Excitons in Semiconductors. – B. Fischer, J. Lagois: Surface Exciton Polaritons. – *P. Y. Yu:* Study of Excitons and Exciton-Phonon Interactions by Resonant Raman and Brillouin Spectroscopies.

Neutron Diffraction

Editor: H. Dachs
1978. 138 figures, 32 tables. XIII, 357 pages
(Topics in Current Physics, Volume 6)
ISBN 3-540-08710-9

Contents:
H. Dachs: Principles of Neutron Diffraction. – *J. B. Hayter:* Polarized Neutrons. – *P. Coppens:* Combining X-Ray and Neutron Diffraction: The Study of Charge Density Distributions in Solids. – *W. Prandl:* The Determination of Magnetic Structures. – *W. Schmatz:* Disordered Structures. – *P.-A. Lindgård:* Phase-Transitions and Critical Phenomena. – *G. Zaccai:* Application of Neutron Diffraction to Biological Problems. – *P. Chieux:* Liquid Structure Investigation by Neutron Scattering. – *H. Rauch, D. Petraschek:* Dynamical Neutron Diffraction and Its Applications.

Positrons in Solids

Editor: P. Hautojärvi
1979. 66 figures, 25 tables. XIII, 255 pages
(Topics in Current Physics, Volume 12)
ISBN 3-540-09271-4

Contents:
P. Hautojärvi, A. Vehanen: Introduction to Positron Annihilation. – *P. E. Mijnarends:* Electron Momentum Densities in Metals and Alloys. – *R. N. West:* Positron Studies of Lattice Defects in Metals. – *R. M. Nieminen, M. J. Manninen:* Positrons in Imperfect Solids: Theory. – *A. Dupasquier:* Positrons in Ionic Solids.

Springer-Verlag
Berlin
Heidelberg
New York